Recent Trends in Nanomedicine and Tissue Engineering

RIVER PUBLISHERS SERIES IN RESEARCH AND BUSINESS CHRONICLES: BIOTECHNOLOGY AND MEDICINE

Series Editors

PAOLO DI NARDO
University of Rome Tor Vergata
Italy

PRANELA RAMESHWAR
Rutgers University
USA

ALAIN VERTES
London Business School, UK
and
NxR Biotechnologies, Switzerland

Editorial Board

- **Sonia Melino**, University of Rome Tor Vergata, Italy
- **Elisabeth Engel**, IBEC, Barcelona, Spain
- **Carlo Guarnieri**, University of Bologna, Italy
- **Veronique Santran**, ICELTIS Toulouse, France
- **Chandra**, University of Texas, USA
- **Pere Gascon**, University of Barcelona, Spain
- **Banerjee Debabrata**, Rutgers University, USA

Indexing: All books published in this series are submitted to the Web of Science Book Citation Index (BkCI), to CrossRef and to Google Scholar.

Combining a deep and focused exploration of areas of basic and applied science with their fundamental business issues, the series highlights societal benefits, technical and business hurdles, and economic potentials of emerging and new technologies. In combination, the volumes relevant to a particular focus topic cluster analyses of key aspects of each of the elements of the corresponding value chain.

Aiming primarily at providing detailed snapshots of critical issues in biotechnology and medicine that are reaching a tipping point in financial investment or industrial deployment, the scope of the series encompasses various specialty areas including pharmaceutical sciences and healthcare, industrial biotechnology, and biomaterials. Areas of primary interest comprise immunology, virology, microbiology, molecular biology, stem cells, hematopoiesis, oncology, regenerative medicine, biologics, polymer science, formulation and drug delivery, renewable chemicals, manufacturing, and biorefineries.

Each volume presents comprehensive review and opinion articles covering all fundamental aspect of the focus topic. The editors/authors of each volume are experts in their respective fields and publications are peer-reviewed.

For a list of other books in this series, visit www.riverpublishers.com

Recent Trends in Nanomedicine and Tissue Engineering

Editors

Jince Thomas

Mahatma Gandhi University
Kerala, India

Sabu Thomas

Mahatma Gandhi University
Kerala, India

Jiya Jose

Mahatma Gandhi University
Kerala, India

Nandakumar Kalarikkal

Mahatma Gandhi University
Kerala, India

LONDON AND NEW YORK

Published 2017 by River Publishers
River Publishers
Alsbjergvej 10, 9260 Gistrup, Denmark
www.riverpublishers.com

Distributed exclusively by Routledge
4 Park Square, Milton Park, Abingdon, Oxon OX14 4RN
605 Third Avenue, New York, NY 10158

First published in paperback 2024

Recent Trends in Nanomedicine and Tissue Engineering / by Jince Thomas, Sabu Thomas, Jiya Jose, Nandakumar Kalarikkal.

Routledge is an imprint of the Taylor & Francis Group, an informa business

Publisher's Note
The publisher has gone to great lengths to ensure the quality of this reprint but points out that some imperfections in the original copies may be apparent.

While every effort is made to provide dependable information, the publisher, authors, and editors cannot be held responsible for any errors or omissions.

ISBN: 978-87-93609-16-7 (hbk)
ISBN: 978-87-7004-413-4 (pbk)
ISBN: 978-1-003-33923-6 (ebk)

DOI: 10.1201/9781003339236

Contents

Preface   **xix**

Acknowledgement   **xxiii**

List of Contributors   **xxv**

List of Figures   **xxix**

List of Tables   **xxxv**

List of Abbreviations   **xxxvii**

**1   Nanomedicine and Nanotechnology: State of Art,
     New Challenges and Opportunities   1**

Jince Thomas, Jiji Abraham and Sabu Thomas

   1.1   Introduction   1
   1.2   Scope   2
         1.2.1   Drug Delivery   2
         1.2.2   Tissue Engineering and Nanoscaffolding
                 and Wound Healing   3
         1.2.3   Diagnostics   4
   1.3   History   4
   1.4   Commercial Significance and Current Challenges   7
         References   12

2   Novel Approaches to Nanomedicine and Nanotechnology   19

Apparao Gudimalla, Raghvendra K. Mishra and Prerna Arora

   2.1   Introduction   19
   2.2   Nanomedicine   21
   2.3   History of Nanomedicine   25

        2.3.1   Present Facing Challenges; Translation
                of Nanotechnology to Nanomedicine   29
    2.4 Nanoparticles in Cancer Therapy   30
        2.4.1   Liposomes   30
        2.4.2   The Solid Lipid Nanoparticles   31
        2.4.3   Carbon Nanotubes   32
        2.4.4   Carbon Nanotubes in Cancer Treatment   33
                2.4.4.1   Single-walled carbon nanotubes in the
                          treatment of cancer   33
                2.4.4.2   Multi-walled carbon nanotubes in the
                          treatment of cancer   34
        2.4.5   Drug Delivery   34
        2.4.6   The Applications of Nanoparticles in Drug
                Delivery   35
    2.5 Nanoparticles Anti-Oxidative Role in Diabetes   37
        2.5.1   Nanomedicine in Management of Diabetes   38
    2.6 Nanorobotics   39
        2.6.1   Types of Nanorobots   39
                2.6.1.1   In surgery   39
                2.6.1.2   Drug delivery   40
                2.6.1.3   In diagnosis   40
        2.6.2   Applications of Nanorobots in Medicine   40
        2.6.3   Advantages of Nanorobots   42
        2.6.4   Disadvantages of Nanorobots   42
    2.7 Future Development of Nanomedicine   42
    2.8 Conclusion   45
        Acknowledgment   45
        References   46

3   **Chitosan and Its Derivatives as a Potential Nanobiomaterial:**
    Drug Delivery and Biomedical Application   **57**
    Abhay Raizaday, Hemant K. S. Yadav and Susmitha Kasina

    3.1 History   58
    3.2 Chemistry   58
    3.3 Advantages   60
    3.4 Disadvantages   61
    3.5 Properties of Chitosan   61
        3.5.1   Physicochemical   61
                3.5.1.1   Crystalline structure   61

		3.5.1.2	Degree of *N*-Acetylation	62
		3.5.1.3	Molecular weight	62
		3.5.1.4	Solubility	63
		3.5.1.5	Viscosity	63
	3.5.2		Biological Properties	64
		3.5.2.1	Mucoadhesive properties	64
		3.5.2.2	Permeation enhancing properties	65
		3.5.2.3	Haemostatic activity	65
		3.5.2.4	Antimicrobial activity	65
		3.5.2.5	Analgesic effect	66
		3.5.2.6	Biodegradability	66
3.6	Extraction of Chitosan			66
	3.6.1		Preparation of Chitosan and Water Soluble Chitosan	68
		3.6.1.1	Extraction of Chitin from the Beetle	69
		3.6.1.2	Extraction of collagen from squid	70
		3.6.1.3	Extraction of chitosan from fungi cell wall	70
		3.6.1.4	Extraction of Chitin, Chitosan, from Shrimp by biological method	73
3.7	Chitosan Derivatives			75
	3.7.1		Carboxymethylchitosan	75
	3.7.2		Mono-Carboxymethylated Chitosan	75
	3.7.3		*N*-Succinyl Chitosan	76
	3.7.4		*N*-Acetylated Chitosan	76
	3.7.5		*N*-Trimethyl Chitosan	76
	3.7.6		*N*-Trimethylchitosan Chloride	77
	3.7.7		Succinate and Chitosan Phthalate	77
	3.7.8		Amphiphilic Chitosan Derivatives	78
	3.7.9		Graft-Copolymerization of Chitosan	78
	3.7.10		Thiolated Chitosan Conjugate	78
	3.7.11		Cyclodextrin (CD)-Chitosan Derivative	79
3.8	Applications of Chitosan as Nanobiomaterial			79
	3.8.1		Mucoadhesive Property	79
	3.8.2		Permeation Enhancement	82
	3.8.3		Wound Healing	83
	3.8.4		Gene Delivery	86
	3.8.5		Vaccine Delivery	88
3.9	Conclusion			89
	References			89

**4 Design and Characterization of Lipid Mediated
Nanoparticles Containing an Anti-Psychotic Drug
for Enhanced Bio-Availability   95**

*Jawahar Natarajan, Gowtham Reddy Naredla and
Veera Venkata Satyanarayana Reddy Karri*

4.1   Introduction   96
4.2   Experimental Part   96
      4.2.1   Preformulation Studies   96
              4.2.1.1   Solubility studies   97
              4.2.1.2   Compatibility study   97
              4.2.1.3   Development of calibration curve   97
              4.2.1.4   Partition coefficient studies   98
      4.2.2   Preparation of Solid Lipid Nanoparticles (SLN) by
              Microemulsion Technique   98
              4.2.2.1   Optimization of lipid quantity   99
              4.2.2.2   Study on the effect of formulation process
                        variables   99
              4.2.2.3   Preparation of drug loaded batches . . . 100
      4.2.3   Evaluation of Solid Lipid Nanoparticles 100
              4.2.3.1   Particle size, zeta potential and
                        polydispersity index 100
              4.2.3.2   Entrapment efficiency and drug loading . . 100
              4.2.3.3   Differential scanning calorimetry 101
      4.2.4   *In vitro* Release Studies 101
      4.2.5   *In vivo* Oral Bioavailability Studies 101
      4.2.6   Bioanalytical Method Development and Analysis . . 102
              4.2.6.1   Chromatographic conditions 102
              4.2.6.2   Preparation of olanzapine standard stock
                        solution 103
                        4.2.6.2.1   *Standard stock solution of IS
                                    (Internal standard)* 103
              4.2.6.3   Preparation of analytical calibration curve
                        solutions 103
              4.2.6.4   Preparation of blank plasma 103
              4.2.6.5   Preparation of bio-analytical calibration
                        curve samples 103
              4.2.6.6   Preparation of plasma samples 103
              4.2.6.7   Method of analysis 103

4.3 Results and Discussion 104
    4.3.1 Preformulation Studies 104
        4.3.1.1 Solubility studies 104
        4.3.1.2 Compatibility Studies 104
        4.3.1.3 Development of calibration curve 105
        4.3.1.4 Partition coefficient studies 105
    4.3.2 Effect of Formulation Process Variables 106
    4.3.3 Evaluation of Solid Lipid Nanoparticles 107
        4.3.3.1 Zeta potential 107
        4.3.3.2 Polydispersity 108
        4.3.3.3 Entrapment efficiency and drug loading . . 108
        4.3.3.4 Differential scanning colorimetry 110
        4.3.3.5 *In vitro* release studies 110
        4.3.3.6 Release kinetics 112
    4.3.4 Bioanalytical Method Development and Analysis . . 112
4.4 Conclusion 118
    References 118

5 Nanogels: The Emerging Carrier in Drug Delivery System   **121**
Prashant Sahu, Samaresh Sau, Arun K. Iyer and Sushil K. Kashaw

5.1 Introduction 122
5.2 Properties of Nanogels 123
    5.2.1 Good Drug Loading Capacity 123
    5.2.2 Solubility 124
    5.2.3 Colloidal Stability 124
    5.2.4 Particle Size 124
    5.2.5 Biocompatibility and Degradability 124
    5.2.6 Electro Mobility 124
    5.2.7 Non-Immunologic Response 124
    5.2.8 Others 125
5.3 Classification of Nanogels 125
    5.3.1 Responsive Type 125
        5.3.1.1 Non-responsive nanogels 125
        5.3.1.2 Stimuli-responsive nanogels 126
    5.3.2 Linkage Type 126
        5.3.2.1 Physical cross-linked gels 126
        5.3.2.2 Liposomes modified nanogels 126
        5.3.2.3 Micellar nanogels 126

| | | 5.3.2.4 | Hybrid nanogel . . . . | 127 |

5.3.2.4   Hybrid nanogel 127
5.3.2.5   Chemically cross-linked nanogels 127
5.4   Method of Preparation of Nanogel 128
5.4.1   Photolithographic Technique 128
5.4.2   Micro-Moulding Method 128
5.4.3   Bi-Polymers Synthesis Technique 129
5.4.4   Water in Oil (W/O) Heterogeneous Emulsion
Method 129
5.4.5   Inverse Mini Emulsion Method 129
5.4.6   Reverse Micellar Method 130
5.4.7   Membrane Emulsification Method 130
5.4.8   Heterogeneous Free Radical Polymerization 131
5.4.8.1   Inverse micro emulsion 131
5.4.8.2   Inverse mini-emulsion polymerization . . 131
5.4.8.3   Precipitation polymerization 132
5.4.8.4   Dispersion polymerization 132
5.4.8.5   Heterogeneous controlled/living radical
polymerization 132
5.4.9   Conversion of Macrogels to Nanogels 133
5.4.10   Chemical Cross-Linking Method 133
5.5   Characterization of Nanogel 133
5.5.1   Morphological Analysis 134
5.5.1.1   Scanning Electron Microscopy (SEM) . . 135
5.5.1.2   Transmission Electron
Microscopy (TEM) 135
5.5.2   Size and Shape 135
5.5.3   Viscosity 135
5.5.4   Phase Behaviour 136
5.5.5   Optical Transparency 136
5.5.6   Spectroscopic Analysis 136
5.5.7   *pH* 137
5.6   Routes of Administration of Nanogel 137
5.6.1   Parenteral Drug Delivery System 138
5.6.2   Oral Drug Delivery System 138
5.6.3   Transdermal Drug Delivery 140
5.6.4   Ocular Drug Delivery System 140
5.6.5   Pulmonary or Intranasal Drug Delivery System . . . 141
5.7   Application of Nanogels 142
5.7.1   Nano-Sized Drug Delivery System 144
5.7.2   Peptide and Protein Delivery 144

5.7.3   Vaccine Delivery 145

5.7.4   Gene Delivery 145

5.7.5   Antiviral Nanogel Delivery 146

5.7.6   Antifungal Nanogel Delivery 147

5.7.7   In Autoimmune Diseases 148

5.7.8   Ophthalmic Delivery 148

5.7.9   Diabetes 148

5.7.10 Coagulating Agent 149

5.7.11 Anti-Inflammatory Agent 149

5.8   Disadvantages of Nanogel 149

5.9   Conclusion 149

     References 150

6 Fe, Co Based Bio-Magnetic Nanoparticles (BMNPs): Synthesis, Characterization, and Biomedical Application   157

Amirsadegh Rezazadeh Nochehdehi, Sabu Thomas, Minoo Sadri, S. M. Mehdi Hadavi, Yves Grohens, Nandakumar Kalarikkal and Neerish Revaprasdu

6.1   Introduction 158

6.1.1   Magnetic Properties 161

6.1.2   Magnetic Nanoparticles 162

    6.1.2.1   Iron and iron oxide nanoparticles 164

    6.1.2.2   Cobalt-based nanoparticles 165

6.2   Synthesis and Characterization of Magnetic Nanoparticles (MNPs) 166

6.2.1   Iron Oxide (Fe$_3$O$_4$) Nanoparticles (ION) 166

6.2.2   Cobalt-Based (FeCo) Nanoparticles (CBN) 168

6.3   Synthesis and Characterization of Core/Shell Magnetic Nanoparticles (CS-MNPs) 169

6.3.1   Iron Oxide Core/Shell Nanoparticles (IOCSN) . . . 169

    6.3.1.1   Fe$_3$O$_4$@Ag core/shell nanoparticles . . . 169

    6.3.1.2   Fe$_3$O$_4$@Chitosan core/shell nanoparticles 171

6.3.2   Cobalt-Based Core/Shell Nanoparticles (CBCSN) . . 173

    6.3.2.1   FeCo@C core/shell nanoparticles 173

    6.3.2.2   FeCo@PEG core/shell nanoparticles . . . 175

6.4   Biomedical Application of Magnetic Nanoparticles (MNPs) 176

6.4.1   Bioimaging Application of MNPs 180
6.4.2   Controlled Drug Delivery (TDD)
        Applications of MNPs 181
6.4.3   Cancer Diagnosis and Treatment via Hyperthermia
        Method (CDT) Using MNPs 182
6.5   Conclusion 183
      Acknowledgement 185
      References 185

**7   Comparative Study on Cytotoxic and Bactericidal Effect
of Nanoscale Zero Valent Iron Synthesized through Chemical
and Biological Methods   197**

*Sharath R., Harish B. G., Chandraprabha M. N., Samrat K.,
Nagaraju Kottam, Hari Krishna R., Rakesh G. Kashyap
and Muktha H.*

7.1   Introduction 198
7.2   Materials and Methods 206
      7.2.1   Materials 206
      7.2.2   Methods 206
              7.2.2.1   Synthesis of nanoscale Zero Valent Iron
                        (nZVI) by chemical and biological
                        methods 206
      7.2.3   Characterization Studies 207
      7.2.4   Screening of Bactericidal and Cytotoxic Activity . . 207
7.3   Results and Discussion 208
      7.3.1   Synthesis of Nanoscale Zero Valent Iron (nZVI)
              by Chemical and Biological Methods 208
      7.3.2   Characterization of nZVI Particles 208
      7.3.3   Screening of Bactericidal and Cytotoxic Activity . . 211
7.4   Conclusion 213
      References 213

**8   Simulation Studies of Nanomotors Based on Carbon
Nanotubes for Nanodelivery Systems   219**

Sunita Negi

8.1   Introduction 219
8.2   Nanomotor 220
8.3   Protein Structure 221

8.4 Simulation Method 224
8.5 Summary and Future Scope 225
    References 225

**9 Synthesis and Characterization of Lipid-Conjugated Carbon
  Nanotubes for Targeted Drug Delivery to Human Breast
  Cancer Cells** **227**

*Jawahar Natarajan, Surendra Ekkuluri
and Veera Venkata Satyanarayana Reddy Karri*

9.1 Introduction 228
9.2 Experimental Part 228
    9.2.1 Preformulation Studies 228
          9.2.1.1 Solubility studies 229
          9.2.1.2 Standard calibration curve 229
          9.2.1.3 Compatibility study 229
    9.2.2 Preparation of Carbon Nanotubes 229
          9.2.2.1 Study on the effect of formulation process
                  variables 230
    9.2.3 Purification, Cutting, and Oxidation of CNTs 230
    9.2.4 Particle Size, Zeta Potential, and Polydispersity
          Index (PDI) 231
    9.2.5 Surface Morphology by Scanning Electron
          Microscopy (SEM) 231
    9.2.6 Preparation of CNTs-RH-Folic Acid
          (CNTs-RH-FA) 231
    9.2.7 Characterization of CNT-RH-FA 231
          9.2.7.1 Size distribution and zeta potential 231
    9.2.8 Scanning Electron Microscopy (SEM) 232
    9.2.9 Determination of Loading Efficiency 232
    9.2.10 *In Vitro* Drug-Release Studies 232
    9.2.11 Determination of Mitochondrial Synthesis
           by MTT Assay 233
9.3 Results and Discussion 233
    9.3.1 Preformulation Studies 233
          9.3.1.1 Solubility studies 233
          9.3.1.2 Development of calibration curve 234
          9.3.1.3 Crystallinity study by using DSC 234
          9.3.1.4 Compatibility studies using FT-IR 234

9.3.2   Preparation of Carbon Nanotubes   237
9.3.3   Particle Size Distribution and Zeta Potential   239
9.3.4   Scanning Electron Microscopy (SEM)   240
9.3.5   Drug Loading Efficiency   240
9.3.6   *In Vitro* Drug-Release Studies   241
9.3.7   *In Vitro* Cytotoxicity Studies   244
9.4   Conclusion   246
References   247

10 Phytosynthesis of Silver Nanoparticles and Its Potent Antimicrobial Efficacy   249

Soumya Soman and Joseph George Ray

10.1   Introduction   250
10.1.1   Advantages of Phytosynthesis   251
10.1.2   Major Objective   252
10.2   Phytosynthesis of AgNPs   253
10.2.1   Extracellular Synthesis of AgNPs Using Plant
Extracts: A Few Case Studies   254
10.2.2   Effect of Environment Parameters Influencing
Phytosynthesis   257
10.3   Probable Mechanism for AgNP Formation   260
10.4   Importance of Antibacterial Activity
of Phytosynthesized AgNPs   261
10.4.1   Antibacterial Activity of Phytosynthesized AgNPs:
Case Studies   263
10.5   Mechanism of Action of AgNPs   266
10.5.1   Different Postulates of Mechanism of AgNPs
Toxicity to Bacteria   266
10.6   Antibacterial Applications of AgNPs   270
10.7   Conclusion   272
References   273

11 Recreation of Turmeric Matrix with Enhanced Curcuminoids— Enhances the Bioavailability and Bioefficacy   289

Augustine Amalraj and Sreeraj Gopi

11.1   Introduction   290
11.2   Discovery of Curcumin   290
11.3   Isolation of Curcumin   291

11.4 Physical, Chemical, and Molecular Properties
of Curcuminoids 291
11.5 Experimental Part 294
    11.5.1 Preparation Method of Cureit^TM 294
    11.5.2 Analytical Method for Analysis of Plasma Curcumin
        Level in Cureit^TM and Standard Curcumin 296
    11.5.3 Statistical Analysis 296
11.6 Results and Discussion 297
    11.6.1 Chemical Analysis of Cureit^TM 297
    11.6.2 Characterization of Cureit^TM 297
        11.6.2.1 NMR studies of cureit^TM 297
        11.6.2.2 FT-IR studies of cureit^TM 299
        11.6.2.3 XRD studies of cureit^TM 302
        11.6.2.4 TGA/DTA studies of cureit^TM 303
        11.6.2.5 SEM analysis of cureit^TM 303
        11.6.2.6 I–V studies of cureit^TM 304
11.7 Bioavailability and Bioefficacy Studies of Cureit^TM 305
    11.7.1 Comparative Oral Bioavailability Study of Cureit^TM
        with Standard Curcumin 305
    11.7.2 Recent Studies on the Cureit^TM 307
11.8 Conclusions 307
    References 308

**12 The Good Tooth, The Bad Influence of Aciduric Germs
and The Ugly Stench of Decay**   **315**
T. Jesse Joel and P. W. Ramteke

12.1 The Good, The Bad, and The Ugly Microbe – *Streptococcus
Mutans* 316
    12.1.1 Introduction 317
        12.1.1.1 Streptococcus mutans: Isolation
            and identification 317
        12.1.1.2 Habitat and nature of source 320
        12.1.1.3 Taxonomy 321
    12.1.2 Dangerous Etiology or a Farce 321
        12.1.2.1 Anaerobiosis 321
        12.1.2.2 Pathogenecity 322
        12.1.2.3 Virulence 323
        12.1.2.4 Transmissibility 328

12.1.2.5 Risk factors 332
12.2 Other Organisms Associated with Caries 333
References 335

13 Therapeutic Angiogenesis in Cardiovascular Diseases, Tissue Engineering, and Wound Healing   343

*K. R. Rakhimol, Robin Augustine, Sabu Thomas
and Nandakumar Kalarikkal*

13.1 Introduction 343
13.2 Therapeutic Angiogenesis: Concept, Approaches,
and Applications 345
13.2.1 Approaches 347
13.2.1.1 Direct VEGF administration 347
13.2.1.2 Cell-based therapy 347
13.2.1.3 Regulation at genomic/molecular level . . 348
13.2.1.4 Hypoxia-induced angiogenesis 348
13.2.2 Applications of Therapeutic Angiogenesis 350
13.2.2.1 Wound healing 350
13.2.2.2 Bone development 350
13.2.2.3 Cardiac diseases 351
13.3 Growth Factors Needed for Angiogenesis 351
13.3.1 Fibroblast Growth Factor 351
13.3.2 Vascular Endothelial Growth Factor 352
13.3.3 Platelet-Derived Growth Factor 352
13.4 Reactive Oxygen Species-Dependent Angiogenesis 353
13.5 Metal Nanoparticle-Based Angiogenesis 354
13.6 Stimulating Angiogenesis in Scaffolds by Therapeutic
Angiogenesis 355
13.7 Challenges and Risks 357
13.8 Conclusion 358
References 358

14 Toxicity of Nanomaterials Used in Nanomedicine   365

*Parvathy Prasad, Sunija Sukumaran, Nitheesha Shaji,
V. K. Yadunath, Jiya Jose, Nandakumar Kalarikkal
and Sabu Thomas*

14.1 Introduction 365
14.2 Nanomedicine 367

14.3 Nanomaterials Used for Nanomedicine 368
14.4 Toxicokinetics of Nanoparticles 370
14.5 Toxicity of Nanoparticles 372
14.6 Effect of Nanoparticles in Some Aquatic Organisms 378
14.7 Discussion 379
14.8 Conclusion 379
       References 380

Index **385**

About the Editors **387**

Preface

The book entitled "Recent trends in Nanomedicine and Tissue Engineering" precises many of the recent technological and research accomplishments in the area of Nanomedicine and Tissue Engineering. Comprised in the book are presentations of state of art in the area, novel approaches of nanomedicine and nanotechnology, new challenges, and opportunities. Also discussed are the different types of nanomedicine drugs, their production, and commercial significance. Other topics enclosed are the use of natural and synthetic nanoparticles for the production nanomedicine drugs, different types of drug delivery systems, drug carriers, wound-healing antimicrobial activity, effect of natural materials in nanomedicine, toxicity of nanoparticles. As the heading specifies the book highlights various facets of nanomedicine and nanotechnology and their outputs for all community.

This novel book serves as an up to date record on the key findings, observations and fabrication of drugs related to nanomedicine and nanotechnology. It is intended to assist as a pioneer reference resource for all medical research fields as well as in science fields. The various chapters in this book are contributed by prominent researchers from academe, industry, and government–private research laboratories across the world. This book is a very valuable reference source for university and college faculties, professionals, post-doctoral research fellows, senior graduate students and researchers from R&D laboratories working in the area of health care and medical fields.

Chapter 1 gives a brief outline and an overview of the state of art in the area and also presents new challenges, opportunities, and commercial significance of this area. Chapter 2 discussed the detailed studies on nanomedicine and some technological approaches of this field. The authors explain the history of nanomedicine and nanoparticles used as in cancer therapy and also mention other topics such as drug delivery, nanorobotics, and future of nanomedicine. Chapter 3 provides the benefits of chitosan and their derivatives used in drug delivery systems. And also mention the physio-chemical and biological properties of chitosan and their derivatives.

Chapter 4 focuses the design and characterization of lipid mediated nanoparticles for different biomedical applications. *In vitro* and *In vivo* studies of newly formulated lipid nanoparticles is also reported and important characterizations such as zeta potential, partition coefficient studies, differential scanning calorimetry, and release kinetics results are discussed. Chapter 5 presents the applications of nanogels for drug delivery systems. Properties, preparation, benefits and disadvantages of nanogels are also correlated in this chapter. In Chapter 6, biomedical applications of bio magnetic nanoparticles are presented. In this chapter, the authors reviews the effect of FeCo-based biomagnetic nanoparticles for enhance the biomedical and tissue engineering applications. Synthesis of Nanoscale zero valent iron (nZVI) by chemical and biological methods are discussed in Chapter 7. This chapter mainly focuses the characterization of nZVI particles and their Bactericidal and Cytotoxic Activity. Chapter 8 gives the idea about Nanomotors-based carbon Nanotubes for Nano-delivery Systems. Chapter 9 highlights the drugs against human breast cancer cells. The authors explain the working principle of CNT, used against breast cancer cells. Preparation, characterization, and compatibility of CNT's are also discussed.

Phytosynthesis of silver nanoparticles and their antimicrobial efficacy are discussed in Chapter 10. This chapter includes various discussions about phytosynthesis, effect of environment parameters influencing phytosynthesis, probable mechanism for AgNP formation, different postulates of mechanism of AgNPs toxicity to bacteria. Chapter 11 is about the recreation of turmeric matrix with enhanced curcuminoids. This chapter includes preparation method of Cureit^{TM}, isolation of curcumin, comparison of plasma curcumin level in Cureit^{TM} and standard curcumin and following analysis such as NMR, XRD, FT-IR, SEM are also included. Chapter 12 is about the protection of human teeth. Taxonomy, pathogenecity, virulence, risk factors, and other organisms associated with caries are correlated with this chapter. Chapter 13 is focusing the Tissue engineering and wound healing applications. The last chapter discussed the toxicity of nanoparticles. In this chapter, authors review the effect of toxicity of nanomaterials in nanomedicines and how aquatic organisms are affected by nanoparticles.

Finally, Thanks to God for the successful completion of this book and the editors would like to express their honest gratitude to all the contributors of this book who provided excellent support for the fruitful completion of this endeavor. We appreciate them for their commitment and the genuineness for their contributions to this book. We would like to acknowledge all the reviewers who have taken their valuable time to make critical comments on each chapter. We thank 'River Publishers' who recognized the demand for this book and also acknowledge their support.

Jince Thomas
Sabu Thomas

Acknowledgement

Thanks to Divinity for the successful completion of this book. After fruitful completion of this book, it is a pleasant task to express the honest gratitude to all the contributors of this book who provided excellent support for the productive achievement of this endeavor. We appreciate them for their commitment and the genuineness for their contributions to this book. We also thank the publisher 'River Publishers' who recognized the demand for such a book and for realizing the increasing importance of the area of nanomedicine and tissue engineering.

List of Contributors

Abhay Raizaday, *Department of Pharmaceutics, JSS College of Pharmacy, JSS University, Mysore, Karnataka, India*

Amirsadegh Rezazadeh Nochehdehi, *Researcher, Biomaterials research group, Biomedical Engineering Department, Materials and Biomaterials Research Group (MBMRC), Tehran, Iran*

Apparao Gudimalla, *International and Inter University Centre for Nanoscience and Nanotechnology, M G University, Kottayam, Kerala, India*

Arun K. Iyer, *Use-inspired Biomaterials & Integrated Nano Delivery (U-BiND) Systems Laboratory, Department of Pharmaceutical Sciences, Wayne State University, Detroit, Michigan, USA*

Augustine Amalraj, *R&D Centre, Aurea Biolabs Pvt. Ltd, Kolenchery, Cochin, India*

Chandraprabha M. N., *Department of Biotechnology, M. S. Ramaiah Institute of Technology, Bangalore, 560054, India*

Gowtham Reddy Naredla, *Department of Pharmaceutics, JSS College of Pharmacy, Ootacamund, Jagadguru Sri Shivarathreeswara University, Mysuru, India*

Hari Krishna R., *Department of Chemistry M. S. Ramaiah Institute of Technology, Bangalore, 560054, India*

Harish B. G., *Department of MCA, Visvesvaraya Institute of Advanced Technology, Muddenahalli, Chikkaballapur, 562101, India*

Hemant K. S. Yadav, *Department of pharmaceutics, RAK Medical & Health Sciences University, Ras Al Khaimah, UAE*

Jawahar Natarajan, *Department of Pharmaceutics, JSS College of Pharmacy, Ootacamund, Jagadguru Sri Shivarathreeswara University, Mysuru, India*

Jiji Abraham, *International and Inter University Centre for Nanoscience and Nanotechnology, Mahatma Gandhi University, Kerala, India*

Jince Thomas, *International and Inter University Centre for Nanoscience and Nanotechnology, Mahatma Gandhi University, Kerala, India*

Jiya Jose, *International and Inter University Centre for Nanoscience and Nanotechnology, Mahatma Gandhi University, Kottayam, Kerala, India*

Joseph George Ray, *Laboratory of Ecology and Eco-technology, School of Biosciences, Mahatma Gandhi University, Kottayam, Kerala, India*

K. R. Rakhimol, *International and Inter University Centre for Nanoscience and Nanotechnology, Mahatma Gandhi University, Kottayam, Kerala 686560, India*

Minoo Sadri, *Department of Biochemistry and Biophysics, Education and Research Center of Science and Biotechnology, Malek Ashtar University of Technology, Tehran, Iran*

Muktha H., *Department of Biotechnology, M. S. Ramaiah Institute of Technology, Bangalore, 560054, India*

Nagaraju Kottam, *Department of Chemistry M. S. Ramaiah Institute of Technology, Bangalore, 560054, India*

Nandakumar Kalarikkal, *Professor, School of Pure and Applied Physics, Mahatma Gandhi University, Kottayam 686 560, Kerala, India*

Neerish Revaprasdu, *Professor, SARCHI Chair in Nanotechnology, Department of Chemistry, Faculty of Science and Agriculture, University of Zululand, KwaZulu-Natal, South Africa*

Nitheesha Shaji, *International and Inter University Centre for Nanoscience and Nanotechnology, Mahatma Gandhi University, Kottayam, Kerala, India*

P. W. Ramteke, *Head, Department of Biological Sciences, Sam Higginbottom University of Agriculture, Technology and Sciences, Allahabad, India*

Parvathy Prasad, *International and Inter University Centre for Nanoscience and Nanotechnology, Mahatma Gandhi University, Kottayam, Kerala, India*

Prashant Sahu, *Department of Pharmaceutical Sciences, Dr. Hari Singh Gour University, Sagar [MP], India*

Prerna Arora, *Department of Biotechnology, Thapar University, Patiala, Punjab, India*

Raghvendra K. Mishra, *International and Inter University Centre for Nanoscience and Nanotechnology, M G University, Kottayam, Kerala, India*

Rakesh G. Kashyap, *Department of Nanotechnology, Visvesvaraya Institute of Advanced Technology, Muddenahalli, Chikkaballapur, 562101, India*

Robin Augustine, *School of Nano Science and Technology, National Institute of Technology Calicut, Kozhikode, Kerala 673601, India*

S. M. Mehdi Hadavi, *Professor, President of Materials and Biomaterials Research Center (MBMRC) and Iran National Institute of Materials and Energy (MERC), Tehran, Iran*

Sabu Thomas, *Professor, Polymer Science & Engineering, School of Chemical Sciences and Hon. Director of International and Inter University Centre for Nanoscience and Nanotechnology (IIUCNN), Mahatma Gandhi University (MGU), Kottayam, Kerala, India*

Samrat K., *Department of Biotechnology, M. S. Ramaiah Institute of Technology, Bangalore, 560054, India*

Samaresh Sau, *Use-inspired Biomaterials & Integrated Nano Delivery (U-BiND) Systems Laboratory, Department of Pharmaceutical Sciences, Wayne State University, Detroit, Michigan, USA*

Sharath R., *Department of Biotechnology, M. S. Ramaiah Institute of Technology, Bangalore, 560054, India*

Soumya Soman, *Laboratory of Ecology and Eco-technology, School of Biosciences, Mahatma Gandhi University, Kottayam, Kerala, India*
Sreeraj Gopi, *R&D Centre, Aurea Biolabs Pvt. Ltd, Kolenchery, Cochin, India*

Sunija Sukumaran, *International and Inter University Centre for Nanoscience and Nanotechnology, Mahatma Gandhi University, Kottayam, Kerala, India*

Sunita Negi, *Amity University, Manesar, Haryana, India*

Surendra Ekkuluri, *Department of Pharmaceutics, JSS College of Pharmacy, Ootacamund, Jagadguru Sri Shivarathreeswara University, Mysuru, India*

Sushil K. Kashaw, *Department of Pharmaceutical Sciences, Dr. Hari Singh Gour University, Sagar [MP], India*

Susmitha Kasina, *Department of Pharmaceutics, JSS College of Pharmacy, JSS University, Mysore, Karnataka, India*

T. Jesse Joel, *Department of Biotechnology, School of Agriculture and Biosciences, Karunya Institute of Technology and Sciences, Tamil Nadu, India*

V. K. Yadunath, *International and Inter University Centre for Nanoscience and Nanotechnology, Mahatma Gandhi University, Kottayam, Kerala, India*

Veera Venkata Satyanarayana Reddy Karri, *Department of Pharmaceutics, JSS College of Pharmacy, Ootacamund, Jagadguru Sri Shivarathreeswara University, Mysuru, India*

Yves Grohens, *Université de Bretagne Sud, Laboratoire Ingénierie des Matériaux de Bretagne, BP 92116, 56321 Lorient Cedex, France*

List of Figures

Figure 1.1   Indicates the recent drug delivery system.   3

Figure 1.2   **i.** Indicates the application of electro spun membranes, **ii.** Schematic representation of electro spinning unit for the preparation of electro spun membranes, **iii.** Scanning electron microscopy image of spun membrane, **iv.** Wound healing mechanism of electro spun membranes: **(a).** In open wound area bacteria will colonize and hinder the wound healing process, **(b).** Bacterial entry is prevented and cells guided towards the center of the wound in presence of electro spun membrane, **(c).** Completion of wound healing process without any complications.   5

Figure 1.3   Shows the X-ray scanning of hand.   6

Figure 1.4   Pie chart on nanomedicine market by region in 2016.   10

Figure 2.1   Various types of pharmaceutical nanomaterials.   . .   20

Figure 2.2   Application of nanotechnology in the clinical field.   21

Figure 2.3   Example of various types of nanomaterials.   24

Figure 2.4   Future development of Nanomedicine.   43

Figure 3.1   Structure of chitin.   59

Figure 3.2   Structure of chitosan.   60

Figure 3.3   Procedure for extraction of chitosan by chemical method.   68

Figure 3.4   Procedure for extraction of chitosan by biological method.   74

Figure 3.5   Wound healing mechanisms of chitosan.   84

Figure 4.1   IR Spectra of olanzapine and glyceryl tripalmitate. .   105

Figure 4.2   Calibration curve of Olanzapine in 0.1 N HCl
(λ_{max} = 258 nm).   106

Figure 4.3   Calibration curve of Olanzapine in pH 6.8
(λ_{max} = 258 nm).   106

Figure 4.4   Particle size distribution for NF-2 formulation. . . .   109

Figure 4.5   Zeta potential of NF-2 formulation.   109

Figure 4.6   Differential scanning calorimeter curves of OL and
OL-loaded SLN formulations. OL indicates
Olanzapine; GTP – Glyceryl Tripalmitate.   110

Figure 4.7   Comparative *in vitro* release profile in
0.1 N HCl.   111

Figure 4.8   Comparative *in vitro* release profile in phosphate
buffer 6.8.   112

Figure 4.9   Higuchi's plot for GTP-SLN.   113

Figure 4.10   Korsmeyer Peppa's plot for GTP-SLN.   113

Figure 4.11   Bio-analytical calibration curve of Olanzapine. . .   114

Figure 4.12   Chromatogram of Blank plasma.   114

Figure 4.13   Chromatograph of pure drug and IS in plasma. . . .   115

Figure 4.14   Chromatogram of Plasma and formulation.   115

Figure 4.15   Concentration time profile after oral administration
of Glyceryl tripalmitate solid Lipid nanoparticles
(GTP-SLN) and Olanzapine suspension
(OL-SUSP).   117

Figure 5.1   A simple nanogel system.   122

Figure 5.2   Properties of nanogel.   123

Figure 5.3   Classification of nanogel.   125

Figure 5.4   Different characterization parameters
of nanogel.   134

Figure 5.5   Different routes of administration of nanogel. . . .   137

Figure 5.6   Applications of nanogel.   142

Figure 6.1   The Slater–Pauling curve showing the mean atomic
moment for binary alloys of transition metals
as a function of their composition.   169

Figure 7.1   Mechanism of biosynthesis of nanoparticles.   204

Figure 7.2   Process of biosynthesis of metallic
nanoparticles.   205

Figure 7.3   XRD pattern of nZVI synthesized by (A) chemical
and (B) biological methods.   208

Figure 7.4   FT-IR Spectrum of nZVI synthesized
by (A) chemical and (B) biological
methods. 209

Figure 7.5   Scanning electron micrograph (A) magnification
of 50.00 K X of nZVI particles synthesized
by chemical method and (B) magnification
of 50.00 K X of nZVI synthesized
by biological method.   210

Figure 7.6   Screening of antibacterial activity of synthesized
nZVI against (A) Staphylococcus *aureus,*
(B) *Bacillus subtilis,* (C) *Escherichia coli,*
(D) *Klebsiella pneumonia,* (E) *Salmonella
paratyphi, and* (F) Pseudomonas aeruginosa –
(1) ciprofloxacin, (2) DMSO, (3) biological
method, and (4) chemical method. 212

Figure 7.7   Graph of % viability of cancer cells treated
with (A) chemically synthesized nZVI
and (B) biologically synthesized nZVI.   213

Figure 8.1   A nanomotor consisting of an inner CNT (shaft)
and outer CNT (sleeve). 221

Figure 8.2   Fully loaded calmodulin (CaM) protein with four
calcium ions; N-lobe shown in green, the linker
in purple, and C-lobe in cyan. 222

Figure 8.3   Root mean square deviation (RMSD) of the N-lope,
C-lobe, linker, and protein at a lower pH of 5.0. . . . 222

Figure 8.4   Normalized distance distribution (in Å) measured
between residues 34 and 110, observed at the end
of 100 ns in (a) ionic strength corresponding
to a physiological value (b) ionic strength
equal to 150 mM. 223

Figure 8.5   RMSD of the protein as a function of time at
different temperatures. A significant deviation
can be observed at 500 K as compared to 300 K
implying a much greater flexibility at 500 K. 224

Figure 9.1   Calibration curve of RH in phosphate
buffer pH 4.0. 234

Figure 9.2   Calibration curve of RH in phosphate
buffer pH 7.4. 235

Figure 9.3   DSC thermogram for formulation F3
and excipients. 235

Figure 9.4   DSC thermogram for formulation F4
and excipients. 236

Figure 9.5   FT-IR spectrum of physical mixture for F3
formulation. 236

Figure 9.6   FT-IR spectrum of physical mixture for F4
formulation. 237

Figure 9.7   FT-IR spectrum of formulation F3. 237

Figure 9.8   FT-IR spectrum of formulation F4. 238

Figure 9.9   Particle size distribution of formulation F3. 239

Figure 9.10   Zeta potential of formulation F3. 240

Figure 9.11   Particle size distribution of formulation F4. 241

Figure 9.12   Zeta potential of formulation F4. 242

Figure 9.13   Scanning electron micrograph of purified CNTs. . 243

Figure 9.14   Scanning electron micrograph
of drug-loaded CNTs. 243

Figure 9.15   *In vitro* drug-release profile. 244

Figure 9.16   Cytotoxicity studies of formulation F3. 244

Figure 9.17   Cytotoxicity studies of formulation F4. 245

Figure 9.18   Cytotoxicity studies of pure drug. 245

Figure 9.19   Cell control MCF-7. 245

Figure 9.20   100% cytotoxic culture at highest concentration
250 μg/ml. 246

Figure 9.21   100% cell protection at lowest concentration
(15.625 μg/ml). 246

Figure 10.1   Biological synthesis and applications of metal NPs
in different fields. 251

Figure 10.2   Different parameters which are involved
in the production of homogenous NPs. 257

Figure 10.3   Steps involved in nanoparticle synthesis. 261

Figure 10.4   TEM images of *E. coli* treated with AgNPs. 269

Figure 11.1   Chemical structures of important constituents
present in turmeric. 292

Figure 11.2   Schematic representation of the PNS technology
design of Cureit^{TM}. 296

Figure 11.3   Q-TOF of (a) curcuminoid and (b) noncurcuminoid
fraction present in Cureit^{TM}. 298

Figure 11.4 ^{1}H NMR spectra of (a) curcuminoid
and (b) Cureit^{TM}. 299

Figure 11.5 ^{1}H NMR (a) and ^{13}C NMR (b) spectra
of curcuminoid in DMSO-d$_6$. 300

Figure 11.6 ^{1}H NMR (a) and ^{13}C NMR (b) spectra of Cureit^{TM}
in DMSO-d$_6$. 301

Figure 11.7 FT-IR spectra of (a) curcumin, (b) demethoxycurcumin (DMC), (c) bisdemethoxy-curcumin (BDMC),
(d) curcuminoid, and (e) Cureit^{TM} 302

Figure 11.8 XRD pattern of (a) curcuminoid
and (b) Cureit^{TM}. 302

Figure 11.9 TGA curve of (a) curcuminoid
and (b) Cureit^{TM}. 303

Figure 11.10 SEM images of Cureit^{TM} with different cross
sections (a) single granule, (b) cross section
of semilayer form, and (c) three layers
of PNS form. 304

Figure 11.11 Current–voltage study of (a) curcuminoid
and Cureit^{TM} [insert: equivalent current]
(b) conductivity and percentage of increase
in conductivity of curcuminoid and Cureit^{TM}
[insert: SEM micrograph of Cureit^{TM}]. 304

Figure 11.12 Mean curcumin concentration in plasma after
treatment with Cureit^{TM} and standard curcumin
at different time intervals. 306

Figure 13.1 Process of angiogenesis. 345

Figure 13.2 Hypoxia-induced angiogenesis. Hypoxia results
in the expression of VEGF and this VEGF
helps to stimulate angiogenesis. 349

Figure 13.3 Role of reactive oxygen species
in angiogenesis. 353

Figure 13.4 Mechanism of angiogenesis in PCL scaffolds
containing ZnO nanoparticles. 354

Figure 13.5 Metal nanoparticle induced angiogenesis.
Anti-angiogenic nanoparticle inhibits
the angiogenesis, and proangiogenic
nanoparticle promotes the angiogenesis. 355

Figure 13.6 Triad of tissue vascularization. 356

Figure 13.7   Stimulation of angiogenesis by using ZnO
nanoparticle-incorporated PCL scaffold.
(a) Initiation of sprouting of blood vessels through
the scaffold, (b) blood vessel formation through
the scaffold having different concentration of ZnO
nanoparticle, (c) matured blood vessel through
the scaffold after subcutaneous implantation. 357
Figure 14.1   The different types of engineered nanoparticles. . . 369
Figure 14.2   Important recorded toxic effects of therapeutically
used nanoparticles. 372
Figure 14.3   The effects of nanoparticles into the organisms. . . 373

List of Tables

Table 1.1   Nanomaterials used in clinical purposes   8

Table 1.2   Important patents in the field of nanomedicine
and nanotechnology   9

Table 2.1   The development of Nanotechnology   29

Table 2.2   Nanoparticles-based delivery systems with their
therapeutic and diagnostic uses in cancer therapy . .   31

Table 2.3   Nanodrugs approved and there indications   33

Table 3.1   Chitosan derivatives and their applications   80

Table 4.1   Solubility profile of olanzapine in different media . .   104

Table 4.2   Functional groups present in the IR spectrum   104

Table 4.3   Composition of drug-loaded batches   107

Table 4.4   Influence of stirring time on particle size   107

Table 4.5   Influence of stirring speed on particle size   107

Table 4.6   Influence of Stearyl amine on zeta potential   108

Table 4.7   Entrapment efficiency for different batches   109

Table 4.8   Pharmacokinetic parameters of olanzapine after oral
administration (mean $\pm$ S.D.) (*$p < 0.05$)   116

Table 5.1   Applications of nanogel as drug delivery system . .   143

Table 6.1   The types of magnetism seen in materials:
Blue arrows signify the direction of the applied
field and blue arrows in the black circle signify
the direction of the electron spin   162

Table 6.2   Saturation magnetization of samples synthesized
using different methods   167

Table 6.3   Biomedical application of different nanoparticles . .   184

Table 7.1   Zone of inhibition (mm) of nZVI synthesized
by chemical and biological methods   211

Table 9.1   Formulation of RH-CNTs-FA   232

Table 9.2   Zeta potential, particle size, and PDI   238

Table 9.3   Drug loading efficiency   243

Table 10.1 Lists summary of a few case studies done in the past
decade toward the antimicrobial activity
of phytosynthesized AgNPs 264

Table 10.2 List of clinical trials of AgNPs 271

Table 11.1 The average pharmacokinetic variables from plasma
curcumin levels after administration of the Cureit^{TM}
and standard curcumin at different time intervals . . 307

Table 13.1 Different types of angiogenic growth factors and their
role in angiogenesis 346

List of Abbreviations

&	And
16S	16 Svedberg Unit
Ab	Antibiotics
ACE	Angiotensin-Converting Enzyme
ACN	Acetonitrile
ADA	American Dental Association
AEP	Acquired Enamel Pellicle
AFM	Atomic force microscopy
Ag NPs	Silver Nanoparticles
$AgNO_3$	Silver nitrate
AgNPs	Silver Nanoparticles
AIDS	Acquired Immuno deficiency virus
APCs	Antigen presenting cell
AT	Analytical Tools
ATP	Adenosine triphosphate
ATRP	Atom transfer radical polymerization
Au NPs	Gold nanoparticles
AUC	Area under the curve
AUMC	Area Under mean curve
AuNPs	Gold nanoparticles
BBB	Blood Brain Barriers
BHI	Brain-Heart Infusion
BMMSCs	Bone Marrow Mesenchymal Stem Cells
BMNPs	Biomagnetic Nanoparticles
BMPs	Bone morphogenetic protein
BOP	Bleeding on Probing
Ca	Calcium
CBN	Cobalt-based Nanoparticles
CD	Cyclodextrins
CDA	Chitin deacetylase
CDC	Centre for Disease Control and Prevention

CDD	Controlled Drug Delivery
CdSe	Cadmium selenium
Ch-HCl	Chitosan Hydrochloride
CHP	Cholesterol bearing Pullulan
CIS	Cell Immunomagnetic Separation
ClpL	Cytoplasmic Protease
CMC	Carboxy methyl cellulose
CMCS	Carboxymethyl-chitosan
CNS	Central nervous system
CNT	Carbon nanotube
COMSTAT	COMPARE STATistics
CRP	Controlled radical polymerization
CS-MNPs	Core/Shell Magnetic Nanoparticles
CT	Computed Tomography
DA	Degree of acetylation
DDS	Drug Delivery Systems
DLS	Dynamic Light Scattering
DMEM	Dulbecco's Modified Eagle's medium
DMFS	Decay Missing Filled Surface
DMFT	Decay Missing Filled Teeth
DMSO	Dimethyl sulfoxide
DNA	Deoxyribonucleic acid
DSC	Differential scanning calorimetry
DWNTs	Double Walled Nanotubes
EC	Endothelial Cells
ECC	Early Childhood Caries
ECM	Extracellular matrix
EDS	Energy Dispersive Spectroscopy
EGF	Epithelial growth factor
ELISA	Enzyme Linked Immunosorbent Assay
EPCs	Endothelial Progenitor Cells
FA	Folic acid
FBS	Fetal Bovine serum
FESEM	Field emission scanning electron microscopy
FGF	Fibroblast growth factor
FT-IR	Fourier Transform Infra Red microscopy
FT-IR	Fourier Transform Infra Red Spectroscopy
FTIR	Fourier transform infrared spectrometer
FT-IR	Fourier Transform infrared spectroscopy

GFP	Green fluorescent proteins
GIT	Gastrointestinal tract
GMS	Glyceryl monostearate
GNSs	Gold nanoshells
GPC	Gel permeation chromatography
GreA	Transcription elongation factor
GSH	Chitosan–Glutathione
GTF	Glucosyltransferases
GTP	Glyceryl Tripalmitate
H_2O	Water
H_2O_2	Hydrogen peroxide
HDMR	High Density Magnetic Recording
HIF	Hypoxia-Inducible Factor
HIV	Human Immuno deficiency virus
HPLC	High Performance Liquid Chromatography
HR-TEM	High Resolution TEM
HSV	Herpes Simplex Virus
ION	Iron Oxide Nanoparticles
IPM	Isopropyl myristate
IS	Internal standard
K	Potassium
kDa	Kilodalton
LMWH	Low molecular weight heparin
MBMRC	Materials and Biomaterials Research Center
MCC	Mono-carboxymethylated chitosan
MCF-7	Michigan Cancer Foundation-7
MD	Medical Diagnosis
MeHA	Metha-acrylated hyaluronic acid
MEM	Minimum Essential Medium
MEM-PR	MEM without phenol red
MES	Mercaptoethanesulfonate
MFH	Magnetic Fluids Hyperthermia
Mm	Millimetre
MMP-9	Metalloproteinase-9
MNPs	Magnetic Nanoparticles
MOD	Magneto-optics Devices
MOS	Magneto Optical Switches
MRI	Magnetic Resonance Imaging
mRNA	Messenger Ribonucleic acid

MRSA	Methicillin resistant *staphylococcus aureus*
MRT	Mean residence time
MS	Mutans Streptococci
MSA	Mannitol Salt Agar
MSB	Mitis-Salivarius Bacitracin
MWCNTs	Multi Walled Carbon Nanotubes
MWNTs	Multi Walled Nanotubes
Na	Sodium
NADPH	Nicotinamide Adenine Dinucleotide Phosphate
NBCS	New Born Calf Serum
N-CMCh	N Carboxymethylchitosan
NF	Nano formulation
NI	Nanoimaging
NIR	Near Infra red radiation
nm	nanometer
NM/ND	Nanomaterial and Nanodevices
NMR	Nuclear Magnetic Resonance
NPs	Nanoparticles
NSAIDs	Non-steroidal anti inflammatory drugs
NTMC	N-Trimethylchitosan chloride
nZVI	nanoscale zero valent iron
O/W	Oil in water
O_2	Oxygen
ODNs	Oligo deoxy Nucleotide
OI	Optical Imaging
OL	Olanzapine
OL-SOL	Plain olanzapine solution
PAAs	Poly acrylic acids
PB	Phosphate buffer
PBS	Fhosphate-buffered saline
PC	Phosphatidylcholine
PCL	Polycaprolactone
PCR	Polymerase Chain Reaction
PCRs	Polymerase Chain Reactions
PDGF	Platelet-Derived Growth Factors
PDI	Polydispersity Index
PDMS	Polydimethylsiloxane
PEC	Polyelectrolyte
PEG	Poly ethylene glycol

PEI	Polyethylene amine
PEO	Polyethylene oxide
PepD	Dipeptidase
PET	Positron Emission Tomography
PHD2	Prolyl hydroxylase domain-2
PHEMA	Polyhydroxyethylmethacrylate
PI	Polydispersity index
PLGA	Poly lactic-glycolic acid
PMMA	Polymethyl methacrylate
PNIPAm	Poly (N-isopropylacrylamide)
PnpA	Polyribonucleotide Nucleotidyltransferase
PRINT	Particles replication in non-wetting templates
PVA	Polyvinyl alcohol
PVDF	Polyvinylidene fluoride
PVP	Polyvinyl pyrrolidone
PXRD	X-ray powder diffractometer
QD	Quantum dots
RAFT	Reversible addition fragmentation chain transfer
RBC	Red Blood Corpuscles
RGD	Arginine-Glycine-Aspartic acid
RH	Raloxifene hydrochloride
RNA	Ribonucleic acid
RNAi	Ribonucleic acid interference
ROS	Reactive Oxygen Species
RSV	Respiratory syncytial virus
RTI	Regulatory and Toxicological Issues
RTKs	Receptor Tyrosine Kinases
SAED	Selected Area Electron Diffraction
SCC	Squamous cell carcinoma
SCDF	Stromal Cell-Derived Factor
SEM	Scanning electron microscopy
SiRNA	Small (or short) interfering RNA
si-RNA	small interfering Ribonucleic acid
SLN	Solid Lipid Nanoparticles
SOD	Superoxide dismutase
SPD	Sensory Processing Disorder
SPE	Solid phase extraction
SPF	Shirasu porous glass
SPIO	Super Paramagnetic Iron Oxide Particles

Spx	Spexin Hormone
Ssb	Single-Stranded DNA-Binding Protein
SSB	Sugar Sweetened Beverages
SUSP	Suspension
SWCNTs	Single Walled Carbon Nanotubes
SWNTs	Single walled Nanotubes
SWPH	Shrimp waste protein hydrolysate
TDD	Targeted Drug Delivery
TEM	Transmission Electron Microscopy
TGA	Thermo gravimetric Analysis
TGF	Transforming Growth Factors
TGT	Targeted Gene Therapy
TMC	N-trimethyl chitosan
TP	Tripalmitate
TSP	Thrombosporins
USA	United States of America
USDA	United States Department of Agriculture
US-FDA	United States Food and Drug Administration
USPIO	Ultra-Small Super Paramagnetic Iron Oxide Particles
UV	Ultraviolet
VEGF	Vascular Endothelial Growth Factor
W/O	Water in oil
WHO	World health organization
XRD	X-Ray Diffraction
ZNO	Zinc oxide

1

Nanomedicine and Nanotechnology: State of Art, New Challenges and Opportunities

Jince Thomas, Jiji Abraham and Sabu Thomas

International and Inter University Centre for Nanoscience
and Nanotechnology, Mahatma Gandhi University,
Kerala, India

Abstract

Nanomedicine and Nanotechnology is a key science of the 21st century because it finds great role in human health care systems. Nanomedicine has a vital role in various medical fields such as drug delivery, tissue engineering, nano-scaffolding, wound healing, and diagnostics. Introduction of nanomaterials to the medical field leads to various commercial applications. Different nanomaterials in clinical uses and the various patents filed in the area of nanomedicine are discussed in this chapter. The final part of this chapter addresses the toxicity aspects of various nanomaterials.

1.1 Introduction

Nanotechnology is directly pertinent to medicine due to the significance of nanoscale phenomena to the cell cycle and enzyme action, cellular signalling, whereas nanomedicine is merely the utilization of nanotechnology in the field of all living things, healthcare settings. The popular benefits involve the use of nanoparticles to advance the behaviour and delivery of the drug substances [1, 2]. By scrutinizing the structures of tissue in the scale of atomic and the cellular levels using nanotechnology, it is possible to design and create synthetic biomaterials in the nanoscale level as substitutes for damaged organs and therapies. The important benefits of nanotechnology is

1

in biological pathways, that means it is used for the production of precisely battered nanodrugs for the protein and nucleic acid sites allied with disease and syndromes. More over transport of delicate organic macromolecules and peptides to their active sites of action without troubling other parts of the body is possible with the help of nanotechnology. Currently, nanomedicines are widely used to augment the cures and lifetime of patient who is suffering from a range of disorders including kidney disease, fungal infections, elevated cholesterol, menopausal symptoms, multiple sclerosis, ovarian and breast cancer, chronic pain, asthma and emphysema. The key benefits of nanomedicine is confirming the proper entry of sufficient amount of drugs in the body of living systems, stays in the body for long periods and effective functioning on to the zones which need treatment. Since nanomedicine is considered to be the extension of molecular medicine, its size is in the scale of 100 nanometres or less than the biological structures and molecules inside living cells. The prime target of nanomedicine is to employ nanoscale machine systems to address medical problems and other related issues, and hence to retain and improve people's health [3–6].

1.2 Scope

Scientists are forced to improve the current technologies used in medical field to develop new effective curatives because of the introduction new diseases and increasing the number of patients day by day as the effect of some drugs and modern life style. In this occasion, nanomedicine and nanotechnology had great openings in the following fields.

1.2.1 Drug Delivery

Drug releasing is an important area in medical field where nanomedicine has a significant role. Nanoparticle can help doctors to target drugs at the origin of diseases with increased efficacy, minimum toxicity and high bioavailability [4, 7–9]. Nanotechnology based targeted drug transfer systems are already available in the market and a large variety of nanoscale drug delivery systems are now under process. Some examples for commercially available products in the market include Doxil (antineoplastic), Abraxane (metastatic breast cancer), Emend (antiemetic), Abelcet (for fungal infection), etc. Recently, very few of nano formulations have been sanctioned by US-FDA (United States Food and Drug Administration) and will launch on coming years.

Various nanoscale drug delivery systems such as nanoparticles, nanoliposomes [10], dendrimers [11], fullerence, nanopores [12], nanotubes [13], nanoshells [14, 15], quantum dots, nanocapsule, nanosphere, nanovaccines, nanocrystals, green nanoparticulate [16–18] etc. have abilities to revolutionize drug delivery systems in present and future [19, 20]. Other than the above-mentioned systems nanorobotics, nanomaterials on chips, magnetic nanoparticles connected to specific antibody, nanosize empty virus capsids and magnetic immunoassay are found to be innovative materials in drug delivery applications [21–23]. Nanoliposomes are well established and possessed reasonably good number of clinical trials. Even though polymer-based nanomaterial; carbon nanotubes, gold nanoparticles etc. [24, 25] have less importance because of the fewer number of clinical trials, research is going on in this area and in the coming years it will take the major portion in the market. Nanomachines are also in the research-and-development stage, but at the same time some basic molecular tackles have been tested in this area. Nanorobot is the example for nanomachine which is capable of penetrating the many biological barriers of human body to recognize the cancer cells. So nanodrug delivery systems have an important role in nanomedicine in near imminent [15, 26, 27] (Figure 1.1).

1.2.2 Tissue Engineering and Nanoscaffolding and Wound Healing

The new idea of nanotechnology in regenerative medicine provides hope to various patients with organ failure and severe injuries [29–33]. American army researchers have developed a man's fingertip and the internal organs from polymer fibres using "nanoscaffolding" technology. Nanoscaffolds are

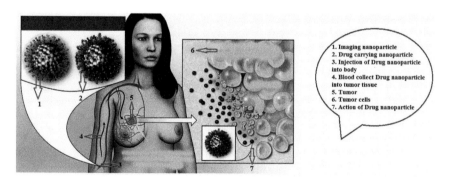

Figure 1.1 Indicates the recent drug delivery system [28].

hundreds times finer than a human hair and can be used as a substitute for missing limb or damaged organ. The nanoscaffold monitors adhesion of cells onto it so that they can start to reconstruct missing bones and tissue. Finally, the scaffold breaks down and is removed from the body naturally. Researchers from the City University of Hong Kong reported the revolutions of nanoscaffoldings in bone grafts and implants. Electrospun membranes emerged as novel materials for skin substitutes and wound healing. This membrane prevents the entry of bacteria to the wound and boost the healing process [34, 35]. At present, research is still persistent for a better alternative for this field. Figure 1.2 shows the function and mechanism of polymer fibres in wound healing and skin substitute's zone [36]. In future, we can hope the introduction of self-healing scaffold systems.

1.2.3 Diagnostics

Diagnosis at nanoscale is found to be developing a branch of nanomedicine. The ultimate aim of this area is to identify a disease at the earliest possible period. More specifically a single cell with ill actions would be identified and cured or eliminated [37–40]. Other main scopes involved in this field are Cancer [41, 42], Dentistry [43, 44], Gene delivery [45], magnetofection [46, 47] etc.

1.3 History

Nanomedicine is an emerging field of science, the prefix "nano" comes from the Greek word and its meaning is "dwarf". Mathematically one nanometer (nm) is equal to one-billionth of a meter, or about the width of six carbon atoms or ten water molecules [48, 49]. The American theoretical physicist Richard Feynman was the pioneer of nanotechnology and mentioned the importance of this field during his lecture in 1959 entitled as There's plenty of room at the bottom" [50]. The nanotechnology exploration could develop advances in various sectors such as communication, engineering, chemistry, physics, robotics, biology, and medicine. Moreover Nanotechnology has been exploited in medicine for therapeutic drug delivery and the advancement of treatments for a range of disorders and diseases.

Nanomedicine was born as a focused effort of applying nanotechnology in various medical challenges. Nanomedicine terminology was coined by Drexler and colleagues in the 1980s and the Tokyo university professor Norio Taniguchi introduced the nanotechnology term in 1974. But the nanoscale

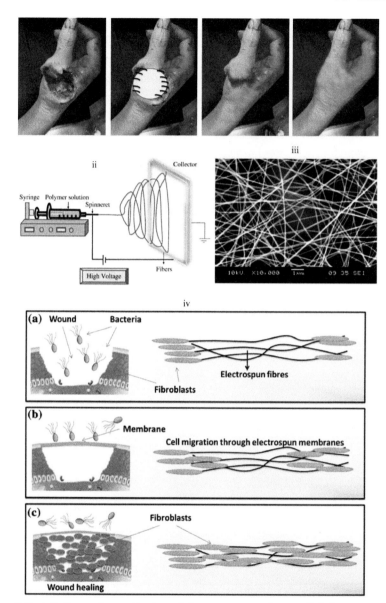

Figure 1.2   i. Indicates the application of electro spun membranes, **ii.** Schematic representation of electro spinning unit for the preparation of electro spun membranes, **iii.** Scanning
electron microscopy image of spun membrane, **iv.** Wound healing mechanism of electro spun
membranes: **(a).** In open wound area bacteria will colonize and hinder the wound healing
process, **(b).** Bacterial entry is prevented and cells guided towards the center of the wound in
presence of electro spun membrane, **(c).** Completion of wound healing process without any
complications [36].

techniques were being used in the medical fields and applied in diagnostics even before the introduction of term nanotechnology [4, 51, 52].

Millennial article of the New England Journal of Medicine reported the past five hundred developments in nanotechnology and nanomedicine, some of the important developments are quoted here,

i. ***Elucidation of Human Anatomy and Physiology and Discovery of Cells and their Substructures:*** It begins from sixteenth century to give exact anatomies, circulation, pulse, blood pressure, electrical nerve stimulus etc. During these centuries invention microscope helps to discover the bacteria and protozoa, complex inner structures of plant and animal tissues, plant cells, animal cells, cell division, nucleus, etc.

ii. ***Development of Biochemistry:*** This includes discovery of oxygen, other gases, development of organic and physical chemistry, enzyme chemistry, pathways of metabolism, Krebs cycle, role of calcium, sodium, potassium, magnesium, etc., chlorophyll, hemoglobin, hormones, neurotransmitters, cell signaling.

iii. ***Medical Imaging and Biomarkers for Diagnosis and Research:*** Introduction of X-rays helps the imaging of bone and hard tissue structure. Photochromic markers and bioluminescent markers are used to find molecular activity within the cells (Figure 1.3).

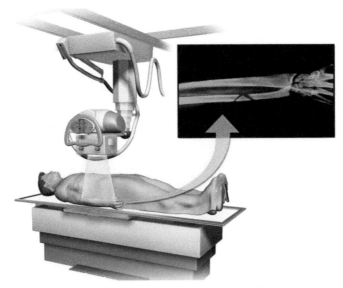

Figure 1.3   Shows the X-ray scanning of hand.

Source: https://en.wikipedia.org/wiki/X-ray

In the opening of 20th era Paul Ehrlich tried to develop "magic bullets" [53] to which drugs were added and this could be used to target diseases and would eliminate all pathogens after a single treatment. Several historical milestones were behind on this nanomedicine and nanotechnology fields and these were thoroughly explained in different books.

1.4 Commercial Significance and Current Challenges

Introduction of nanomaterials enrich the developments in medical fields and also these materials have the numerous commercial applications. Nano-materials consist of metal atoms, non-metal atoms or a mixture of both. Nanomaterials are classified as follows: (a) Nanotubes: Hollow cylinders made of carbon atoms or any other elements (b) Nanowires: Around five times smaller than virus and its uses include early sensing of ovarian malig-nancies and breast cancer (c) Nanoshells: Hollow silica spheres covered with gold and the attachment of antibodies to their surface helps in targeting cancer cells. Another important scope of nanoshells is their ability to carry drug containing polymer for drug delivery system, (d) Quantum dots: Tiny semiconductor particles which can serve as sign pots of certain type of molecules or cells in the body, (e) Nano pores: These have applications in cancer research and treatments, (f) Gold nanoparticles: Development of the ultra-sensitive detection systems for DNA and protein markers associated with many forms of cancer including breast, prostate cancer, (g) Bucky balls: Made of 60 carbon atoms formed in the shape of hollow ball. It can be used for the construction of drug delivery vehicles for cancer ther-apy [54–59]. Table 1.1 displays the list of nanomaterials used in clinical purpose.

United States of America is the leading nation in the field of nanomedicine with one-third of all publications and half of patent filings compared to other nations. Comparison between Europe and the USA shows that Europe is the vanguard of research publications whereas the USA tops in the number of patent filings which indicated the more progressive com-mercialization status of USA. Some of the important patents are revealed in Table 1.2.

The global nanomedicine market size was projected at USD 138.8 billion in 2016 and it is expected to reach USD 350.8 billion by 2025 as per the recent report by Grand View Research, Inc. BCC Research stated that the global market for nanoparticles in the life sciences is estimated at over \$29.6 billion for 2014 and this market is predicted to attain \$79.8 billion by 2019. In the

Table 1.1   Nanomaterials used in clinical purposes

Nanomaterial	Trade Name	Application	Target	Adverse Effects
Protein	Abraxane	Cancer therapy	Breast	Cytopenia
Polymer	Oncaspar	Cancer therapy	Acute lymphoblastic leukemia	Urticaria, rash
	CALAA-01	Cancer therapy	Various forms	Mild renal toxicity
Micelle	Genexol-PM	Cancer therapy	Various forms	Peripheral sensory neuropathy, neutropenia
Dendrimer	VivaGel	Microbicide	Cervicovaginal	Abdominal pain, dysuria
Liposome	Doxil/Caelyx	Cancer therapy	Various forms	Hand–foot syndrome, stomatitis
Quantum dot	Qdots, EviTags, Semiconductor nanocrystals	Fluorescent contrast, *in vitro* diagnostics	Tumors, cells, tissues, and molecular sensing structures	Not applicable
Gold	Verigene	Vitro diagnostics	Genetic	Not applicable
	Aurimmune	Cancer therapy	Various forms	Fever
Nanoshells	Auroshell	Cancer therapy	Head and neck	Under examination
Iron oxide	Feridex	Magnetic resonance imaging contrast	Liver	Vasodilatation, Back pain
	Resovist	Magnetic resonance imaging contrast	Liver	None
	Combidex	Magnetic resonance imaging contrast	Lymph nodes	None
	NanoTherm	Magnetic resonance imaging contrast	Various forms	Acute urinary retention

Table 1.2 Important patents in the field of nanomedicine and nanotechnology

Patent No.	Assignee/Inventors	Filed On	Title
US20100284982 [60]	Yang; Victor; et al.	December 22, 2012	Erythrocyate-encapsulated L-asparaginase For Enhanced Acute Lymphoblastic Leukemia Therapy
US20110151004	Wu; Daging; et al.	January 27, 2011	Injectable Microspheres
US20110028431	Zerbe; Horst; et al.	July 15, 2010	Oral Mucoadhesive Dosage Form
US20120141540	Magnani; Mauro; et al.	June 7, 2010	Drug Delivery System
US20100278920 [61]	University of South Florida, Tampa, FL	April 26, 2010	Polyacrylate Nanoparticle Drug Delivery
US20100209492	SDG, Inc, Cleveland, OH	January 14, 2010	Targeted Liposomal Drug Delivery System
US20090155374 [62]	Sung; Hsing-Wen; et al.	January 15, 2009	Nanoparticle For Protein Drug Delivery
US20110190716 [63]	Easterbrook; Timothy; et al.	June 2, 2009	Transdermal Drug Delivery Device
US20080248114 [64]	Reliant pharmaceuticals INC. Liberty corner, NJ	June 11, 2008	Oral Osmotic Drug Delivery System
US20100100064 [65]	Convatec technologies INC, Reno, NV	March 6, 2008	Ostomy Devices With mucoadhesives
US20090004281 [66]	Biovail Laboratories International S.R.L. St. Michael, BB	June 26, 2007	Multiparticulate Osmotic Delivery System
US20070104777 [67]	Lau; John; et al.	December 21, 2006	Targeted Liposomal Drug Delivery System

coming years, growth of nanotechnology in the area of medical field will be very rapid because of the benefits of nanotechnology-based contrast reagents for diagnosis and monitoring of the effects of drugs with short span of time. Moreover, tissue renovation is anticipated to be an attractive zone because of the demand for biodegradable implants with higher lifetimes. As per the WHO booklet, cancer is one of the main reasons of mortality and illness in worldwide, with approximately 14 million new cases reported in 2012 and 8.2 million peoples lost their life by cancer-related diseases. Therefore, the demand for nanomedicine in order to control such high incidence rate is anticipated to increase market progress during the forecast period.

Nano based medical products are accessible in the market over a decade ago and some of them become top sellers in their therapeutic categories. The main areas of nanomedical products are cancer, central nervous system diseases (CNS), cardiovascular disease, and infection control.

From Figure 1.4 for it is clear that about half of the market of nano-medicine products is with USA which shows the good commercialization of medical products there. US companies produce 45–50% of nanomedicine products, European companies had 35% share and other regions had less contributions. Various biopharmaceutical and medical devices companies are enthusiastically involved in the development of novel products for getting the income and developing partnership between nanomedicine startups top enterprises [68].

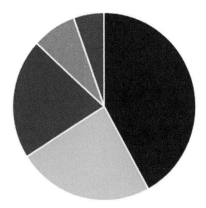

■ North America ■ Europe ■ Asia Pacific ■ Latin America ■ MEA

Figure 1.4   Pie chart on nanomedicine market by region in 2016.

Source: http://www.grandviewresearch.com

In November 2015, Ablynx and Novo Nordisk made a universal collaboration and an authorizing contract for the discovery and improvement of novel drugs with multi-specific nanobodies. This deliberate partnership is projected to raise the net yearly sales of the nanomedicine products and uplifting the market growth. In February 2017, Celgene International Sàrl received approval from the European Commission for their product-REVLIMID. This medicine is meant for the treatment of adult patients with multiple myeloma who have undergone autologous stem cell transplantation. This agreement allows the company to distribute its products in European countries, thereby enhancing company's presence in Europe market. The main players who are operating the nanomedicine and nanotechnology market include Ablynx NV, Merck & Co. Inc., Arrowhead Research, Abraxis Bioscience Inc., GE Healthcare, Combimatrix Corp, Teva Pharmaceutical Industries Ltd, Celgene Corporation, Pfizer Inc., and Nanosphere Inc [69–72].

Cancer is considered as one of the main troubling area in which nano-enabled products have made chief contributions; Nanomaterials used for cancer treatment include Oncospar, Depocyt, Abraxane, Doxil, and Neulasta. Various nanopharmaceutical R&D and companies include Celgene, Access, Camurus, Cytimmune etc. give special emphasis on cancer treatment [73–76]. Nanotherapeutic research is going on to treat the CNS disorders including Alzheimer's and stroke disease. Some products such as Tysabri, Copazone, and Diprivan were already developed. This area is very hungry for successful therapeutic advancements and annual growth from current and innovative products is anticipated to reach 16% over the coming five years [77, 78]. Nanotechnology had contributed novel products such as Humira and Remicade for the autoimmune-related inflammatory disease. Enzon enterprises forcefully pursuing new product development and these novel products are expected to improve the annual growth rate as around 15%. Another fascinating contribution of nanotechnology is in the area of anti-infective products, from PEGylated interferons (used for the viral disease) to nanocrystalline silver (used for wound healing applications). Companies like Biosanté and NanoBio were actively involved in this field.

Current medical nanotech applications is based on single nanoparticles and simple assemblies, future promises will involve combining such single elements into hybrid structures which can solve more complicated tasks than the existing technologies, for example, releasing drug payloads. Subsequently, nanostructures may be industrialized which can insert probes into chosen cells and inject DNA or protein to precise genetic abnormalities. Another chance is the designing of nanostructures which monitor the renewal of nerve cells; these would be used in the treatment of stroke and trauma victims, and feasibly for the renovation of lost function in Alzheimer's disease. However, the expect timing of these changes to become reality until 10–20 years from now [79–81].

Toxicity of nanomaterials is the major challenge involved in this nanomedicine and nanotechnology field. It is identified by *in vitro* cell-culture and *in vivo* animal studies. For example, metabolism of CdSe quantum dots leads to cadmium toxicity, with the negative effects on the function, feasibility, and morphological structures of freshly isolated rat hepatocytes. Carbon nanotubes can cause the asbestos-like inflammation and granulomas in female mice. Capability of heavy metals to enter into the vital organs leads to unique toxic reactions [82–84]. There is no specific evidence or reports for the human toxic response which is specifically originated by nanomaterials. Some findings showed that high dosage of nanoparticles based therapeutic agents directed to reversible kidney toxicity and also magnetic nanoparticles used for thermal ablation found to be retained in the urinary tract [85]. However abundant studies reported that certain nanomaterials do not encourage toxic responses in animals and it is determined by histopathological studies and analyses of hepatic and renal markers [86, 87]. The problem of nanomaterial toxicity remains contentious and wants more specific investigations [88, 89].

References

[1] Whitman, A. G., et al. (2008). "Applications of nanotechnology in the biomedical sciences: small materials, big impacts, and unknown consequences," in *Emerging Conceptual, Ethical and Policy Issues in Bionanotechnology*, F. Jotterand (Berlin: Springer), 117–130.

[2] Sami, A., et al. (2012). Nanotechnology in human health care system: a review. *J. Public Health Biol. Sci.* 1, 121–126.

[3] Taniguchi, N. (1974). "On the Basic Concept of Nano-Technology," in *Proceedings of the International Conference on Production Engineering Tokyo Part II Japan Society of Precision Engineering.*

[4] Krukemeyer, M., et al. (2015). History and Possible Uses of Nano-medicine Based on Nanoparticles and Nanotechnological Progress. *J. Nanomed. Nanotechnol.* **6**:1.

[5] Mayer, C. (2005). Nanocapsules as drug delivery systems. *Int. J. Artif. Org.* 28, 1163–1171.

[6] Jain, K. (2009). Role of nanobiotechnology in the development of personalized medicine. *Nanomedicine* 4, 249–52.

[7] Jain, K. K. (2008). "Future of nanomedicine," in *The Handbook of Nanomedicine*, ed. K. K Jain (Berlin: Springer), 369–376.

[8] Mukherjee, B., et al. (2014). "Current status and future scope for nano-materials in drug delivery," in *Application of Nanotechnology in Drug Delivery*, eds J. Silva, A. R. Fernandes, and P. V. Baptista (Rijeka: InTech).

[9] Emerich, D. F. and Thanos, C. G. (2007). Targeted nanoparticle-based drug delivery and diagnosis. *J. Drug Targeting* 15, 163–183.

[10] Yousefi, A., et al. (2009). Preparation and in vitro evaluation of a pegy-lated nano-liposomal formulation containing docetaxel. *Sci. Pharm.* 77, 453–464.

[11] Dhanikula, R. S., Hammady, T., and Hildgen, P. (2009). On the mecha-nism and dynamics of uptake and permeation of polyether – copolyester dendrimers across an in vitro blood–brain barrier model. *Journal of pharmaceutical sciences* 98, 3748–3760.

[12] Yadav, A., Ghune, M., and Jain, D. K. (2011). Nano-medicine based drug delivery system. *J. Adv. Pharm. Educ. Res.* 1, 201–213.

[13] Reilly, R. M. (2007). Carbon nanotubes: potential benefits and risks of nanotechnology in nuclear medicine. *J. Nuclear Med.* 48, 1039–1042.

[14] Kherlopian, A. R., et al. (2008). A review of imaging techniques for systems biology. *BMC Syst. Biol.* 2:74.

[15] Hirsch, L. R., et al. (2003). Nanoshell-mediated near-infrared thermal therapy of tumors under magnetic resonance guidance. *Proc. Natl. Acad. Sci.* 100, 13549–13554.

[16] Mandal, G., and Ganguly, T. (2011). Applications of nanomaterials in the different fields of photosciences. *Indian J. Phys.* 85, 1229–1245.

[17] Lam, P.-L., et al. (2017). Recent advances in green nanoparticu-late systems for drug delivery: efficient delivery and safety concern. *Nanomedicine* 12, 357–385.

[18] Wilner, O. I., et al. (2011). Self-assembly of DNA nanotubes with controllable diameters. *Nat. Commun.* 2:540.

[19] Nam, J.-M., Thaxton, C. S., and. Mirkin, C. A. (2003). Nanoparticle-based bio-bar codes for the ultrasensitive detection of proteins. *Science* 301, 1884–1886.

[20] Chattopadhyay, P. K., et al. (2006). Quantum dot semiconductor nanocrystals for immunophenotyping by polychromatic flow cytometry. *Nat. Med.* 12, 972–977.

[21] Allen, T. M., and Cullis, P. R. (2004). Drug delivery systems: entering the mainstream. *Science* 303, 1818–1822.

[22] Verma, R. K., and Garg, S. (2001). Drug delivery technologies and future directions. *Pharm. Technol.* 25, 1–14.

[23] Singh, S., and Singh, A. (2013). Current Status of Nanomedicine and Nanosurgery. *Anesth. Essays Res.* 7, 237–242.

[24] Choi, H. S., and Frangioni, J. V. (2010). Nanoparticles for biomedical imaging: fundamentals of clinical translation. *Mol. Imaging* 9, 291–310.

[25] Wang, Y., et al. (2006). Co-delivery of drugs and DNA from cationic core–shell nanoparticles self-assembled from a biodegradable copolymer. *Nat. Mater.* 5, 791–796.

[26] Zhang, L., et al. (2010). Development of nanoparticles for antimicrobial drug delivery. *Curr. Med. Chem.* 17, 585–594.

[27] Sajja, H. K., et al. (2009). Development of multifunctional nanoparticles for targeted drug delivery and noninvasive imaging of therapeutic effect. *Curr. Drug Discov. Technol.* 6, 43–51.

[28] Kim, B., Rutka, J. T., and Chan, W. C. (2010). Nanomedicine. *N. Engl. J. Med.* 363, 2434–2443.

[29] Lanza, R., Langer, R., and Vacanti, J. P. (2011). *Principles of Tissue Engineering*. Cambridge, MA: Academic Press.

[30] Greenberg, S., Margulis, A., and Garlick, J. A. (2005). In vivo transplantation of engineered human skin. *Methods Mol. Biol.* 289, 425–429.

[31] Enis, D. R., et al. (2005). Induction, differentiation, and remodeling of blood vessels after transplantation of Bcl-2-transduced endothelial cells. *Proc. Natl. Acad. Sci. U.S.A.* 102, 425–430.

[32] Augustine, R., et al. (2013). "Biopolymers for health, food, and cosmetic applications," in *Handbook of Biopolymer-Based Materials: From Blends and Composites to Gels and Complex Networks*, eds S. Thomas, D. Durand, C. Chassenieux, and P. Jyotishkumar (Weinheim: Wiley-VCH Verlag GmbH & Co. KGaA), 801–849.

[33] Augustine, R., et al. (2014). Investigation of angiogenesis and its mechanism using zinc oxide nanoparticle-loaded electrospun tissue engineering scaffolds. *RSC Adv.* 4, 51528–51536.

[34] Augustine, R., et al. (2015). Dose-dependent effects of gamma irradiation on the materials properties and cell proliferation of electrospun polycaprolactone tissue engineering scaffolds. *Int. J. Polym. Mater. Polym. Biomater.* 64, 526–533.

[35] Boateng, J. S., et al. (2008). Wound healing dressings and drug delivery systems: a review. *J. Pharm. Sci.* 97, 2892–2923.

[36] Augustine, R., Kalarikkal, N., and Thomas, S. (2014). Advancement of wound care from grafts to bioengineered smart skin substitutes. *Prog. Biomater.* 3, 103–113.

[37] Meetoo, D. and Lappin, M. (2009). Nanotechnology and the future of diabetes management. *J. Diabetes Nurs.* 13, 288–297.

[38] Stiriba, S. E., Frey, H., and Haag, R. (2002). Dendritic polymers in biomedical applications: from potential to clinical use in diagnostics and therapy. *Angew. Chem. Int. Ed.* 41, 1329–1334.

[39] Kalloo, A. N., et al. (2004). Flexible transgastric peritoneoscopy: a novel approach to diagnostic and therapeutic interventions in the peritoneal cavity. *Gastrointest. Endosc.* 60, 114–117.

[40] Geeraedts, L., et al. (2009). Exsanguination in trauma: A review of diagnostics and treatment options. *Injury* 40, 11–20.

[41] Patel, S., et al. (2011). Nano delivers big: designing molecular missiles for cancer therapeutics. *Pharmaceutics* 3, 34–52.

[42] Das, M., Mohanty, C., and Sahoo, S. K. (2009). Ligand-based targeted therapy for cancer tissue. *Expert Opin. Drug Deliv.* 6, 285–304.

[43] Bhardwaj, A., et al. (2014). Nanotechnology in dentistry: present and future. *J. Int. Oral Health* 6, 121–126.

[44] Xu, H. (2014). Dental composites comprising nanoparticles of amorphous calcium phosphate. U.S. Patent No. 8, 889, 196. Washington, DC: U.S. Patent and Trademark Office.

[45] Pack, D. W., et al. (2005). Design and development of polymers for gene delivery. *Nat. Rev. Drug Discov.* **4**, 581–593.

[46] Holzbach, T., et al. (2010). Non-viral VEGF165 gene therapy–magnetofection of acoustically active magnetic lipospheres ('magneto-bubbles') increases tissue survival in an oversized skin flap model. *J. Cell. Mol. Med.* **14**, 587–599.

[47] Schillinger, U., et al. (2005). Advances in magnetofection—magnetically guided nucleic acid delivery. *J. Magn. Magnet. Mater.* **293**, 501–508.

[48] Whitesides, G. M. (2003). The 'right' size in nanobiotechnology. *Nat. Biotechnol.* **21**, 1161–1165.

[49] Whitesides, G. M. (2005). Nanoscience, nanotechnology, and chemistry. *Small* **1**, 172–179.

[50] Feynman, R. P. (1960). There's plenty of room at the bottom. *Eng. Sci.* **23**, 22–36.

[51] Tibbals, H. F. (2010). *Medical Nanotechnology and Nanomedicine.* Boca Raton, FL: CRC Press.

[52] Webster, T. J. (2007). IJN's second year is now a part of nanomedicine history! *Int. J. Nanomed.* **2**, 1.

[53] Kreuter, J. (2007). Nanoparticles—a historical perspective. *Int. J. Pharm.* **331**, 1–10.

[54] Singh, M., Manikandan, S., and Kumaraguru, A. (2011). Nanoparticles: a new technology with wide applications. *Res. J. Nanosci. Nanotechnol.* **1**, 1–11.

[55] Jokerst, J. V., and Gambhir, S. S. (2011). Molecular imaging with theranostic nanoparticles. *Account. Chem. Res.* **44**, 1050–1060.

[56] West, J. L. and Halas, N. J. (). Engineered nanomaterials for biophotonics applications: improving sensing, imaging, and therapeutics. *Annu. Rev. Biomed. Eng.* **5**, 285–292.

[57] Prabha, S., et al. (2002). Size-dependency of nanoparticle-mediated gene transfection: studies with fractionated nanoparticles. *Int. J. Pharm.* **244**, 105–115.

[58] Roohani-Esfahani, S.-I., and Zreiqat, H. (2017). Nanoparticles: a promising new therapeutic platform for bone regeneration? *Fut. Med.* 12, 419–422.

[59] Zhao, Y., et al. (2017). Targeted nanoparticle for head and neck cancers: overview and perspectives. Hoboken, NJ: John Wiley & Sons.

[60] Yang, V. C., et al. (2008). Erythrocyte-encapsulated L-asparaginase for enhanced acute lymphoblastic leukemia therapy. Google Patents. US 20100284982 A1.

[61] Turos, E., Cormier, R., and Kyle, D. E. (2010). Polyacrylate nanoparticle drug delivery. Google Patents. WO 2009055650 A2.

[62] Heppe, K., Heppe, A., and Schliebs, R. (2005). Chitosan-based transport system. Google Patents. US 20060051423 A1.

[63] Easterbrook, T. J., Gosden, E., and Meyer, E. (2009). Transdermal drug delivery device. Google Patents. WO 2014047191 A1.

[64] Patel, H. B. (2008). Oral osmotic drug delivery system. *Int. J. Pharm. Sci. Res.* 7, 2302–2312.

[65] Sambasivam, M. (2008). *Ostomy devices with mucoadhesives.* 2008, Google Patents. US 20100100064 A1.

[66] Nghiem, T., and Jackson, G. (2007). Multiparticulate osmotic delivery system. 2007, Google Patents. US 20090004281 A1.

[67] Lau, J. R., Geho, W. B., and Snedeker, G. H. (2007). Targeted liposomal drug delivery system. U.S. Patent No. 8303983 B2, Washington, DC: U.S. Patent and Trademark Office.

[68] Würmseher, M. and Firmin, L. (2017). Nanobiotech in big pharma: a business perspective. *Nanomedicine.* 12, 535–543.

[69] Rösslein, M., Liptrott, N. J., Owen, A., Boisseau, P., Wick, P., and Herrmann, I. K. (2017). Sound understanding of environmental, health and safety, clinical, and market aspects is imperative to clinical translation of nanomedicines. *Nanotoxicology*, 11, 147–149.

[70] Agrahari, V., and Hiremath, P. (2017). Challenges associated and approaches for successful translation of nanomedicines into commercial products. *Nanomedicines* 12, 819–823.

[71] Webster, T. J. (2016). *15 Years of Commercializing Nanomedicine into Real Medical Products.* Available at: http://dc.engconfintl.org/nanotech_med/32016

[72] Hobson, D. W. (2016). "The commercialization of medical nanotechnology for medical applications," in *Intracellular Delivery III*, eds A. Prokop, and V. Weissig (Berlin: Springer), 405–449.

[73] Wang, A. Z., Langer, R., and Farokhzad, O. C. (2012). Nanoparticle delivery of cancer drugs. *Annu. Rev. Med.* 63, 185–198.

[74] Kantarjian, H. M., Fojo, T., Mathisen, M., Zwelling, L. A. (2013). Cancer drugs in the United States: justum pretium—the just price. *J. Clin. Oncol.* 31, 3600–3604.

[75] Peters, W. P. and Rogers, M. C. (1994). Variation in approval by insurance companies of coverage for autologous bone marrow transplantation for breast cancer. *N. Engl. J. Med.* 330, 473–477.

[76] Raaschou-Nielsen, O., et al. (2003). Cancer risk among workers at Danish companies using trichloroethylene: a cohort study. *Am. J. Epidemiol.* 158, 1182–1192.

[77] Kola, I. and Landis, J. (2004). Can the pharmaceutical industry reduce attrition rates? *Nat. Rev. Drug Discov.* 3, 711–716.

[78] Trouiller, P., et al. (2002). Drug development for neglected diseases: a deficient market and a public-health policy failure. *Lancet* 359, 2188–2194.

[79] Shoseyov, O. and Levy, I. (2008). *NanoBioTechnology: Bioinspired Devices and Materials of the Future.* Berlin: Springer.

[80] Etheridge, M. L., et al. (2013). The big picture on nanomedicine: the state of investigational and approved nanomedicine products. *Nanomedicine* 9, 1–14.

[81] Halappanavar, S., et al. (2017) Promise and peril in nanomedicine: the challenges and needs for integrated systems biology approaches to define health risk. *Wiley Interdiscip. Rev. Nanomed. Nanobiotechnol.* doi: 10.1002/wnan.1465 [Epub ahead of print].

[82] Derfus, A. M., Chan, W. C., and Bhatia, S. N. (2004). Probing the cytotoxicity of semiconductor quantum dots. *Nano Lett.* 4, 11–18.

[83] Poland, C. A., et al. (2008). Carbon nanotubes introduced into the abdominal cavity of mice show asbestos-like pathogenicity in a pilot study. *Nat. Nanotechnol.* 3, 423–428.

[84] Heidel, J. D., et al. (2007). Administration in non-human primates of escalating intravenous doses of targeted nanoparticles containing ribonucleotide reductase subunit M2 siRNA. *Proc. Natl. Acad. Sci. U.S.A* 104, 5715–5721.

[85] Johannsen, M., et al. (2007). Morbidity and quality of life during thermotherapy using magnetic nanoparticles in locally recurrent prostate cancer: results of a prospective phase I trial. *Int. J. Hyperthermia* 23, 315–323.

[86] Schipper, M. L., et al. (2008). A pilot toxicology study of single-walled carbon nanotubes in a small sample of mice. *Nat. Nanotechnol.* 3, 216–221.

[87] Hauck, T. S., et al. (2010). In vivo quantum-dot toxicity assessment. *Small* 6, 138–144.

[88] Mirshafiee, V., et al. (2017). Facilitating translational nanomedicine via predictive safety assessment. *Mol. Ther.* 25, 1522–1530.

[89] Chan, W. C. (2017). Nanomedicine 2.0. *Acc. Chem. Res.* 50, 627–632.

2

Novel Approaches to Nanomedicine and Nanotechnology

Apparao Gudimalla[1], Raghvendra K. Mishra[1] and Prerna Arora[2]

[1]International and Inter University Centre for Nanoscience and Nanotechnology, M G University, Kottayam, Kerala, India
[2]Department of Biotechnology, Thapar University, Patiala, Punjab, India

Abstract

Nanotechnology today has fetched importance in almost all the domains of science as it involves the materials manoeuvred at atomic or molecular level. The nanomaterials behave entirely different in physicochemical and biological properties when compared to bulk counterparts. The production of nanomaterials was initiated in the ancient era, but a modern interdisciplinary branch of nanomedicine was established in the late nineties. There is rapid growth in the area of nanomedicine and has progressed tremendously. The advancement opened novel treatment strategies for curing degenerative disorders such as cancer, diabetes, and nanoparticle-mediated targeted drug delivery system. The potential applications of various nano-sized materials such as carbon nanotubes, dendrimers, nanocrystals, nanoshells, nanowires, quantum dots etc include robust drug delivery systems and encapsulating drugs. The progress, challenges, and opportunities in cancer nanomedicine and diabetes along with novel engineering approaches such as nanorobots, liposomes are main highlights of this article.

2.1 Introduction

Nanotechnology is the art of extremely tiny structures. The title "nano" is a Greek word that suggests "dwarf", it signifies "one billionth". And it's the treatment of specific compounds, atoms, or molecules into structures to fabricate materials and devices with unique properties. Nanotechnology includes from the top down, i.e., decreasing the size of large structures to smallest structure [1]. It trades with the material size of 0.1–100 nm.

Nanotechnology has established itself as a central feature intended to enhance product performance in many applications, including food technology, cosmetics, surfaces, new materials, and medical products. The term "nanomedicine" it is broadly used to refer to the collective medical applications emerging for nanotechnology. It is currently employed as a tool to explore the darkest avenues of medical sciences in several ways like drug delivery, sensing, imaging, gene delivery, implant technology, tissue engineering, and artificial implants [2–4]. The new ages of drugs are nanoparticles of ceramics, metals, polymers, which can combat conditions like cancer and fight human pathogenic such as microorganisms, bacteria, and fungus [5–7]. Up to date, around 250+ nanomedicine products are either approved for human use or still in human testing seeking such regulatory approvals [8]. Nanotechnology also offers potential applications of pharmaceuticals [9], medical imaging and diagnosis [10], cancer treatment [11], implantable materials [12], tissue regeneration [13], and even multifunctional platforms combining several of these modes of action into packages a fraction the size of a cell [14]. Figure 2.1 show the various types of pharmaceutical nanomaterials.

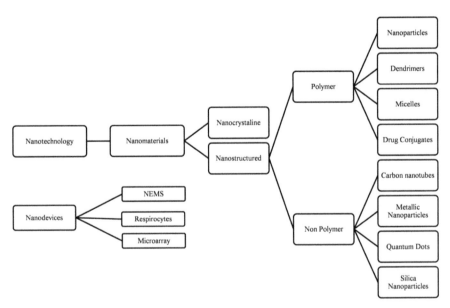

Figure 2.1 Various types of pharmaceutical nanomaterials [15].

Modality		Potential Applications
Cantilevers		• High-throughput screening • Disease protein biomarker detection • DNA mutation detection (SNPs) • Gene expression detection
Carbon Nanotubes		• DNA mutation detection • Disease protein biomarker detection
Dendrimers		• Image contrast agents
Nanocrystals		• Improved formulation for poorly soluble drugs
Nanoparticles		• Targeted drug delivery, permeation enhancers • MRI and ultrasound image contrast agents • Reporters of apoptosis, angiogenesis, etc.
Nanoshells		• Tumor-specific imaging
Nanowires		• High-throughput screening • Disease protein biomarker detection • DNA mutation detection (SNPs) • Gene expression detection
Quantum Dots		• Optical detection of genes and proteins in animal models and cell assays • Tumor and lymph node visualization

Figure 2.2 Application of nanotechnology in the clinical field.

Wagner et al. summarized the funding of 2005 educational commissioned by the European Science and Technology Observatory (ETSO) [16], including a list of approved products and data on rising applications and the companies involved, but with more importance on the economic potential than on trends in the technology. A number of articles have analyzed specific sectors of nanomedicine, including liposomes [17], nanoparticles for drug delivery [18], emulsions [19], biomaterials [20], and *in vitro* diagnostics [21], but such focused discussions don't provide insight into the overall trajectory of nanomaterials in medicine. The application of nanotechnology in various clinical applications is mentioned in Figure 2.2 [22].

2.2 Nanomedicine

Nanomedicine is the application of nanotechnology to accomplish innovation in healthcare. It uses to developing the properties by nanoscale materials at the range is 10^{-9} m. The nanometric size is also the scale of many biomechanisms in the human body allowing nanoparticles and materials to potentially cross-natural barriers to access new sites of delivery and interact with DNA, RNA or small proteins at different levels, in blood or within the tissues, cells or organs. Nanomedicine ranges from the medical applications of nanomaterials and biological devices to biosensors, and even possible future applications of molecular nanotechnology such as biological mechanics. Nanomaterials can be used both *in vivo* and *in vitro* biomedical research and applications. The addition of nanomaterials with biology has led to the improvement of diagnostic devices, analytical tools, contrast agents, drug delivery vehicles, and physical therapy applications. Today, nanomedicine are used globally to improve the treatments and lives of patients suffering from a range of disorders including ovarian and breast cancer, fungal infections, kidney disease, menopausal symptoms, elevated cholesterol, asthma, chronic pain, emphysema, and multiple sclerosis. However, nanomedicine has under more than 70 clinical trial products, covering all most important diseases including cardiovascular, inflammatory, musculoskeletal, and neurodegenerative. Enabling technologies in all healthcare areas, nanomedicine is already accounting for 77 marked products, ranging from nano-delivery and pharmaceutical to imaging, diagnostics, and biomaterial. Investigating the tumor sites and hyperthermia treatment of cancer have been accomplished through magnetic resonance and fluorescence imaging and doxorubicin, Fe_3O_4 magnetic nanocrystals uniformly decorated on the surface dye-doped mesoporous silica nanoparticles have been employed to enhance the visibility of internal body structures in magnetic resonance imaging due to potential biomedical applications [23].

Furthermore, protein-capped multifunctional nanoplatforms (histidine-tagged cyan fluorescent protein-capped magnetic mesoporous silica nanoparticles) have been used as a highly biocompatible system for intracellular drug delivery and fluorescent imaging, where fluorescent imaging agent and capping agent was cyan fluorescent protein [24].

A solid tumor is an organ made of cancer and host cells enclosed in an extracellular matrix and supported by blood vessels. The quantum dots (fluorescent semiconductor nanocrystals) can be used for tumor analysis and cellular imaging, the quantum dots are nanomaterials, which having size range 1–10 nm. [25, 26].

Zinc sulfide–capped cadmium selenide quantum dots can be used for ultrasensitive biological detection because these types of quantum dots covalently coupled to biomolecules. These types of fluorescent probes are 20 times brighter, more stable against photobleaching than organic dyes such as rhodamine, and they are water soluble and biocompatible [27].

Encapsulated individual nanocrystals in phospholipid block–copolymer micelles can be used for imaging both *in vitro* and *in vivo* [28]. Quantum dots are small-sized (1–10 nm) semiconductor nanocrystals composed of the inorganic elemental core (e.g., Cd and Se) surrounded by a metallic shell (ZnS). Inorganic elemental core (e.g., Cd and Se) surrounded by a metallic shell (ZnS) nanomaterials have been widely used as drug carriers or simply as a fluorescent probe for other drug carriers [29].

Surface Plasmons characteristics of gold, copper, and silver made them well suitable for size-dependent light absorption. The gold nanoparticles can be used for a favorable carrier for delivery of drugs due to its biocompatibility and non-cytotoxicity, and the functionalized such as amino acid, proteins gold nanoparticles are used as drug delivery as biomarkers of the drug resistance cancer cell. The anti-cancer chemotherapy drug (Paclitaxel)-conjugated Fe_3O_4 and gold core nanoparticles have been used as anticancer drugs [30, 31].

Nowadays, polymeric nanoparticles are another option for promising drug carrier, gelatin, albumin, polylactides, polyalkylcyanoacrylates, poly-caprolactone. The nanoparticles are used to design as a drug carrier for supplying active molecules. The main aim of this carrier is to supply the active compound to the target, and adsorption capacity of the drug depends on polymer hydrophobicity, nanoparticle area, and monomer concentration. Polylactide–polyglycolide copolymers, polyacrylates and polycaprolactones, albumin, gelatin, alginate, collagen and chitosan, Polylactides and poly (DL-lactide-co-glycolide) polymers can be used for drug carrier; drugs are loaded within the nanoparticle polymer network. However, biodegradability and drug realizing ability are the key parameters, which have to be optimized and decide the proper selection of polymer [32–35].

For targeted and selective drugs delivery, various nanomaterials such as metal, organic, polymeric, liposomes have been used; schematic images of these nanomaterials are mentioned in Figure 2.3.

The aim of nanomedicine may be broadly defined as the comprehensive monitoring, repair, construction, control, resistance and improvement of all the human biological systems, working from the molecular level, using engineered devices and nanostructures, finally to achieve medical benefits. In this context, nanoscale must be taken to include active components or objects in the size range from 1–100 nm. These may be included in a microdevice or in a biological environment. The focus, however, is always on nano interactions within the framework of a larger device or directly within a subcellular system [36].

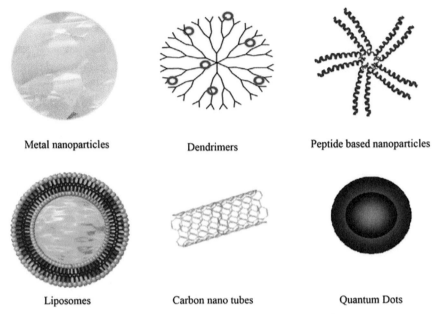

Figure 2.3   Example of various types of nanomaterials [29].

Humans have always tried to improve their health condition and lifestyle. Nowadays, there are several drugs and medical technologies that can treat conditions that only a few decades ago were deadly, like:

1. Nanoparticles that kill cancer cells
2. Targeted cancer therapy
3. ID Tumors noninvasively
4. Nanoparticles for help regenerate bones
5. Fluorescent nanoprobe
6. Implants that minimized the risk of adverse events
7. Ultrasound to penetrate bone
8. Nanoparticles to monitor cancer and other diseases
9. Imaging ID's receptors
10. Technology to improve lung cancer detection
11. Imaging tricks to observe Alzheimer's development
12. New methods to treat tumors with antennas
13. Genomics and truly personalized medicine
14. Body sensors
15. Medical recorders and portable diagnostics

Nanoscience and Nanotechnology have an enormous potential, and a bright future with multiple applications in many areas like engineering, optics, energy, consumer products, nanomedicine (superior diagnostic, therapeutic & preventive measures). Nanomedicine is already a reality that is producing advances in diagnosis, prevention, and treatment of diseases because, among other reasons, to interact with the biomolecules (proteins & nucleic acids). In addition, this capability will enable a better understanding of the complex regulatory and signaling pathways that direct the behavior of normal and transformed cells. Cells within tissue derive mechanical anchorage and particular molecular signals from the unsolvable extracellular matrix that surroundings them. Understanding the role of different cues that extracellular matrices provide is critical for controlling and predicting called responses to scaffolding materials, these complex systems present multiple kinds of cues including mechanical and topographic features, and multiple adhesive ligands on the same molecule [36–37]. Hence understanding these cues are important to design new medical applications and to understand cellular behavior not only for tissue engineering or implants.

Certain fields are particularly interested in nanotechnology, especially:

1. Monitoring
2. Treatment
3. Diagnosis
4. Protection
5. Prevention
6. Tissue repair
7. Evolution control of diseases
8. Applying drugs directly to the cells
9. Improvement of human biological systems

Beyond that, nanomedicine provides important new tools to deal with the grand challenge of an ageing population and is thought to be instrumental for improved and cost effective healthcare, one crucial factor for making medicines and treatments available and affordable to all. Over the coming years, the benefits of nanomedicines and new diagnostic tools will be felt by an increasing number of patients with considerable impact on global health.

2.3 History of Nanomedicine

How nanotechnology could be of use to medicine, pharmacology and medical technology has been researched since the 1990s. After the invention of high-resolution microscopy, it evolved simultaneously in biology, physics, and chemistry in the course of the 20th century and spawned new disciplines such as microelectronics, biochemistry and molecular biology. For nanomedicine, a nanobiotechnology fact which investigates the structure and function of cells as well as Intra- and intercellular processes and cell communication is of prime importance [38].

This research possible at the 20th century when the door to the nano cosmos was burst open with the invention of innovative microscopes. Nanoporous ceramic filters were indeed already being used in the 19th century to separate viruses, and around 1900 Max Planck and Albert Einstein produced theoretical evidence that there must be a range of tiny particles which obeyed their own laws. These particles could not be made visible however the necessary instruments for this had yet to be invented. In 1902 structures smaller than 4 nm were successfully detected in ruby glasses using the ultra-microscope developed by Richard Zsigmondy and Henry Siedentopf [39].

From 1931 onwards significantly better resolutions were achieved with the Transmission Electron Microscope (TEM) developed by Max Knoll and Ernst Ruska than with the light microscopes conventionally used up until then [38]. Insight into the atomic range, however, first became possible with the field electron microscope developed by Erwin Müller in 1936 and its further development to the Field Ion Microscope (FIM), with which in 1951 physicists were able to see individual atoms and their arrangement on a surface [40].

With the aid of further inventions, such as the voltage clamp, understanding of the structure and function of the cell membrane, diffusion processes and systematic cell communication by means of receptors and antibodies according to fixed rules became ever better in the following decades [41]. The mechanisms of maintaining and modifiable metabolism, the position of enzymes and proteins and the functioning of the immune system were also researched and effective vaccines developed. The description and understanding of DNA and RNA in the 1950s and 1960s [41, 42] led to the concept of genetic diseases and to the vision of cures at the molecular level tailor-made for patients. Finally, direct viewing in the nano range became possible at the start of the 1980s with scanning probe microscopy: Gerd Binnig and Heinrich Rohrer developed the Scanning Tunneling Microscope (STM), with which an individual atom was successfully shown graphically in 1981. The first Atomic Force Microscope (AFM) was commissioned in 1986 [38].

Using the different methods of scanning probe microscopy, it became possible not only to reveal nanoscale structures precisely but also to the arrangement and manipulate them in a controlled approach. This opened up diverse promising uses, and new scientific disciplines tailor-made to the nano assortment, including nanomedicine, arose. The term "nanotechnology" was coined in 1974 by Norio Taniguchi, and its definition is still valid even today. Nanotechnology mostly consists of the processing of division, consolidation, and deformation of materials by one atom or one molecule [43]. That there would one day be nanotechnologies and the related possibilities was predicted by the physicist and Nobel prizewinner "Richard P Feynman" as early as 1959 in his paper "There's Plenty of Room at the Bottom". An invitation to come into a new field of physics [44]. And although the term "nano" does not occur a single time in it, this paper is regarded as the founding text of nanotechnology. Feynman invited us to consider the invention and control of tiny machines on the source of quantum mechanics and predicted that the development of more precise microscopes would open up way into the field of individual atoms and that it would be possible to arrange atoms as desired. He even mentioned the use of tiny machines in medicine. It would be fascinating in surgery if you could swallow the surgeon [44]. Keep the mechanical surgeon inside the blood vessel and it goes into the heart and 'looks' around. It finds out which valve is the faulty one and takes a little knife and slices it out. Other small machines might be permanently incorporated in the body to assist some inadequately functioning organ [44]. After Feynman staked out the new field of research and awakened the attention of many scientists, two directions of thought arose describing the different possibilities for producing nanostructures. The top-down approach largely corresponds to Feynman's comments on stepwise reduction in the size of already existing machines and instruments. The bottom-up approach revolves around the construction of nanostructures atom for atom by physical and chemical methods and by using and controlled manipulation of the self-organizing forces of atoms and molecules. This theory of "molecular engineering" became popular in 1986 when Engines of Creation [45]. 'The Coming Era of Nanotechnology' [45] was published, the first and controversially discussed book on nanotechnology in which the author K. Eric Drexler described the construction of complex machines from individual atoms, which can independently manipulate molecules and atoms and thereby produce things and self-replicate. The possible uses of such "nanobots" or "assemblers" in medicine are described by K. Eric Drexler, Chris Peterson and Gayle Pergamit in their book- 'Unbounding the Future'. The nanotechnology revolution [46] published in 1991, in which the term "nanomedicine" was allegedly used for the first time. The term became recognized with the book 'Nanomedicine' [47] by Robert A Freitas published in 1999 and has been used since then in the technical literature. Because the conversion of the visions of Feynman and Drexler of nanoscale robots which patrol the body, render disease foci harmless and detect and repair organs and cells of impaired function is still in the distant future, nanomedicine is concentrated on research into the possibilities of controlling and manipulating cell processes, for example by targeted transport of active substances [38].

At the beginning of the 20th century, Paul Ehrlich attempted to develop "magic bullets" to which drugs were added and which could be used to target diseases and would kill all pathogens after only a single treatment. The Salvarsan he developed is regarded as the first specifically acting therapeutic of this type and marks the start of chemotherapy. The knowledge gained in the course of the 20th century on cells and their constituents and on Intra- and intercellular processes and cell communication, as well as advances in biochemistry and biotechnology, made the production of ever more sophisticated "magic bullets" possible. At the end of the 1960s Peter Paul Speiser developed the first nanoparticles which can be used for targeted drug therapy [48], and in the 1970s Georges Jean Franz Köhler and César Milstein succeeded in producing monoclonal antibodies. Since then there has been intensive research into the possible synthesis and uses of several carrier systems and physicochemical functionalization of their surface structure.

At the beginning of the 1990s nanoparticles were personalized for the first time for transport of DNA fragments and genes and were sluiced into cells with the aid of antibodies [49]. At present biocompatible polymers, liposomes and micelles above all are being researched as carriers for drugs, vaccines, and genes. Because of their small size (<200 nm), nanomaterials are not filtered out of the blood and can travel in the organism until they get in touch with their target. Active substances can be encapsulated in their hollow interiors and their surface can be modified so that they overcome natural barriers such as cell membranes like "Trojan horses", and with the aid of biosensors recognize particular cells and tissue attach themselves to these and release the active substances to the target over a relatively long period of time [50]. These mechanisms are of interest above all for cancer treatment since by the controlled release of the cytostatics exclusively in the tumor tissue the side effects can be reduced and at the same time, higher doses of active substance than hitherto arrive at the tissue affected. Cancer treatment based on targeted transport of active substances can moreover take advantage of the EPR (enhanced permeability and retention) effect described in 1986 by Yasuhiro Matsumura and Hiroshi Maeda: the fact that nanoparticles are

Table 2.1 The development of Nanotechnology [15]

Year	Development of Nanotechnology
1959	R. Feynman initiated thought process
1974	The term nanotechnology was used by Taniguchi for the first time
1981	IBM Scanning Tunneling Microscope
1985	"Bucky Ball"
1986	First book on nanotechnology Engines of Creation published by K. Eric Drexler, Atomic Force Microscope
1989	IBM logo was made with individual atoms
1991	S. Iijima discovered Carbon Nanotube for the first time
1999	1st nanomedicine book by R. Freitas "Nanomedicine" was published
2000	For the first time, National Nanotechnology Initiative was launched
2001	For developing theory of nanometer-scale electronic devices and for synthesis and characterization of carbon nanotubes and nanowires, Feynman Prize in Nanotechnology was awarded
2002	Feynman Prize in Nanotechnology was awarded for using DNA to enable the self-assembly of new structures and for advancing our ability to model molecular machine systems
2003	Feynman Prize in Nanotechnology was awarded for modeling the molecular and electronic structures of new materials and for integrating single molecule biological motors with nano-scale silicon devices
2004	First policy conference on advanced nanotech was held. The first center for nanomechanical systems was established, Feynman Prize in Nanotechnology was awarded for designing stable protein structures and for constructing a novel enzyme with an altered function
2005–2010	3D Nanosystems like robotics, 3D networking, and active nano-products that change their state during use were prepared
2011	Era of molecular nanotechnology started

deposited in tumors to a greater degree than in healthy tissue [38]. The development of nanotechnology is mentioned in Table 2.1.

2.3.1 Present Facing Challenges; Translation of Nanotechnology to Nanomedicine

Publicity, more than used terminology, and inflated prospect for nano-medicine capabilities aside, the practical realities for advancing *in vivo* applications of nanotechnology toward clinical use face a set of technical and scientific challenges [51, 52]. Despite the fascinating science, translating *in vitro* technologies to applications must assert clear performance advantages to the adoption of select nano-component or miniature designs. By contrast, all *in vivo* biomedical applications of nanoparticles require direct absorption, inhalation, and injection into host tissues or circulation. Both require reagent stability, namely resistance to particle aggregation in complex biological media [52]. Both require compatibility assessments as to acceptable toxicity profiles in their context of use: *in vivo* assays with biological components have vastly different requirements than nanomaterials *in vivo*. Possible dose-response and toxicity effects from these new nanomaterials on both isolated biological elements (e.g., enzymes, cells, proteins), tissues, and to human health is a critical prerequisite to translation [53]. Currently, there is no successful predictive test strategy for *in vitro-in vivo* correlations to provide such risk assessments. Each nanomaterial must be evaluated separately under the conditions of use after validating materials purity and particle surface analytics [54]. Full *in vivo* profiling would then follow with Absorption, Distribution, Metabolism and Excretion (ADME) test and physicochemical and characterization, involving both *in vitro* tests and *in vivo* animal studies. Figure shows this conceptual design and performance evaluation. How these methods are performed is critical to understanding both risk and functional assessments of each nanomaterials design. Few nanomaterials to date subscribe to a thought, rigorous analytical protocol across *in vitro* and *in vivo* test bends [55, 56]. The result is a frustrating and persistent gap between *in vitro* result and *in vivo* performance for poorly characterized nanomaterials that stymies ready translation [57, 58].

2.4 Nanoparticles in Cancer Therapy

Nanoparticles used for anticancer drug delivery can be prepared from a diversity of materials, including polymers, dendrimers, liposomes, viruses, carbon nanotubes and metals such as iron oxide and gold [59]. So far almost all the nanoparticles delivery systems which have been approved by the FDA or are currently in clinic trials are based polymers or liposomes [60].

2.4.1 Liposomes

Liposomes are artificial phospholipid vesicles made up of an aqueous interior surrounded by a lipid bilayer. They are biologically inert and completely biocompatible without any toxic reactions [61]. The inner aqueous compartment loaded with water-soluble drugs is used to increase the bioavailability of poorly absorbed drugs like amphotericin B. This has revolutionized the treatment of visceral leishmaniasis and life-threatening fungal infections in immunocompromised patients [62]. Similarly, in cancer chemotherapy,

Table 2.2 Nanoparticles-based delivery systems with their therapeutic and diagnostic uses in cancer therapy [60]

S. No.	Nanoparticle-Based Delivery Systems	Therapeutic and Diagnostic Use
1	Liposomes	Controlled and targeted drug delivery; Targeted gene delivery
2	Nanoshells	Tumor targeting
3	Fullerene-based derivatives	As targeting and imaging agent
4	Carbon nanotubes	Drug-gene and DNA delivery; Tumor targeting
5	Dendrimers	Targeting drug delivery
6	Quantum dots	Targeted delivery and imaging agent
7	Solid lipid nanoparticles	Controlled and targeted drug delivery
8	Nanowires	As targeting and imaging agent
9	Paramagnetic nanoparticles	As targeting and imaging agent

liposomal doxorubicin formulations were shown to be more potent with fewer side effects [62]. These conventional liposomes termed as 'non-stealth' liposomes have high affinity for the reticuloendothelial system and are rapidly removed from circulation. However, it has been overcome by newer formulations of liposomes like stealth liposomes and acoustically active liposomes [63–65]. Of these stealth liposomes have a coating with PEG which significantly decreases their uptake by macrophages and allows them to concentrate in tumors by a phenomenon known as Enhanced Permeation and Retention (EPR) effect [66, 67]. EPR effect is attributed to the enhanced permeability of tumor endothelium and lack of lymphatic drainage in the tumor cells. As a result, there is increased extravasation of the drug from tumor vasculature into the tumor cells and gets accumulated due to lack of lymphatic drainage [68]. These improved liposomal formulations are expected to be of use as sustained drug release systems.

2.4.2 The Solid Lipid Nanoparticles

The solid lipid nanoparticles owing to their lesser size and the ability to transport the drug to the target site, has potential applications in clinical medicine, drug delivery, research and pharmaceutical products. The solid lipid nanoparticles are at present acting as a substitute to liposomes due to its distinctive properties [69]. The different methods used for the preparation of the solid lipid nanoparticles are high shear homogenization, ultrasonication, microemulsion based preparation, spray drying, solvent emulsification, supercritical fluid technology and the double emulsion techniques [70, 71]. The various advantages of the solid lipid nanoparticles are bioavailability of the poorly soluble drugs, non-toxic, site-specific delivery, increased penetration, and stability [72, 73]. The solid lipid nanoparticles loaded with doxorubicin, paclitaxel and cholesterol butyrate are developed for treating colorectal cancer. The *in vitro* studies done in HT-29 Colorectal Cancer (CRC) cell line proved that the solid lipid nanoparticles loaded with doxorubicin had increased anti-cancer effect than the commercially available cancer drugs [74]. The solid lipid nanoparticles are used for the controlled release of 5-fluorouracil in a colonic fluid, and the loading efficiency of the drug was found to be 69% [75]. The solid lipid nanoparticles proved to be a better delivery system for the cholesteryl butyrate which in turn down-regulated the ERK and p38 phosphorylation along with the inhibition of claudin-1 expression. The cholesteryl butyrate nanoparticles also prevent the adhesion of the polymorphonuclear cells to the endothelium by acting as an effective antimetastatic and anti-inflammatory agent [74, 76].

2.4.3 Carbon Nanotubes

Carbon nanotubes discovered in 1991 [77] are tubular structures like a sheet of graphite rolled into a cylinder capped at one or both ends by a buckyball. Nanotubes can be Single-Walled Carbon Nanotubes (SWCNTs) and Multiwalled Carbon Nanotubes (MWCNTs) in a concentric fashion. The microphysiometer is built from MWCNT, which are like several flat sheets of carbon atoms rolled and stacked into miniature tubes. Nanotubes are electrically conductive and the concentration of insulin in the chamber can be directly related to the current at the electrode and the nanotubes operate reliably at pH levels characteristic of living cells. Present exposure methods measure insulin creation at intervals by periodically collecting tiny samples and measuring their insulin levels. The new sensor detects insulin levels continuously by measuring the transfer of electrons produced when insulin molecules oxidize in the presence of glucose. While the cells have produced more insulin molecules, the present in the sensor increases and vice versa, allowing monitoring insulin concentrations in real time [78].

SWCNTs fluorescence in the NIR spectral region and since they also suffer no photobleaching, SWCNTs are thus particularly suitable as fluorophore probes in glucose sensors designed for eventual *in vivo* use [79]. SWCNTs have been engaged in a fluorescence-based aggressive glucose sensing approach where dextran is bound to the carbon nanotubes (CNTs), and binding of concanavalin A or apo-glucose oxidase to the dextran–SWCNT attenuates the fluorescence, which is reversed by the addition of glucose [78]. Apo-glucose oxidase has also been covalently attached to polyvinyl alcohol to make a glucose responsive hydrogel that can be monitored by the fluorescence of SWCNT embedded in the hydrogel [80].

2.4.4 Carbon Nanotubes in Cancer Treatment

Carbon nanotubes are allotropes of carbon and structure of graphene sheets rolled at specific angles called chiral angles to generate a cylindrical structure called Carbon Nanotubes (CNT) [81–87]. They are black in color and can have very high aspect ratio and show inert nature to most chemicals. They have a hollow inner structure which acts as a container to take drugs to specific sites for cancer treatment [88] (Table 2.3).

2.4.4.1 Single-walled carbon nanotubes in the treatment of cancer

The CNT along with etoposide and dexamethasone (anti-cancer drugs) was being used on HeLa and Panc1 cells and results were calculated, which show that at high concentrations CNT with anti-cancer drugs were more effective to cause an increase in etoposide than the drug without CNT was used. The etoposide produces sDNA and dsDNA breakages, which causes delay progression through late S or early G2 phase of the malignant cell cycle. The etoposide induces redox reactions with the production of derivatives that binds with the DNA, thus DNA damage is caused by these anti-cancer agents with the drug carrying ability of CNT [89].

Table 2.3 Nanodrugs approved and there indications [68]

Nanodrugs	Indications
Abraxane	Breast cancer
Amphotericin B	Visceral leishmaniasis
All-trans retinoic acid	Acute promyelocytic leukemia; non-Hodgkin's lymphoma; renal cell carcinoma; Kaposi's sarcoma
Aprepitant	Cancer chemotherapy-induced emesis
Daunorubicin	Kaposi's sarcoma
Doxorubicin	Combinational therapy of recurrent breast cancer
Doxorubicin in polyethylene glycol liposomes	Refractory Kaposi's sarcoma; cancer, ovarian recurrent breast cancer
Vincristine	Non-Hodgkin's lymphoma

2.4.4.2   Multi-walled carbon nanotubes in the treatment of cancer

CNTs are able to absorb light of Near Infrared Region (NIR), results in heating of tubes. This ability of CNT is used to kill cancerous cells by supplying thermal energy to them, similar to that of magnetic hyperthermia. The length of the nanotube is such that, is it half of the wavelength of the light being used in the treatment (according to the principle of antenna theory). MWCNTs are being used as they have high chances of having defects and have a high number of electrons ultimately will generate higher thermal energy [82, 90]. The dopants of group III and can be used. N-doped MWCNT is being used to kill cancerous cells in kidney In the presence of near-infrared-light [88].

2.4.5 Drug Delivery

Nanoparticles drug delivery systems appear to be a feasible and promising strategy for the pharmaceutical industry. They have advantages over conventional drug delivery systems. They can raise the bioavailability, solubility, and permeability of several potent drugs which are otherwise difficult to deliver orally. Nanoparticles drug delivery systems will also decrease the drug dosage, frequency and it will potentially enlarge the patient fulfillment. In near future, nanoparticles drug delivery systems can be used for exploring various biological drugs which have a poor aqueous solubility, permeability and fewer bioavailabilities [36]. Nanoparticles can minimize some of these drugs unusual problems by safeguarding stability and preserving their structure. In addition, nanoparticles provide original treatment by enabling target delivery and controlled release [91].

To successfully integrate a drug into a nanoparticle, some design strategies can be explored, including physical complexion with hydrophobic drugs, or covalent bonding with cleavable linkages for intracellular release. Drugs loaded through hydrophobic connections are typically encapsulated within the nanoparticle coating, limiting nonspecific cell interaction. This approach is advantageous in applications where a drug being delivery and could sincerely harm non-targeted tissue [92]. Process in understanding the nanoparticle internalization by a variety of mammalian cells has already allowed the design of effective nanomedicines, especially for the treatment of infectious diseases and some cancers [93].

Drug delivery strategies are a rising part of research, relatively focusing on finding new molecular targets and pathways in autoimmune diseases, drug delivery strategy can provide the tissue selectivity with current therapies by altering their pharmacokinetics and biodistribution. Nano-carries have been confirmed to have potential in civilizing the protection profile and therapeutic efficiency of the recent therapies for autoimmune diseases, mostly for those with potent but toxic compounds [94]. Inorganic nanomedicine holds vast assure in diagnostics, drug and gene delivery, sensing and biosensing, and *in vivo* imaging under the present scenario elegantly engineered inorganic nanoparticles can enhance drug efficacy and can improve drug targeting to exact areas in the body, therefore making treatment less toxic and less invasive [95, 96].

Nanomedicines used for drug delivery are made up of nanoscale particles or molecules which can improve drug bioavailability. For maximizing bioavailability mutually at particular places in the body and over a period of time, molecular targeting is done by nano-engineered devices such as nanorobots [97]. The molecules are targeted and delivering of drugs is done with cell precision. *In vivo* imaging is one more area where nanotools and devices are individually developed for *in vivo* imaging. Using nanoparticle images such as in ultrasound and MRI, nanoparticles are used as contrast. The nano-engineered materials have been developed for successfully treating illnesses and diseases like cancer. With the advancement of nanotechnology, self-assembled biocompatible nanodevices can be created which will detect the cancerous cells and automatically evaluate the disease, will cure and prepare reports [15].

2.4.6 The Applications of Nanoparticles in Drug Delivery

Abraxane is albumin-bound paclitaxel, a nanoparticle used for the treatment of breast cancer and non-small-cell lung cancer (NSCLC). Nanoparticles are used to deliver the drug with enhanced effectiveness for treatment for head and neck cancer, in mice model study, which was carried out at from Rice University and University of Texas MD Anderson Cancer Center. The reported treatment uses Cremophor EL which allows the hydrophobic paclitaxel to be delivered intravenously. When the toxic Cremophor is replaced with carbon nanoparticles its side effects diminished and drug targeting was much improved and needs a lower dose of the toxic paclitaxel [98]. Nanoparticle chain was used to deliver the drug doxorubicin to breast cancer cells in a mice study at Case Western Reserve University. The scientists prepared a 100 nm long nanoparticle chain by chemically linking three magnetic, iron oxide nanospheres, to one doxorubicin in a loaded liposome. After penetration of the nano-chains inside the tumor magnetic nanoparticles were made to vibrate by generating, radio-frequency field which resulted in the rupture of the liposome, thereby dispersing the drug in its free form throughout the tumor. Tumor growth was halted more effectively by nanotechnology than the standard treatment with doxorubicin and is less harmful to healthy cells as very fewer doses of doxorubicin were used [99, 100]. Polyethylene glycol (PEG) nanoparticles carrying payload of antibiotics at its core were used to target bacterial infection more precisely inside the body, as reported by scientists at MIT. The nano delivery of particles, containing a sub-layer of pH sensitive chains of the amino acid histidine, is used to destroy bacteria that have developed resistance to antibiotics because of the targeted high dose and prolonged release of the drug. Nanotechnology can be efficiently used to treat various infectious diseases [101]. Researchers in the Harvard University Wyss Institute have used the biomimetic strategy in a mouse model. Drug-coated nanoparticles were used to dissolve blood clots by selectively binding to the narrowed regions in the blood vessels as the platelets do. Biodegradable nanoparticle aggregates were coated with tissue plasminogen activator, tPA was injected intravenously which bind and degrade the blood clots. Due to shear stresses in the vessel narrowing region dissociation of the aggregates occurs and releases the tPA-coated nanoparticles. The nanotherapeutics can be applied greatly to reduce the bleeding, commonly found in standard thrombosis treatment [15]. The researchers at the University of Kentucky have created X-shaped RNA nanoparticles, which can carry four functional modules. These chemically and thermodynamically stable RNA molecules are able of remaining intact in the mouse body for more than 8 hours and to resist degradation by RNAs in the blood stream. These X-shaped RNA can be effectively performing therapeutic and diagnostic functions. They regulate gene expression and cellular function, and are capable of binding to cancer cells with precision, due to its design [102]. 'Minicell' nanoparticle is used in early phase clinical trial for drug delivery for treatment of patients with advanced and untreatable cancer. The minicells are built from the membranes of mutant bacteria and were loaded with paclitaxel and coated with cetuximab, antibodies and used for the treatment of a variety of cancers. The tumor cells engulf the mini cells. Once inside the tumor, the anti-cancer drug destroys the tumor cells. The larger size of mini cells plays a better profile in side effects. The mini cell drug delivery system uses a lower dose of the drug and has less side-effects can be used to treat a number of different cancers with different anti-cancer drugs [15]. Nanosponges are important tools [103] in drug delivery, due to their small size and porous nature they can bind poorly-soluble drugs within their matrix and improve their bioavailability. They can be made to carry drugs to specific sites, thus help to prevent drug and protein degradation and can prolong drug release in a controlled manner [15].

2.5 Nanoparticles Anti-Oxidative Role in Diabetes

Oxidative stress plays a foremost role in etiology of several diabetic complications [104, 105]. The major problem with diabetes patients is delayed healing of the wound and still challenging its complete cure. It is a complex programmed sequence of cellular and molecular processes. Healing impairment is characterized by delayed cellular infiltration and granulation tissue formation, reduced angiogenesis, decreased collagen, and its organization [106, 107]. The mechanism of this alteration is thought to result from the production of a high level of ROS production that leads to premature apoptosis of inflammatory cells, which in turn impairs keratinocyte endothelial cells, fibroblasts, and collagen metabolism. Some of the nanoparticles now a day are prepared in a manner that acts as free radical scavenger [108].

Nanoparticles made of other metal oxides were also considered for its potential scavenger behavior. These included particles aluminum oxide (Al_2O_3), Cerium oxide (CeO_2), Yttrium oxide (Y_2O_3), and Silver nitrate ($AgNO_3$). Silver nitrate has been used in the clinical setting as an antimicrobial agent for the treatment of treatment of chronic wounds [78]. It is effective against a broad range of aerobic, anaerobic, Gram-negative and Gram-positive bacteria, yeast, filamentous fungi and viruses [109–111]. It has been reported a novel Ag^+ loaded zirconium phosphate nanoparticle plays a crucial role in diabetic wound healing.

Cerium oxide (CeO_2) plays a major active role due to its excellent free radical scavenging potentials [112]. This metal oxide is a monodisperse particle with a single crystal and few twin boundaries [113] with expanded lattice parameter. Cerium atom characterized by both +4 and +3 oxidation states. This dual oxidation state means that these nanoparticles have oxygen vacancies [114]. The loss of oxygen and the reduction of Ce^{4+} to Ce^{3+} are accompanied by the creation of an oxygen vacancy. This property enhances CeO_2 nanoparticle attractive for wound healing process followed by scavenging properties.

Yttrium oxide (Y_2O_3) is now a day considered most significant due to its highest free energy of oxide formation from elemental yttrium among known metal oxides **[115]**. It is characterized by only small changes from stoichiometry under normal conditions of temperature and pressure and by atmospheric absorption of H_2O and CO_2. These groups of nanoparticles are relatively non-toxic to neutrophils and macrophages, where CeO_2 and Y_2O_3 particles protect cells from death due to oxidative stress. Cerium and yttrium oxide nanoparticles are able to rescue cells from oxidative stress-induced cell death in a manner that appears to be dependent upon the structure of the particle but independent of its size within the range of 6–1000 nm. This might be useful for the diabetic wound healing. There are three alternative explanations for the cerium oxide and yttrium oxide particles protection against oxidative stress [78].

- They may act as direct antioxidants, block ROS production, which inhibits programmed cell death pathway.
- They may directly cause a low level of ROS production, which rapidly induces an ROS defense system before the glutamate-induced cell death program is complete.
- The latter is a form of preconditioning that could be caused by the exposure of cells to particulate material known to induce low levels of ROS [116].

The gold nanoparticles (AuNPs) are known for their tremendous applications in the field of therapeutics and diagnosis. AuNPs are emerging nanomedicine which is renowned for its promising therapeutic possibilities, due to its significant properties such as biocompatibility, high surface reactivity, resistance to oxidation and plasmon resonance [117]. The ability of gold nanoparticles in inhibiting the lipid from peroxidation thereby preventing the ROS generation has restored the imbalances in the antioxidants [118]. Selvaraj Barath Mani Kanth et al. have reported the anti-oxidative and antihyperglycemic activities of gold nanoparticles, whereas the mechanism of the delayed wound healing in diabetes is still unclear [119].

2.5.1 Nanomedicine in Management of Diabetes

Worldwide 285 million peoples are suffering from a pervasive, chronic and often insidious diabetes is caused by the inability of the pancreas to control the blood glucose concentration [120]. The preferred approach of insulin intake since the past decades is via subcutaneous route, which, nonetheless, often fails to mimic the glucose homeostasis observed in normal subjects because in this approach insulin delivered to the peripheral circulation rather than to the portal circulation and directly into the liver, which is the physiological route in normal individuals [121]. Furthermore, multiple daily injections of insulin referred for poor patient compliance are associated with subcutaneous route treatment. Therefore, many studies were done to find out the better and safer route of insulin administration, in this regards application of nanotechnology in medicine revealed a solution to overcome this problem [78].

2.6 Nanorobotics

Nanorobotics, sometimes referred to as molecular robotics, is an emerging research area which can be generally divided into two main focus areas. The first area deals with the design, simulation, control and coordination of robots with nanoscale dimensions. Nanorobots, nanomachines and other nanosystems are objects with overall dimensions at or below the micrometer range and are made of assemblies of nanoscale components with individual dimensions ranging approximately between 1 to 100 nm. Much of the research conducted in this area remains highly theoretical at the present, primarily because of the difficulties in fabricating such devices. The second area deals with the manipulation and/or assembly of nanoscale components with macroscale instruments or robots (i.e., nanomanipulators). A much greater number of research papers have been generated in this area. Due to the advances in nanotechnology and its rapidly growing number of potential applications, it is evident that practical technologies for the manipulation and assembly of nanoscale structures into functional nanodevices need to be developed. Nano-manipulation and nano-assembly may also play a crucial role in the development of artificial nanorobots themselves [122].

2.6.1 Types of Nanorobots

The types of nanorobots designed by Robert A. Freitas as artificial blood are [Freitas 1999]:

1. Respirocytes
2. Microbivores
3. Clottocytes

2.6.1.1 In surgery

Robotic surgery can accomplish what doctors cannot because of precision and repeatability of robotic systems. Besides, robots are able to operate in a contained space inside the human body. All these make robots especially suitable for non-invasive or minimally invasive surgery and for better out-comes of surgery. Today, robots have been demonstrated or routinely used for heart, brain, spinal cord, throat, and knee surgeries at many hospitals. Since robotic surgery improves consistency and quality, it is becoming more and more popular.

2.6.1.2 Drug delivery

Nanorobots carry out a very specific function and are just several nanometers wide. They can be used very effectively for drug delivery. Normally, drugs work through the entire body before they reach the disease-affected area. Using nanotechnology, the drug can be targeted to a precise location which would make the drug much more effective and reduce the chances of possible side-effects [123].

2.6.1.3 In diagnosis

Robotic diagnosis reduces invasiveness to the human body and improves the accuracy and scope of the diagnosis. One example is the robotic capsular endoscope that has been developed for the non-invasive diagnosis of the gastrointestinal tract.

2.6.2 Applications of Nanorobots in Medicine

Medical nanorobots can perform a wide range of tasks in diagnosis, moni-toring and treating vital diseases. These nanorobots are capable of delivering medicine or drugs into specific sites/targets in the human body [124]. The potential applications of nanorobots include:

Drug delivery: Pharmacytes are the nanorobots designed for the action of drug delivery. The dosage of the drug will be loaded into the payload of the pharmacyte. The pharmacyte will be capable of precise transport and targeted delivery of drug to specific cellular targets. The pharmacytes upon arriving in the vicinity of tumor or any 46 Biomedical Science and Engineering target cell would release the drug via nano injection or by progressive cytopenetra-tion until the payload delivery is reached [125].

Body surveillance: Monitoring continuously of vitals and wireless trans-mission could be possible using nanorobots, leading to a quantum leap in diagnostics. This would also help in a quick response in case of sudden change in vitals or could warn against a possible risk, such as high blood glucose in case of diabetics [126].

Dentistry: The nanorobots designed for dental treatment are referred to as dentifrobots. These nanorobots can induce oral analgesia, desensitize tooth, manipulate the tissues to realign and straighten irregular set of teeth [127].

In surgery: The surgical program nanorobot can act as a semi-autonomous onsite surgeon inside the body. It would perform various functions such as detection of pathology, diagnosing, correcting lesions by nanomanipulation coordinated by an on-board computer [124].

Cancer detection and treatment: The nanorobots are made with a mixture of polymer and a protein known as transferrin which is capable of detecting tumor cells. The nanorobots would consist of embedded chemical biosensor that can be used for the detection of the tumor. The medical nanorobots with chemical biosensors can be programmed to detect different levels of E-cadherin and beta-catenin, aiding in the target identification and drug delivery. The nanorobot could also carry the chemicals employed in chemotherapy to treat cancer at the site [128]. The robots could either attack tumors directly using lasers, microwaves or ultrasonic signals or as a part of a chemotherapy treatment, delivering medication to the cancer site.

Diagnosis and treatment of diabetes: The glucose molecules are carried through the bloodstream to maintain the human metabolism. The hSGLT3 molecule can define the glucose levels for diabetes patients. The glucose monitoring nanorobot uses the chemosensor which involves in the modulation of hSGLT3 protein gluco-sensor activity [127]. These chemical sensors can effectively determine the need of insulin in the body and inject.

Delicate surgeries: Nanorobots could be soon used for performing micro-surgery of the eye as well as surgeries of the retina and surrounding membranes. In addition, instead of injecting directly into the eye, nanorobots could be injected elsewhere in the body and delivery of the drug can be guided to the eye. Fetal surgery, one of the riskiest surgeries today because of the high mortality rate of either the baby or the mother, could soon have a 100% success rate, due to the fact that nanorobots can provide better access to the required area inducing minimal trauma. Similarly, other difficult surgeries could also benefit from advances in nanorobotics [126].

Gene therapy: The medical nanorobot can treat genetic diseases by comparing the molecular structure of both DNA and proteins found in the cell. The chromosome replacement therapy can carry out using chromallocytes [129].

2.6.3 Advantages of Nanorobots

A major advantage of nanorobots is thought to be their durability. In theory, they can remain operational for years, decades, or centuries. Nanoscale systems can also operate much faster than their larger counterparts because displacements are smaller; this allows mechanical and electrical events to occur in less time at a given speed. Nanorobots might also produce copies of themselves to replace worn-out units, a process called self-replication. A computer instead of giving a pill or a shot, the doctor refers to a special medical team which implants a tiny robot into our bloodstream. The robot detects the cause of fever, travels to the appropriate system and provides a dose of medication directly to the infected area [122, 124 and 130].

2.6.4 Disadvantages of Nanorobots

The initial design cost is very high. The design of the nanorobot is a very complicated one. Electrical systems can create stray fields which may activate bioelectric-based molecular recognition systems in biology. Electrical nanorobots are susceptible to electrical interference from external sources. Like electric fields, EMP pulses, and stray fields from other *in vivo* electrical devices. Hard to interface, customize and design, complex in nature. Nanorobots can cause a brutal risk in the field of terrorism. The terrorism and anti-groups can make use of nanorobots as a new form of torturing the communities as nanotechnology also has the capability of destructing the human body at the molecular level. Privacy is another potential risk involved with nanorobots. As nanorobots deal with the designing of compact and minute devices, there are chances for more eavesdropping than that which already exists [122, 128 and 130].

2.7 Future Development of Nanomedicine

In the coming decade, nanotechnology and nanobiotechnology applications will gain importance in medicine and medical technology. This trend is already clearly detectable at present. For the past half of this decade 2010–2014 the Web of Knowledge records the titles of 3,438 publications under the keyword "nanomedicine" as previous decade of 2000–2009 entire 857 entries. Nanomedicine has the potential to significantly improve the quality of life of patients. Nevertheless, the new possibilities also involve risks and raise sociological and ethical questions which must be analyzed and debated [38]. Figure 2.4 shows the four research and development areas which will probably receive the greatest impetus from nanomedicine in the coming decades.

The future of nano-therapy vision is the treatment of patients with individually tailor-made medicine at the molecular level as soon as the disease is in the development stage [131]. The preparation of nano drugs and the various methods of targeted transport of active substance will play a prominent role here. With these, it could become possible to develop effective and well-tolerated treatments for hitherto incurable diseases.

Nanotechnology provides methods by which biological information can be acquired easily, quickly and inexpensively and analyzed, and thus enormously increase the possibilities of preventive medicine. Therapy and diagnostics are increasingly becoming fused into the new specialist medical field of theranostics because the nanotechnology methods and medicines serve diagnostic and therapeutic purpose simultaneously. Examples are the contrast medium which brings with it directly the active substance in the event of a pathological tissue change [132] and carrier systems which circulate preventively in the organism and react to endogenous signals and automatically secrete active substances if needed [133]. The production of nanomaterials

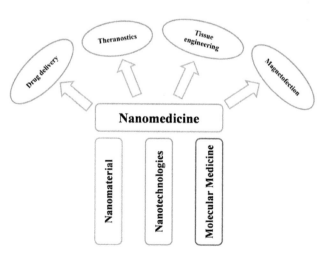

Figure 2.4 Future development of Nanomedicine.

which recognize cells and cell constituents, including individual genes, of impaired function and repair them of their own accord in the organism is also being researched [134–136].

Recently, a promising theranostic application has been addressed in modulating an on switch Chimeric Antigen Receptor (CAR) against chemotherapy-resistant forms of B-cell cancer [137]. By introducing small molecules a switch-on-switch-off mechanism could successfully be established in a draconic engineered T cell therapeutic approach, thus avoiding toxic effects of the cell killer function.

As nanobiotechnology opens up new possibilities above all in the field of regenerative medicine. By stimulation and targeted control of cell growth, damaged or absent tissue- from hair, cartilage and bone, via muscles and organs, through to never cells- could be regenerated or produced artificially with the aid of nanomaterials. Nanoporous carrier materials are already now being used in wound healing and in plastic surgery as matrices along which controlled cell growth takes place [138]. If targeted growth of nerve cells were also to be successful, new possible treatments for incurable neurological diseases such as Alzheimer's, Parkinson's, epilepsy and multiple sclerosis could be developed. And should target manipulation of adult stem cells also be successful, endogenous tissue which causes no rejections could be cultured, and the use of embryonic stem cells could be abandoned.

The fourth area in which nanomedicine will gain importance in the coming years is gene therapy. Intensive research is being conducted into whether and how the various mechanisms of targeted transport of active substances can be used to introduce nucleic acids, DNA fragments, and individual gens into tissue and cells by means of nano-viral nanoparticles. In a cutting-edge animal study using rat's biodegradable, polymeric gene delivery nanoparticles have been shown to effectively kill glioma cells in the brain and extend the survival of the animals [139].

One further possibility is magneto-fiction in which as inactive MDT positively charged nanoparticles containing iron are loaded with negatively charged nucleic acids, brought to the target cells with the aid of a magnetic field and due to their appropriately prepared surface structure are sluiced into these. The nucleic acids are released there and the particles are transferred into cell iron metabolism. One advantage of this method, which is also called "magnetic gene targeting" (MGT) and is of interest above all for the treatment of cancer and neurodegenerative diseases as well as myocardial infarction [140], is that the target cells are not damaged. An example of this strategy is an innovative gene delivery method that may be applicable in stent angioplasty and thereby contribute to bridging the gap between research and clinical application [141]. The approach established in a rat model uses stents as a platform for magnetically targeted gene delivery, where genes are moved to cells at arterial injury locations avoiding adverse effects to other organs. Novel magnetic nanoparticles protect the transferred genes and help them reach their target in active form.

Nano-based gene therapy is aimed at the addition of missing and replacement of defective DNA and therefore intervenes directly in cell process at the molecular level. Because the possibility of replacing DNA sooner or later will also result in the capability of targeted manipulation of DNA, areas are stumbled into here which make confrontation with ethical and social matters unavoidable. Is only the repair of defective or damaged DNA allowed, or does it also desirable for human sensory and physical capabilities to be optimized without a therapeutic indication exist? For whom under what circumstances and by whom will the possibility of genetic modification be generated? How will deal with the body's inadequacies in the future?

2.8 Conclusion

The implementation of nanoparticles in biological sciences is swiftly growing and the scientific community is aiming at evolving from the conventional medicine to nano-based drugs. The advancement in the therapeutics has led to the development of efficient techniques for early detection of diseases, minimum invasive treatment to uncurable diseases, tailoring of medicines without side effects there targeted delivery all with the aid of nanoscience is a new beginning for better life. The public health, safety, and environmental protection are the major concerns while dealing with the nanoparticles. Considerable technological success has been achieved in this field of nanomedicine but the main hurdles to nanomedicine becoming a new paradigm in theranostics is the complexities and heterogeneity of biology, an incomplete understanding of nano-bio interactions and the challenges regarding chemistry, manufacturing, and controls required for clinical translation and commercialization.

Acknowledgment

All the authors' great fully acknowledges and express thanks to the International and Inter University Center for Nanoscience and Nanotechnology, M G University, Kerala, India.

References

[1] Yousaf, S. A., and Salamat, A. (2011). "Effect of heating environment on fluorine doped tin oxide (f: SnO/sub 2/) thin films for solar cell applications," in *Proceedings of the International Conference on Power Generation Systems Technologies*, Islamabad.

[2] Vaseashta, A., and Dimova-Malinovska, D. (2005). Nanostructured and nanoscale devices, sensors and detectors. *Sci. Technol. Adv. Mater.* 6, 312–318.

[3] Roy, K., Mao, H. Q., Huang, S. K., and Leong, K. W. (1999). Oral gene delivery with chitosan–DNA nanoparticles generates immunologic protection in a murine model of peanut allergy. *Nat. Med.* 5, 387–391.

[4] Sachlos, E., Gotora, D., and Czernuszka, J. T. (2006). Collagen scaffolds reinforced with biomimetic composite nano-sized carbonate-substituted hydroxyapatite crystals and shaped by rapid prototyping to contain internal microchannels. *Tissue Eng.* 12, 2479–2487.

[5] Farokhzad, O. C., Cheng, J., Teply, B. A., Sherifi, I., Jon, S., Kantoff, P. W., et al. (2006). Targeted nanoparticle-aptamer bioconjugates for cancer chemotherapy in vivo. *Proc. Natl. Acad. Sci. U.S.A.* 103, 6315–6320.

[6] Panáček, A., Kvítek, L., Prucek, R., Kolář, M., Večeřová, R., Pizúrová, N., et al. (2006). Silver colloid nanoparticles: synthesis, characterization, and their antibacterial activity. *J. Phys. Chem. B* 110, 16248–16253.

[7] Morones, J. R., Elechiguerra, J. L., Camacho, A., Holt, K., Kouri, J. B., Ramírez, J. T., et al. (2005). The bactericidal effect of silver nanoparticles. *Nanotechnology* 16, 2346–2353.

[8] Etheridge, M. L., Campbell, S. A., Erdman, A. G., Haynes, C. L., Wolf, S. M., and McCullough, J. (2013). The big picture on nanomedicine: the state of investigational and approved nanomedicine products. *Nanomedicine* 9, 1-14.

[9] Sahoo, S. K., and Labhasetwar, V. (2003). Nanotech approaches to drug delivery and imaging. *Drug Discov. Today* 8, 1112–1120.

[10] Jain, K. K. (2005). Nanotechnology in clinical laboratory diagnostics. *Clin. Chim. Acta* 358, 37–54.

[11] Nie, S., Xing, Y., Kim, G. J., and Simons, J. W. (2007). Nanotechnology applications in cancer. *Annu. Rev. Biomed. Eng.* 9, 257–288.

[12] Liu, H., and Webster, T. J. (2007). Nanomedicine for implants: a review of studies and necessary experimental tools. *Biomaterials* 28, 354–369.

[13] Engel, E., Michiardi, A., Navarro, M., Lacroix, D., and Planell, J. A. (2008). Nanotechnology in regenerative medicine: the materials side. *Trends Biotechnol.* 26, 39–47.

[14] Sumer, B., and Gao, J. (2008). Theranostic nanomedicine for cancer. *Nanomedicine* 3, 137–140.

[15] Nikalje, A. P. (2015). Nanotechnology and its Applications in Medicine. *Med Chem.* 5, 081–089.

[16] Wagner, V., Dullaart, A., Bock, A. K., and Zweck, A. (2006). The emerging nanomedicine landscape. *Nat. Biotechnol.* 24, 1211–1217.

[17] Kirpotin, D. B., et al. (1998). *Medical Applications of Liposomes.* Amsterdam: Elsevier.

[18] Bawa, R. (2008). Nanoparticle-based therapeutics in humans: a survey. *Nanotechnol. Law Bus.* 5, 135–155.

[19] Shah, P., Bhalodia, D., and Shelat, P. (2010). Nanoemulsion: a pharmaceutical review. *Syst. Rev. Pharm.* 1, 24–32.

[20] Hayman, M. L. (2009). The emerging product and patent landscape for nanosilver-containing medical devices. *Nanotechnol. Law Bus.* 6:148.

[21] Dobson, M. G., Galvin, P., and Barton, D. E. (2007). Emerging technologies for point-of-care genetic testing. *Expert Rev. Mol. Diagn.* 7, 359–370.

[22] Alharbi, K. K., and Al-Sheikh, Y. A. (2014). Role and implications of nanodiagnostics in the changing trends of clinical diagnosis. *Saudi J. Biol. Sci.* 21, 109–117.

[23] Lee, J. E., Lee, N., Kim, H., Kim, J., Choi, S. H., Kim, J. H., et al. (2009). Uniform mesoporous dye-doped silica nanoparticles decorated with multiple magnetite nanocrystals for simultaneous enhanced magnetic resonance imaging, fluorescence imaging, and drug delivery. *J. Am. Chem. Soc.* 132, 552–557.

[24] Yang, X., Li, Z., Li, M., Ren, J., and Qu, X. (2013). Fluorescent protein capped mesoporous nanoparticles for intracellular drug delivery and imaging. *Chem. A Eur. J.* 19, 15378–15383.

[25] Stroh, M., Zimmer, J. P., Duda, D. G., Levchenko, T. S., Cohen, K. S., Brown, E. B., et al. (2005). Quantum dots spectrally distinguish multiple species within the tumor milieu in vivo. *Nat. Med.* 11, 678–682.

[26] Michalet, X., Pinaud, F. F., Bentolila, L. A., Tsay, J. M., Doose, S., Li, J. J., et al. (2005). Quantum dots for live cells, in vivo imaging, and diagnostics. *Science* 307, 538–544.

[27] Chan, W. C., and Nie, S. (1998). Quantum dot bioconjugates for ultrasensitive nonisotopic detection. *Science* 281, 2016–2018.

[28] Dubertret, B., Skourides, P., Norris, D. J., Noireaux, V., Brivanlou, A. H., and Libchaber, A. (2002). In vivo imaging of quantum dots encapsulated in phospholipid micelles. *Science* 298, 1759–1762.

[29] Mirza, A. Z., and Siddiqui, F. A. (2014). Nanomedicine and drug delivery: a mini review. *Int. Nano Lett.* 4, 1–7.

[30] Kim, C. K., Kalluru, R. R., Singh, J. P., Fortner, A., Griffin, J., Darbha, G. K., and Ray, P. C. (2006). Gold-nanoparticle-based miniaturized laser-induced fluorescence probe for specific DNA hybridization detection: studies on size-dependent optical properties. *Nanotechnology* 17:3085.

[31] Dhar, S., Reddy, E. M., Shiras, A., Pokharkar, V., and Prasad, B. E. E. (2008). Natural gum reduced/stabilized gold nanoparticles for drug delivery formulations. *Chem. Eur. J.* 14, 10244–10250.

[32] Gelperina, S., Kisich, K., Iseman, M. D., and Heifets, L. (2005). The potential advantages of nanoparticle drug delivery systems in chemotherapy of tuberculosis. *Am. J. Respir. Crit. Care Med.* 172, 1487–1490.

[33] Egea, M. A., Gamisans, F., Valero, J., Garcia, M. E., and Garcia, M. L. (1994). Entrapment of cisplatin into biodegradable polyalkylcyanoacrylate nanoparticles. *Farmaco* 49, 211–217.

[34] Chawla, J. S., and Amiji, M. M. (2002). Biodegradable poly (ε-caprolactone) nanoparticles for tumor-targeted delivery of tamoxifen. *Int. J. Pharmaceut.* 249, 127–138.

[35] Soppimath, K. S., Aminabhavi, T. M., Kulkarni, A. R., and Rudzinski, W. E. (2001). Biodegradable polymeric nanoparticles as drug delivery devices. *J. Control. Release* 70, 1–20.

[36] Solano-Umaña, V., Vega-Baudrit, J. R., and González-Paz, R. (2015). The New Field of the Nanomedicine. *Int. J. Appl.* 5, 79–88.

[37] Bhadriraju, K., Chung, K. H., Spurlin, T. A., Haynes, R. J., Elliott, J. T., and Plant, A. L. (2009). The relative roles of collagen adhesive receptor DDR2 activation and matrix stiffness on the downregulation of focal adhesion kinase in vascular smooth muscle cells. *Biomaterials* 30, 6687–6694.

[38] Krukemeyer, M. G., Krenn, V., Huebner, F., Wagner, W., and Resch, R. (2015). History and possible uses of nanomedicine based on nanoparticles and nanotechnological progress. *J. Nanomed. Nanotechnol.* 6, 336.

[39] Reibold, M., Paufler, P., Levin, A. A., Kochmann, W., Pätzke, N., and Meyer, D. C. (2006). Materials: carbon nanotubes in an ancient Damascus sabre. *Nature* 444, 286–286.

[40] Yamamoto, V., Suffredini, G., Nikzad, S., Hoenk, M. E., Boer, M. S., Teo, C., et al. (2013). "From nanotechnology to nanoneuroscience/nanoneurosurgery and nanobioelectronics," in *The Textbook of Nanoneuroscience and Nanoneurosurgery*, eds B, Kateb and J.D. Heiss (Boca Raton, FL: CRC Press), 1–28.

[41] Nirenberg, M. W., and Matthaei, J. H. (1961). The dependence of cell-free protein synthesis in *E. coli* upon naturally occurring or synthetic polyribonucleotides. *Proc. Natl. Acad. Sci.* 47, 1588–1602.

[42] Watson, J. D., and Crick, F. H. (1974). Molecular structure of nucleic acids: a structure for deoxyribose nucleic acid. *Nature* 248, 765.

[43] Taniguchi, N. (1974). "On the basic concept of nano-technology." in *Proceedings of the International Conference on Production Engineering* (Tokyo: Japan Society of Precision Engineering).

[44] Feynman, R. P. (1960). There's plenty of room at the bottom. *Eng. Sc.* 23, 22–36.

[45] Eric, D. K. (1986). *Engines of Creation. The Coming Era of Nanotechnology*. New York, NY: Anchor books.

[46] Drexler, K. E., Peterson, C., and Pergamit, G. (1991). *Unbounding the Future*. New York, NY: William Morrow, 294.

[47] Freitas, R. A. (1999). *Nanomedicine: Basic Capabilities*, Vol. I, Georgetown, TX: Landes Biosciences.

[48] Kreuter, J. (2007). Nanoparticles—a historical perspective. *Int. J. Pharm.* 331, 1–10.

[49] Kateb, B., and Heiss, J. D. (2013). *The Textbook of Nanoneuroscience and Nanoneurosurgery*. Boca Raton, FL: CRC Press.

[50] Noorlander, C. W., Kooi, M. W., Oomen, A. G., Park, M. V., Vandebriel, R. J., and Geertsma, R. E. (2015). Horizon Scan of Nanomedicinal Products. *Nanomedicine* 10, 1599–1608.

[51] Park, K. (2013). Facing the truth about nanotechnology in drug delivery. *ACS Nano* 7, 7442–7447.

[52] Loeve, S., Vincent, B. B., and Gazeau, F. (2013). Nanomedicine metaphors: from war to care. Emergence of an oecological approach. *Nano Today* 8, 560–565.

[53] Richman, E. K., and Hutchison, J. E. (2009). The nanomaterial characterization bottleneck. *ACS Nano* 3, 2441–2446.

[54] Grainger, D. W., and Castner, D. G. (2008). Nanobiomaterials and nanoanalysis: opportunities for improving the science to benefit biomedical technologies. *Adv. Mater.* 20, 867–877.

[55] Crist, R. M., Grossman, J. H., Patri, A. K., Stern, S. T., Dobrovolskaia, M. A., Adiseshaiah, P. P., et al. (2013). Common pitfalls in nanotechnology: lessons learned from NCI's nanotechnology characterization laboratory. *Integr. Biol.* 5, 66–73.

[56] Baer, D. R., Engelhard, M. H., Johnson, G. E., Laskin, J., Lai, J., Mueller, K., et al. (2013). Surface characterization of nanomaterials and nanoparticles: Important needs and challenging opportunities. *J. Vac. Sci. Technol. A* 31:50820.

[57] Stirland, D. L., Nichols, J. W., Miura, S., and Bae, Y. H. (2013). Mind the gap: a survey of how cancer drug carriers are susceptible to the gap between research and practice. *J. Control. Release* 172, 1045–1064.

[58] Begley, C. G., and Ellis, L. M. (2012). Drug development: raise standards for preclinical cancer research. *Nature* 483, 531–533.

[59] Pal, D., and Nayak, A. K. (2010). Nanotechnology for targeted delivery in cancer therapeutics. *Int. J. Pharm. Sci. Rev. Res.* 1, 1–7.

[60] Anajwala, C. C., Jani, G. K., and Swamy, S. V. (2010). Current trends of nanotechnology for cancer therapy. *Int. J Pharm. Sci. Nanotechnol.* 3, 1043–1056.

[61] Torchilin, V. P. (2007). Targeted pharmaceutical nanocarriers for cancer therapy and imaging. *AAPS J.* 9, E128–E147.

[62] Moghimi, S. M., Hunter, A. C., and Murray, J. C. (2005). Nanomedicine: current status and future prospects. *FASEB J.* 19, 311–330.

[63] Ahmed, F., and Discher, D. E. (2004). Self-porating polymersomes of PEG–PLA and PEG–PCL: hydrolysis-triggered controlled release vesicles. *J. Control. Release* 96, 37–53.

[64] Coussios, C. C., Holland, C. K., Jakubowska, L., Huang, S. L., MacDonald, R. C., Nagaraj, A., et al. (2004). In vitro characterization of liposomes and Optison® by acoustic scattering at 3.5 MHz. *Ultrasound Med. Biol.* 30, 181–190.

[65] Fanciullino, R., Giacometti, S., Aubert, C., Fina, F., Martin, P. M., Piccerelle, P., et al. (2005). Development of stealth liposome formulation of 2'-deoxyinosine as 5-fluorouracil modulator: in vitro and in vivo study. *Pharm. Res.* 22, 2051–2057.

[66] Greish, K., Sawa, T., Fang, J., Akaike, T., and Maeda, H. (2004). SMA–doxorubicin, a new polymeric micellar drug for effective targeting to solid tumours. *J. Control. Release* 97, 219–230.

[67] Maeda, H. (2001). The enhanced permeability and retention (EPR) effect in tumor vasculature: the key role of tumor-selective macromolecular drug targeting. *Adv. Enzyme Regul.* 41, 189–207.

[68] Sandhiya, S., Dkhar, S. A., and Surendiran, A. (2009). Emerging trends of nanomedicine–an overview. *Fundam. Clin. Pharmacol.* 23, 263–269.

[69] Uner, M., and Yener, G. (2007). Importance of solid lipid nanoparticles (SLN) in various administration routes and future perspectives. *Int. J. Nanomed.* 2:289.

[70] Mukherjee, S., Ray, S., and Thakur, R. S. (2009). Solid lipid nanoparticles: a modern formulation approach in drug delivery system. *Indian J. Pharm. Sci.* 71:349.

[71] Battaglia, L., Gallarate, M., Panciani, P. P., Ugazio, E., Sapino, S., Peira, E., et al. (2014). "Techniques for the preparation of solid lipid nano and microparticles," in *Application of Nanotechnology in Drug Delivery*, ed. A. D. Sezer (Rijeka: InTech).

[72] Gokce, E. H., Korkmaz, E., Tuncay-Tanriverdi, S., Dellera, E., Sandri, G., Bonferoni, M. C., et al. (2012). A comparative evaluation of coenzyme Q10-loaded liposomes and solid lipid nanoparticles as dermal antioxidant carriers. *Int. J. Nanomed.* 7, 5109–5117.

[73] Garud, A., Singh, D., and Garud, N. (2012). Solid lipid nanoparticles (SLN): Method, characterization and applications. *Int. Curr. Pharm. J.* 1, 384–393.

[74] Jerobin, J., Haque, M. Z., Uppada, J. B., Mohammad, R. M., Bhat, and A. A. (2015). Recent developments in nano medicine; treatment options for colorectal cancer. Hyderabad: OMICS Group eBooks.

[75] Tummala, S., Kumar, M. S., and Prakash, A. (2015). Formulation and characterization of 5-Fluorouracil enteric coated nanoparticles for sustained and localized release in treating colorectal cancer. *Saudi Pharm. J.* 23, 308–314.

[76] Minelli, R., Serpe, L., Pettazzoni, P., Minero, V., Barrera, G., Gigliotti, C. L., et al. (2012). Cholesteryl butyrate solid lipid nanoparticles inhibit the adhesion and migration of colon cancer cells. *Br. J. Pharmacol.* 166, 587–601.

[77] Iijima, S. (1991). Helical microtubules of graphitic carbon. *Nature* 354:56.

[78] Rahiman, S., and Tantry, B. A. (2012). Nanomedicine current trends in diabetes management. *J. Nanomed. Nanotechol.* 3:5.

[79] Barone, P. W., Parker, R. S., and Strano, M. S. (2005). In vivo fluorescence detection of glucose using a single-walled carbon nanotube optical sensor: design, fluorophore properties, advantages, and disadvantages. *Anal. Chem.* 77, 7556–7562.

[80] Barone, P. W., Yoon, H., Ortiz-García, R., Zhang, J., Ahn, J. H., Kim, J. H., et al. (2009). Modulation of single-walled carbon nanotube photoluminescence by hydrogel swelling. *ACS Nano* 3, 3869–3877.

[81] Rangasamy, M., and Parthiban, K. G. (2010). Recent advances in novel drug delivery systems. *IJRAP* 1, 316–326.

[82] He, H., Pham-Huy, L. A., Dramou, P., Xiao, D., Zuo, P., and Pham-Huy, C. (2013). Carbon nanotubes: applications in pharmacy and medicine. *BioMed Res. Int.* 2013: 578290.

[83] Singh, B. G. P., Baburao, C., Pispati, V., Pathipati, H., Muthy, N., Prassana, S. R. V., et al. (2012). Carbon nanotubes. A novel drug delivery system. *Int. J. Res. Pharm. Chem.* 2, 523–532.

[84] Chen, Z., Pierre, D., He, H., Tan, S., Pham-Huy, C., Hong, H., and Huang, J. (2011). Adsorption behavior of epirubicin hydrochloride on carboxylated carbon nanotubes. *Int. J. Pharm.* 405, 153–161.

[85] Xiao, D., Dramou, P., He, H., Pham-Huy, L. A., Li, H., Yao, Y., et al. (2012). Magnetic carbon nanotubes: synthesis by a simple solvothermal process and application in magnetic targeted drug delivery system. *J. Nanopart. Res.* 14:984.

[86] Lay, C. L., Liu, J., and Liu, Y. (2011). Functionalized carbon nanotubes for anticancer drug delivery. *Expert Rev. Med. Devices* 8, 561–566.

[87] Madani, S. Y., Naderi, N., Dissanayake, O., Tan, A., and Seifalian, A. M. (2011). A new era of cancer treatment: carbon nanotubes as drug delivery tools. *Int. J. Nanomed.* 6, 2963–2979.

[88] Stirland, D. L. (2016). *Analytical Methods for Studying Intratumoral Drug Delivery in Solid Tumors.* Doctoral dissertation, The University of Utah, Salt Lake City, UT.

[89] Elhissi, A., Ahmed, W., Hassan, I. U., Dhanak, V. R., and D'Emanuele, A. (2011). Carbon nanotubes in cancer therapy and drug delivery. *J. Drug Deliv.* 2012:837327.

[90] Li, R., Wu, R. A., Zhao, L., Wu, M., Yang, L., and Zou, H. (2010). P-glycoprotein antibody functionalized carbon nanotube overcomes the multidrug resistance of human leukemia cells. *ACS nano* 4, 1399–1408.

[91] Kumari, A., Yadav, S. K., and Yadav, S. C. (2010). Biodegradable polymeric nanoparticles based drug delivery systems. *Coll. Surf. B* 75, 1–18.

[92] Veiseh, O., Gunn, J. W., and Zhang, M. (2010). Design and fabrication of magnetic nanoparticles for targeted drug delivery and imaging. *Adv. Drug Deliv. Rev.* 62, 284–304.

[93] Hillaireau, H., and Couvreur, P. (2009). Nanocarriers' entry into the cell: relevance to drug delivery. *Cell. Mol. Life Sci.* 66, 2873–2896.

[94] Yuan, F., Thiele, G. M., and Wang, D. (2011). Nanomedicine development for autoimmune diseases. *Drug Dev. Res.* 72, 703–716.

[95] Sekhon, B. S., and Kamboj, S. R. (2010). Inorganic nanomedicine—part 2. *Nanomedicine* 6 (5), 612–618.

[96] Sekhon, B. S., and Kamboj, S. R. (2010). Inorganic nanomedicine—part 1. *Nanomedicine* 6, 516–522.

[97] Cavalcanti, A., Shirinzadeh, B., Freitas, R. A. Jr., and Hogg, T. (2007). Nanorobot architecture for medical target identification. *Nanotechnology*, 19, 015103.

[98] Hollmer, M. (2012). Carbon nanoparticles charge up old cancer treatment to powerful effect. *Fierce Drug Deliv.* http://www.fiercepharma.com/drug-delivery/carbon-nanoparticles-charge-up-old-cancer-treatment-to-powerful-effect

[99] Garde, D. (2012). Chemo bomb nanotechnology effective in halting tumors. Available at: fiercedrugdelivery.com

[100] Peiris, P. M., Bauer, L., Toy, R., Tran, E., Pansky, J., Doolittle, E., and Griswold, M. A. (2012). Enhanced delivery of chemotherapy to tumors using a multi-component nanochain with radiofrequency-tunable drug release. *ACS Nano*, 6, 4157.

[101] Radovic-Moreno, A. F., Lu, T. K., Puscasu, V. A., Yoon, C. J., Langer, R., and Farokhzad, O. C. (2012). Surface charge-switching polymeric nanoparticles for bacterial cell wall-targeted delivery of antibiotics. *ACS Nano*, 6, 4279.

[102] Haque, F., Shu, D., Shu, Y., Shlyakhtenko, L. S., Rychahou, P. G., Evers, B. M., ana Guo, P. (2012). Ultrastable synergistic tetravalent RNA nanoparticles for targeting to cancers. *Nano Today* 7, 245–257.

[103] Ahmed, R. Z., Patil, G., and Zaheer, Z. (2013). Nanosponges – a completely new nano-horizon: pharmaceutical applications and recent advances. *Drug Dev. Ind. Pharm.* 39, 1263–1272.

[104] Mishra, M., Kumar, H., and Tripathi, K. (2008). Diabetic delayed wound healing and the role of silver nanoparticles. *Dig. J. Nanomater. Biosyst.* 3, 49–54.

[105] Feldman, E. L., Stevens, M. J., and Greene, D. A. (1996). Pathogenesis of diabetic neuropathy. *Clin. Neurosci.* 4, 365–370.

[106] Coleman, D. L. (1982). Diabetes-obesity syndromes in mice. *Diabetes*, 31(Supplement 1), 1–6.

[107] Yue, D. K., McLennan, S., Marsh, M., Mai, Y. W., Spaliviero, J., Delbridge, L., and Turtle, J. R. (1987). Effects of experimental diabetes, uremia, and malnutrition on wound healing. *Diabetes* 36, 295–299.

[108] Warheit, D. B. (2004). Nanoparticles: health impacts? *Mater. Today* 7, 32–35.

[109] Wright, J. B., Lam, K., Hansen, D., and Burrell, R. E. (1999). Efficacy of topical silver against fungal burn wound pathogens. *Am. J. Infect. Control* 27, 344–350.

[110] Demling, R. H., and Desanti, L. (2001). Effects of silver on wound management. *Wounds* 13, 4–15.

[111] Yin, H. Q., Langford, R., and Burrell, R. E. (1999). Comparative evaluation of the antimicrobial activity of ACTICOAT antimicrobial barrier dressing. *J. Burn Care Res.* 20, 195–200.

[112] Zhang, F., Chan, S. W., Spanier, J. E., Apak, E., Jin, Q., Robinson, R. D., and Herman, I. P. (2002). Cerium oxide nanoparticles: size-selective formation and structure analysis. *Appl. Phys. Lett.*

[113] Chung, D. (2003). Nanoparticles have health benefits too. *New Sci.* 179, 2410–2416.

[114] Robinson, R. D., Spanier, J. E., Zhang, F., Chan, S. W., and Herman, I. P. (2002). Visible thermal emission from sub-band-gap laser excited cerium dioxide particles. *J. Appl. Phys.* 92, 1936–1941.

[115] Kilbourn, B. T. Yttrium oxide. *Encycl. Adv. Mater.* 4, 2957–2959.

[116] Becker, S., Soukup, J. M., and Gallagher, J. E. (2002). Differential particulate air pollution induced oxidant stress in human granulocytes, monocytes and alveolar macrophages. *Toxicol. in vitro*, 16, 209–218.

[117] Guo, R., Song, Y., Wang, G., and Murray, R. W. (2005). Does core size matter in the kinetics of ligand exchanges of monolayer-protected Au clusters?. *J. Am. Chem. Soc.* 127, 2752–2757.

[118] Yakimovich, N. O., Ezhevskii, A. A., Guseinov, D. V., Smirnova, L. A., Gracheva, T. A., and Klychkov, K. S. (2008). Antioxidant properties of gold nanoparticles studied by ESR spectroscopy. *Russ. Chem. Bull.* 57, 520–523.

[119] BarathManiKanth, S., Kalishwaralal, K., Sriram, M., Pandian, S. R. K., Youn, H. S., Eom, S., and Gurunathan, S. (2010). Anti-oxidant effect of

gold nanoparticles restrains hyperglycemic conditions in diabetic mice. *J. Nanobiotechnol.* 8, 16.

[120] Shaw, J. E., Sicree, R. A., and Zimmet, P. Z. (2010). Global estimates of the prevalence of diabetes for 2010 and 2030. *Diabet. Res. Clin. Pract.* 87, 4–14.

[121] Clore, J. N., and Thurby-Hay, L. (2004). Basal insulin therapy. *Curr. Diab. Rep. 4*, 342–345.

[122] Senthilnathan, B., Bejoy, J., Suruthi, L., Valentina, P., and Robertson, S. (2016). Nanorobots–a hypothetical concept of interest. *Pharma Sci. Monit.* 7, 70–84.

[123] Muthukumaran, G., Ramachandraiah, U., and Samuel, D. G. (2015). Role of nanorobots and their medical applications. *Adv. Mater. Res.* 1086, 61–67.

[124] Manjunath, A., and Kishore, V. (2014). The promising future in medicine: nanorobots. *Biomed. Sci. Eng. 2*, 42–47.

[125] Robert, A. F. J. (2009). "Medical nanorobotics: the long-term goal for nanomedicine," in *Nanomedicine Design of Particles, Sensors, Motors, Implants, Robots, and Devices*, eds J. M. Schulz, V. N. Shanov (Norwood MA: Artech House), 367–392.

[126] Bhat, A. S. (2014). Nanobots: the future of medicine. *Int. J. Manage Eng. Sci. 5*, 44–49.

[127] Abhilash, M. (2010). Nanorobots. *Int. J. Pharma Biol. Sci.* 1, 1–10.

[128] Kshirsagar, N., Patil, S., Kshirsagar, R., Wagh, A., and Bade, A. (2014). Review on application of nanorobots in health care. *World J. Pharm. Pharm. Sci.* 3, 472–80.

[129] Abeer, S. (2012). Future medicine: nanomedicine. *JIMSA* 25, 187–192.

[130] Hussan Reza, K., Asiwarya, G., Radhika, G., and Bardalai, D. (2011). Nanorobots: the future trend of drug delivery and therapeutics. *Int. J. Pharm. Sci. Rev. Res. 10,* 60–68.

[131] McDermott, U., Downing, J. R., and Stratton, M. R. (2011). Genomics and the continuum of cancer care. *N. Engl. J. Med.* 364, 340–350.

[132] Lu, Z. R., Ye, F., and Vaidya, A. (2007). Polymer platforms for drug delivery and biomedical imaging. *J. Control. Release* 122, 269–277.

[133] Patel, K., Angelos, S., Dichtel, W. R., Coskun, A., Yang, Y. W., Zink, J. I., et al. (2008). Enzyme-responsive snap-top covered silica nanocontainers. *J. Am. Chem. Soc.* 130, 2382–2383.

[134] Xue, D., Zheng, Q., Zong, C., Li, Q., Li, H., Qian, S., et al. (2010). Osteochondral repair using porous poly (lactide-co-glycolide)/

nano-hydroxyapatite hybrid scaffolds with undifferentiated mesenchymal stem cells in a rat model. *J. Biomed. Mater. Res. A* 94, 259–270.

[135] Park, J. S., Yang, H. N., Woo, D. G., Jeon, S. Y., Do, H. J., Lim, H. Y., et al. (2011). Chondrogenesis of human mesenchymal stem cells mediated by the combination of SOX trio SOX5, 6, and 9 genes complexed with PEI-modified PLGA nanoparticles. *Biomaterials* 32, 3679–3688.

[136] Perán, M., García, M. A., Lopez-Ruiz, E., Jiménez, G., and Marchal, J. A. (2013). How can nanotechnology help to repair the body? Advances in cardiac, skin, bone, cartilage and nerve tissue regeneration. *Materials* 6, 1333–1359.

[137] Wu, C. Y., Roybal, K. T., Puchner, E. M., Onuffer, J., and Lim, W. A. (2015). Remote control of therapeutic T cells through a small molecule–gated chimeric receptor. *Science* 350:aab4077.

[138] Zhang, L., and Webster, T. J. (2009). Nanotechnology and nanomaterials: promises for improved tissue regeneration. *Nano Today* 4, 66–80.

[139] Mangraviti, A., Tzeng, S. Y., Kozielski, K. L., Wang, Y., Jin, Y., Gullotti, D., et al. (2015). Polymeric nanoparticles for nonviral gene therapy extend brain tumor survival *in vivo*. *ACS Nano* 9, 1236–1249.

[140] Schwerdt, J. I., Goya, G. F., Calatayud, P., Herenu, C. B., Reggiani, P. C., and Goya, R. G. (2012). Magnetic field-assisted gene delivery: achievements and therapeutic potential. *Curr. Gene Ther.* 12, 116–126.

[141] Chorny, M., Fishbein, I., Tengood, J. E., Adamo, R. F., Alferiev, I. S., and Levy, R. J. (2013). Site-specific gene delivery to stented arteries using magnetically guided zinc oleate-based nanoparticles loaded with adenoviral vectors. *FASEB J.* 27, 2198–2206.

3

Chitosan and Its Derivatives as a Potential Nanobiomaterial: Drug Delivery and Biomedical Application

Abhay Raizaday[1], Hemant K. S. Yadav[2]
and Susmitha Kasina[1]

[1]Department of Pharmaceutics, JSS College of Pharmacy,
JSS University, Mysore, Karnataka, India
[2]Department of pharmaceutics, RAK Medical & Health Sciences
University, Ras Al Khaimah, UAE

Abstract

Chitosan is a unique polysaccharide, a copolymer of glucosamine and N-acetyl glucosamine linked by β 1–4 glucosidic bonds obtained by deacetylation of chitin. This has applications in wide variety of pharmaceutical and non-pharmaceutical fields due to its non-toxic, biocompatible and biodegradable properties. But the limitation of this polymer falls with its poor solubility at low pH and many organic solvent, which was overcome by chemical modification of the initial skeleton. Hence, chitosan and its derivatives have been used for mucoadhesion, permeation enhancement, wound healing, vaccine delivery, gene delivery etc. Chitosan and its derivatives have also been used for development of various drug delivery systems like transdermal, ocular, hydrogels, microparticles, targeted delivery based on pH sensitivity to various parts of gastro-intestinal tract like sublingual, buccal, intestine, colon. The applicability of chitosan and its derivatives in drug delivery along with the products marketed, past and current research trends will be discussed in this chapter.

3.1 History

The origin of chitosan can be traced back to 1811 when "chitin", from which it is derived, was first discovered by Henri Braconnot, a professor of the natural history in France. According to some researches, while Braconnot was conducting research on mushrooms, he isolated what was later to be called chitin. Chitin was the first polysaccharide identified by man, preceding cellulose by about 30 years. In the 1830s, there was a man who authored an article on insects in which he noted that similar substance was present in the structure of insects as well as the structure of plants. He then called this amazing substance as "chitin".

Basically, the name chitin is derived from Greek, meaning "tunic" or "envelope". The concept was further known in 1843 when Lassaigne demonstrated the presence of nitrogen in chitin. In 1859, Professor C. Rouget subjected chitin to alkali treatment, which resulted in a substance that could, unlike chitin itself, be dissolved in acids. The term "chitosan" was given to deacetylated chitin by Hoppe–Seiler. While chitin remained an unused natural resource for a long time, interest in this polymer and its derivatives such as chitosan and chito-oligosaccharides has increased in recent years due to their unique properties.

Intense interest applications grew in the 1930s; however, the lack of adequate manufacturing facilities and competition from synthetic polymers hampered the commercial development in this period. Renewed interest in the 1970s was encouraged by the need to better utilize shellfish shells and the scientists worldwide began to chronicle the more distinct properties of chitin and chitosan to understand the potential of these natural polymers. In the early 1960s, chitosan was investigated for its ability to bind with the red blood cells. That time also, it was considered as a hemostatic agent. Then, for the past three decades, chitosan has been used in water purification. Since then, numerous research studies have been undertaken to find ways to use these materials [1].

3.2 Chemistry

Chitosan is a linear, semi-crystalline amino polysaccharide of glucosamine and *N*-acetyl glucosamine units. Chitosan may be considered as a family of linear binary copolymers of (1,4)-2-acetamido-2-deoxy-b-D-glucan (*N*-acetyl D-glucosamine) and (1,4)-2-amino-2-deoxyb-D-glucan (D-glucosamine) units. The term chitosan does not refer to a uniquely defined compound; it merely refers to polysaccharides having different composition of both the units.

As such, chitosan is not extensively present in the environment—however, it can be easily derived from the partial deacetylation of a natural Polymer, Chitin (Figure 3.1). Chitin is a linear polymer of (1→4)-linked 2-acetamido-2-deoxy-β-D-glucopyranose (GlcNAc; A-unit), which is insoluble in aqueous solvents. It also has many structural similarities with cellulose such as conformation of the monomers and diequatorial glycosidic linkages.

The deacetylation degree of chitosan, giving indication of the number of amino groups along the chains, is calculated as the ratio of D-glucosamineto the sum of D-glucosamine and *N*-acetyl D-glucosamine. To be named "chitosan", the deacetylated chitin should contain at least 60% of D-glucosamine residues which means the deacetylation degree must be at least 60. The deacetylation of chitin is conducted by chemical hydrolysis under severe alkaline conditions or by enzymatic hydrolysisin the presence of particular enzymes, among of chitindeacetylase. Various analytical techniques are used to confirm the structure of chitosan. FTIR spectroscopy is considered to be a very attractive technique, as it is non-destructive, fast, sensitive, user-friendly, low-priced, and suitable for both soluble and non-soluble samples.

Proton NMR spectroscopy is a convenient and accurate method for determining the chemical structure of chitosan and its derivatives. NMR measurements of chitosan compounds are, however, limited to samples that

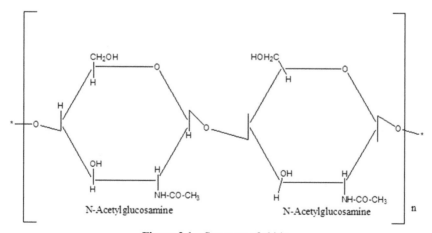

Figure 3.1   Structure of chitin.

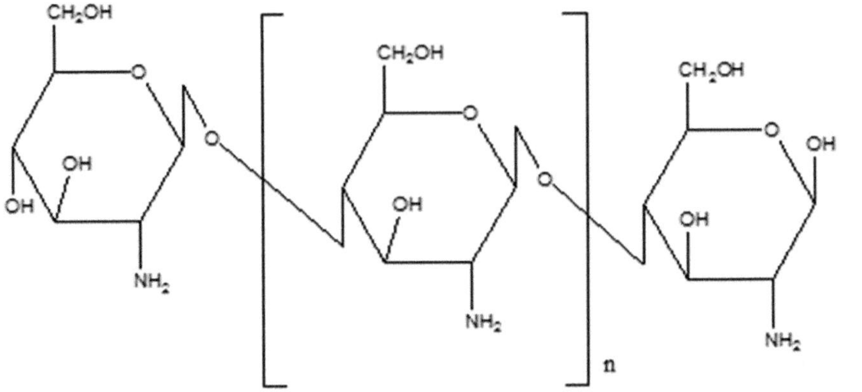

Figure 3.2   Structure of chitosan.

are soluble in the solvent, which limits the analysis of chitosan with DA values lower than 0.3 in aqueous solutions. A typical proton NMR spectrum of chitosan is shown in Figure 3.2. The signal at δ 3.20 ppm was attributed to H-2 of GlcN residue. The intense band at 4.8–5.30 ppm is related to OH groups and HDO (solvent). In this region, as observed more clearly from an extended spectrum, some different anomeric protons (H-1 of GlcN and GlcNAc units) are appeared at 4.88–5.00 ppm [1, 2].

3.3 Advantages

The following major characteristics of chitosan make this polymer advantageous for numerous applications:

1. It has a defined chemical structure;
2. It can be chemically and enzymatically modified;
3. It is physically and biologically functional;
4. It is biodegradable and biocompatible with many organs, tissues, and cells;
5. It can be processed into several products including flakes, fine powders, beads, membranes, sponges, cottons, fibres, and gels. Consequently, chitosan has found considerable.
6. Application in various industrial areas.
7. Non-toxic in nature,
8. Chitosan is widely-used as an antimicrobial agent either alone or blended with other natural polymers.

9. It has some specific properties such as polyoxy salt formation, poly-electrolyte complexation with oppositely charged polymers (guar gum, carragenans, etc.)
10. Film forming property.
11. It has several biological and therapeutic applications such as choles-terol lowering activity, antihypertension activity, fungistatic or bacte-riostatic, anticancerogen, regenerative effect on connective gum tissue, haemostatic, spermicidal, antitumor, and anticholesteramic activity.

3.4 Disadvantages

The main drawback is its poor solubility at physiological pH owing to the partial protonation of the amino groups, thereby causing presystemic metabolism of drugs in intestinal and gastric fluids in the presence of proteolytic enzymes [1].

3.5 Properties of Chitosan

3.5.1 Physicochemical

Since chitosan is a heterogeneous polymer consisting of GlcN and GlcNAc units, its properties depend on the structure and composition [1–5].

3.5.1.1 Crystalline structure

Crystallinity of chitosan depends on the method of its preparation and the conditions employed in the procedure. For example, in case of chitosan samples prepared by two different procedures: (a) the partial deacetyla-tion of chitin, and (b) the partial reacetylation of a pure chitosan, it was observed that the partially reacetylated material was less crystalline than pure chitosan.

Generally, the step of dissolving the polymer results in a decrease in the crystallinity of the material. However, it also depends on the secondary treatment (reprecipitation, drying, and freeze-drying). In addition, the origin may affect the residual crystallinity of chitosan, which in turn controls the accessibility to internal sorption sites and the diffusion properties (rehydra-tion and solute transport) and also deacetylation procedure may affect the solid state structure of chitosan.

3.5.1.2 Degree of *N*-Acetylation

An important parameter to examine closely is the degree of acetylation (DA) in chitin. It is defined as the ratio of Glc-NAc to GlcN structural units. In chitin, the acetylated units prevail (DA $\geq$ 90%), whereas chitosan is fully or partially *N* deacetylated derivative with a DA of less than 30%.

The DA is calculated from the integral ratio between protons of acetyl group and the GlcN protons. The degree of deacetylation (DA) is calculated from the integral ratio between the proton on C-2 and the glucose unit protons.

This ratio has a striking effect on chitin and chitosan solubility and solution properties. To define this ratio, several analytical methods like IR, UV, and NMR spectroscopies, pyrolysis gas, gel permeation chromatography (GPC), thermal analysis, various titration schemes, and acid hydrolysis are employed. The search for quick, user-friendly, low cost, and accurate method to determine the DA has been one of the major concerns over many decades.

The degree of deacetylation of chitin can also be increased by increasing the temperature or strength of the alkaline solution. The degree of deacetylation can also be determined by its ratio of 2-acetamido-2-deoxy-D-glucopyranose to 2-amino-2-deoxy-D-glucopyranose structural units. When the number of 2-amino-2-deoxy-D-glucopyranose units is more than 50%, the biopolymer is said to be Chitosan and when the number of 2-acetamido-2-deoxy-Dglucopyranose units is higher, the polymer is said to be chitin.

3.5.1.3 Molecular weight

The molecular weight of the CS also depends on viscosity, solubility, elasticity and tears strength. Chitosan obtained from deacetylation of crustacean chitin may have a MW over 100,000 Da. In order to evaluate the MW of polymeric chain, various methods can be used extensively. Viscometric and GPC techniques are easy to perform and low time consuming.

Method:

- The viscosity of chitosan solutions is measured by using Ubbelohde Viscometer. The running times of the solution and solvent are used to calculate the relative viscosity (η_{rel}), specific viscosity (η_{sp}), and reduced viscosity (η_{red}) as follows:

$$\eta_{rel} = t_{ch}/t_{sol}$$
$$\eta_{rel} = t_{ch}/t_{sol}$$
$$\eta_{red} = \eta_{sp}/c$$

where,

t_{ch} and t_{sol} are the running times of the chitosan solution and solvent, respectively, c is the chitosan concentration in g/dL.

- The intrinsic viscosity, defined as

$[\eta] = C(\eta_{red})c = 0.$
It is obtained by extrapolating the η_{red} versus concentration data to zero concentration and the intercept on the ordinate.

- Finally, the average molecular weight (M) is calculated based on the MHKS equation

$$[\eta] = KMa,$$

Where K and a are viscometric parameters depending on the solvent.

3.5.1.4 Solubility

Chitosan, being a cationic polysaccharide in neutral or basic pH conditions, contains free amino groups on C-2 of GlcN unit and hence, is insoluble in water. In acidic pH, amino groups can undergo protonation thus, making it soluble in water. Therefore, solubility of chitosan depends upon the distribution of free amino and N-acetyl groups.

Usually 1–3% aqueous acetic acid solutions are used to solubilize chitosan.

The extent of solubility depends on the concentration and type of acid, whereas the solubility decreases with increasing concentration of acid and aqueous solutions of some acids such as phosphoric, sulfuric, and citric acids are not good solvents. Number of solvents for chitin and chitosan can be found in the literature.

Generally, the solubility decreases with an increase in MW. Moreover, few attempts have been made to enhance chitosan's solubility in organic solvents. However, many other attempts have been made to enhance its solubility in water. One major reason is because most biological applications for chemical substances require the material to be processible and functional at neutral pH. Thus, obtaining a water soluble derivative of chitosan is an important step towards the further application as a biofunctional material.

3.5.1.5 Viscosity

Viscosity is an important factor in the conventional determination of chitosan MW and in determining its commercial applications. Higher MW chitosan often render highly viscous solutions, which may not be desirable for industrial handling. However, lower viscosity chitosans may facilitate easy handling. The solution viscosity of chitosan depends on its molecular size, cationic character, and concentration as well as the pH and ionic strength of the solvent. Chitosan is a pseudoplastic material and an excellent viscosity-enhancing agent in acidic environments. The viscosity of CS solution increases with an increase in Chitosan concentration and decreases with increase in temperature. The viscosity of Chitosan also influences the biological properties such as wound-healing properties as well as biodegradation by lysozyme. Since Chitosan is hydrophilic in nature, thereby it has the ability to form gels at acidic pH. This type of gels can be used as a slow-release drug-delivery system.

3.5.2 Biological Properties

CS has been used as a safe excipient in drug formulations. Clinical tests of CS has been carried out in order to promote CS-based biomaterials do not report any inflammatory or allergic reactions following implantation, injection, topical application or ingestion in the human body. It shows a variety of biological activities such as:

- Phytoalexin elicitor activity,
- Activation of immune response,
- Cholesterol lowering activity
- Antihypertension activity
- Fungistatic or bacteriostatic
- Anticancerogen
- Regenerative effect on connective gum tissue
- Accelerates the formation of osteoblast responsible for bone formation
- Haemostatic
- Fungistatic
- Spermicidal
- Antitumor
- Anticholesteramic
- Central nervous system depressant
- Immunoadjuvant

3.5.2.1 Mucoadhesive properties

Due to its bioadhesive property, it can adhere to hard and soft tissues and has been used in dentistry, orthopedics and ophthalmology and in surgical procedures. It adheres to epithelial tissues and to the mucus coat present on the surface of the tissues.

The mucoadhesive properties of chitosan are probably attributable to its cationic character. Furthermore, hydrophobic interactions may help with the mucoadhesive components. Indeed, the mucoadhesion of chitosan for example, can be explained by the presence of negatively charged residues (sialic acid) in the mucin – the glycoprotein that composes the mucus. In acidic medium, chitosan amino groups are positively charged and can thus interact with the mucin.

3.5.2.2 Permeation enhancing properties

Based on the positive charges of chitosan, it was found that these charges are responsible for the mechanism of permeation enhancement, which can interact with the cell membrane of chitosan, resulting in a structural reorganization of tight junction-associated proteins. The permeation enhancing properties and toxicity to a large extent were attributable to the structural properties of chitosan including the degree of deacetylation and molecular mass. As for mucoadhesion, if chitosan degree of deacetylation increases, the permeation ability also increases. Thus, chitosans with high molecular mass and high degree of deacetylation exhibited a comparatively higher increase in epithelial permeability, which could be due to molecular mass and other permeation enhancing polymers such as polyacrylates.

Chitosan can be combined with other permeation enhancers because it acts in a completely different manner from these enhancers, leading to an additive or even a synergistic effect. Using this strategy, the oral bioavailability of ganciclovir could even be improved by 4-fold, using a combination of sodium dodecyl sulfate and chitosan compared with just a 2-fold improvement with sodium dodecyl sulfate alone.

3.5.2.3 Haemostatic activity

The haemostatic activity of chitosan can also be related to the presence of positive charges on chitosan backbone. Indeed, red blood cells membranes are negatively charged, and can thus interact with the positively charged chitosan. Besides, chitin shows less effective haemostatic activity than chitosan, which tends to confirm this explanation.

3.5.2.4 Antimicrobial activity

The case of chitosan antimicrobial activity is slightly more complex; two main mechanisms have been reported in the literature to explain chitosan antibacterial and antifungal activities.

In the first proposed mechanism, positively charged chitosan can interact with negatively charged groups at the surface of cells, and as a consequence, alter its permeability. This would prevent essential materials to enter the cells or/and lead to the leaking of fundamental solutes out of the cell.

The second mechanism involves the binding of chitosan with the cell DNA (still via protonated amino groups), which would lead to the inhibition of the microbial RNA synthesis. Chitosan antimicrobial property might in fact result from a combination of both mechanisms.

3.5.2.5 Analgesic effect

Polycationic nature of chitosan also allows explaining chitosan analgesic effects. Indeed, the amino groups of the D-glucosamine residues can protonate in the presence of proton ions that are released in the inflammatory area, resulting in an analgesic effect.

3.5.2.6 Biodegradability

Chitosan is not only a polymer bearing amino groups, but also a polysaccharide, which consequently contains breakable glycosidic bonds. Chitosan is actually degraded *in vivo* by several proteases, and mainly lysozyme. Till now, eight human chitinases have been identified, three of them possessing enzymatic activity on chitosan. The biodegradation of chitosan leads to the formation of non-toxic oligosaccharides of variable length. These oligosaccharides can be incorporated in metabolic pathways or be further excreted. The degradation rate of chitosan is mainly related its degree of deacetylation, but also to the distribution of *N*-acetyl D-glucosamine residues and the molecular mass of chitosan.

3.6 Extraction of Chitosan

A variety of procedures have been developed and proposed over the years for preparation of pure chitosan. Chitin is the raw material for all commercial production of chitosan with estimated annual production of 2000 tons. Most commonly, chitin forms the skeletal structure of invertebrates. At least 10 Gtons (1×1013 kg) of chitin are constantly present in the biosphere. After cellulose, chitin is the second most abundant biopolymer and is commonly found in invertebrates – ascrustacean shells or insect cuticles – but also in some mushrooms envelopes, green algae cell walls, and yeasts. Insect's wing and fungi's cell wall are also reported to contain chitin. Shrimp consists about 45% of raw material used for processed seafood industry and among them about 30–40% by weight of raw shrimp is discarded as waste which composed of the its exoskeleton (shells) and cephathoraxes [6]. At industrial scale, the two main sources of chitosan are crustaceans and fungal mycelia.

The animal source shows however some drawbacks as seasonal, of limited supplies and with product variability which can lead to inconsistent physicochemical characteristics.

The mushroom source offers the advantage of a controlled production environment all year round that insures a better reproducibility of the resulting chitosan, the physical properties of extracted chitosan being notably related to the growth substrate composition. Moreover, the vegetable source is generally preferred to the crustacean one from anallergenic point of view – the produced chitosan being safer for biomedical and healthcare applications. The mushroom-extracted chitosan typically presents a narrower molecular mass distribution than the chitosan produced from seafood, and may also differ in terms of molecular mass, DD and distribution of deacetylated groups. Production of chitosan from chitin in general involves the following steps.

1. **Demineralization, Deproteinization and Decolourization**

   - Proteins are removed from ground shells by treating them with either sodium hydroxide or by digestion with proteolytic enzymes such as papain, pepsin, trypsin and protease.
   - Minerals such as calcium carbonate and calcium phosphate are extracted with hydrochloric acid.
   - Pigments such as melanin and carotenoids are eliminated with 0.02% potassium permanganate at 60°C or hydrogen peroxide or sodium hypochlorite.

2. **Conversion of chitin to chitosan**

   - It is achieved by hydrolysis of acetamide groups of chitin.
   - This is normally conducted by severe alkaline hydrolysis treatment due to the resistance of such groups imposed by the *trans*-arrangement of the C2-C3 substituents in the sugar ring.
   - Thermal treatments of chitin under strong aqueous alkali are usually needed to give partially deacetylated chitin (degree of acetylation, DA <30%), regarded as chitosan. Usually, this process is achieved by treatment with concentrated sodium or potassium hydroxide solution (40–50%) at 100°C or higher to remove some or all the acetyl groups from the polymer. This process, called

deacetylation, releases amine groups (NH_2) and gives the chitosan a cationic characteristic. This is especially interesting in an acid environment where the majority of polysaccharides are usually neutral or negatively charged.

- The deacetylation process is carried out either at room temperature (homogeneous deacetylation) or at elevated temperature (heterogeneous deacetylation), depending on the nature of the final product desired. However, the later is preferred for industrial purposes.
- In some cases, the deacetylation reaction is carried out in the presence of thiophenol as a scavenger of oxygen or under N_2 atmosphere to prevent chain degradation that invariably occurs due to peeling reaction under strong alkaline conditions (Figure 3.3).

3.6.1 Preparation of Chitosan and Water Soluble Chitosan

The chitin was put into NaOH at 60°C for 8h to prepare crude chitosan. After filtration, the residue was washed with hot distilled water at 60°C for three times. The crude chitosan was obtained by drying in an air oven at 50°C overnight. Crude chitosan was added into 2% acetic acid in a water-bath

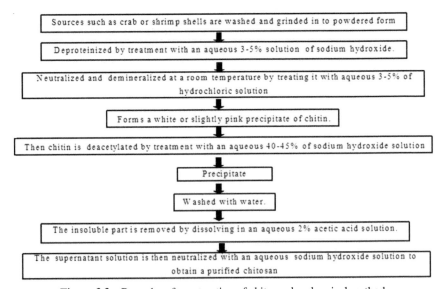

Figure 3.3   Procedure for extraction of chitosan by chemical method.

shaker. The conditions were set as follows: H_2O_2 level (4%), time (4 h) and temperature (60°C). After reaction, NaOH was used to adjust the solution to neutrality. The residue was removed by filtration, while twofold volumes of ethanol were added to the filtrate. The crystal of water-soluble chitosan was liberated after incubation at ambient condition overnight and dried in an air oven at 50°C [7].

3.6.1.1 Extraction of Chitin from the Beetle

Traditionally, chitin is prepared mainly from crab and shrimp shells obtained as by products in the seafood industry. Chitin is also a primary component in insect cuticles. Therefore, insects are an alternative source of chitin and, consequently, of chitosan. Recently, the production of chitin and chitosan from insect sources has drawn increased attention [8].

Advantages:

- First, insects possess enormous biodiversity and represent 95% of the animal kingdom. Therefore, they offer a tremendous potential as a natural resource for chitin and chitosan production.
- Insect cuticles have lower levels of inorganic material compared to crustacean shells, which makes their demineralization treatment more convenient.

The most common method for chitin extraction from insects involves two steps

i. An acidic step to remove catechols and
ii. A basic step to remove the cuticle proteins

- Adult beetles (***Holotrichiaparallela***) were starved for 48 h to eliminate gut contents, washed with water and killed by freezing.
- They were allowed to thaw at room temperature and then air-dried at 50°C for 2 days.
- The dried beetles were milled to a powder to pass through a 20-mesh screen and stored at 4°C in airtight containers.
- The powder was treated with 1 M HCl solution (250 ml) at 100°C for 30 min to remove minerals and catechols.
- The demineralization step was followed by rinsing with distilled water until neutrality was reached.
- Deproteinization was performed using alkaline treatment with 1 M NaOH (250 ml) solution at 80°C for 24 h, and the product was washed with distilled water until the pH became neutral.

- For the purpose of decolorization, the precipitate was treated further with 1% potassium permanganate solution (100 mL) for 1 h.
- Finally, lightly brown chitin was washed with distilled water and dried at 50°C in a dry heat sterilizer.

3.6.1.2 Extraction of collagen from squid

Collagen, the major protein of the connective tissue, exists in a concentration from 3% to 11.1% in the mantle of some squid species like Illex and Loligo, whereas

In Dosidicusgigas mantle, collagen was found in a concentration of 18.33%.

- Mantle was manually skinned, cut in portions of 100 g, packed in polyethylene bags, and stored at freezing temperature for no longer than 35 days until collagen extraction.
- All operations were performed at 4°C. Preparation of Acid–soluble collagen involves a preliminary extraction with 6 M urea in sodium acetate (pH 6.8) solution and neutral buffer (0.05 M Tris + 1 M NaCl, pH 7.2) to remove non-collagenous proteins.
- For salt-soluble extraction, the mantle was homogenized during 2 min with three volumes (v/w) of urea solution, stirred for 24 h, and centrifuged at 10,000 g for 30 min.
- Then, three volumes of neutral buffer are added to the residue, stir for another 24 h, and centrifuge at 10,000 g for 30 min.
- Finally, three volumes of 0.5 M acetic acid are added to the residue, stir for 24 h and centrifuge at 10,000 g for 30 min.
- The supernatant was collected, freeze dried, and used as Acid soluble chitosan fraction.

3.6.1.3 Extraction of chitosan from fungi cell wall

Conventionally, an industrial scale, chitosan is mainly derived from the waste product of crustacean exoskeletons obtained after the industrial processing of seafood, such as shrimp, crabs, squids, and lobsters shell by chemical deacetylation, using a hot concentrated base solution (30–50% w/v) at high temperatures (<100°C) for a prolonged time. However, the chitosan obtained by such treatments suffers some inconsistencies, such as protein contamination, inconsistent levels of deacetylation, and high molecular weight, which results in variable physico-chemical characteristics. There are some additional problems, such as environmental issues due to the large amount of waste concentrated alkaline solution, seasonal limitation of seafood shell supply, and high cost. In this context, production and purification of chitosan from the cell walls of fungi grown under controlled conditions offers the advantage of being environmental-friendly and provides greater potential for a consistent product [12].

Chitosan is a substantial component of the cell wall of certain fungi, particularly those belonging to the class Zygomycetes including *Absidia, Gongronella, Mucor* and *Rhizopus* Fungal cell walls are composed of polysaccharides and glycoproteins. Polysaccharides, such as chitin and glucan are the structural components, whereas the glycoproteins, namely mannoproteins, galactoproteins, xylomanno proteins and glucurono proteins form the interstitial components of fungal cell walls.

There is already evidence that the enzymatic deacetylation takes place in fungi and certain bacteria to form chitosan. The formation of chitosan in cell walls of fungi is a result of the complex synergistic action of two enzymes:

i. Chitin synthase
ii. Chitin deacetylase (CDA).

In the first step, chitin synthase synthesizes a chain of chitin using the chitin precursor, uridino-di phospho *N*-acetylglucosamine.

In the next step, CDA hydrolyzes acetic groups from *N*-acetylglucosamine transforming it into glucosamine (GlcN) and finally forming chitosan.

Advantages:

There are several advantages of obtaining chitosan from fungi:

- Chitosan can be obtained free of allergenic shrimp protein, and the molecular weight and degree of deacetylation of fungal chitosan can be controlled by varying the fermentation conditions.
- Fermentative production of fungi on cheap biowaste is an unlimited and, in principle, a very economic source of chitosan.
- β-glucan, that can be isolated from the myceliar chitosan-glucan complex, has important medical applications.

1. Production of mycelia biomass

- *Gongronellabutleri*, was grown on peeled sweet potato cut into $1-1.5 \times 4-6$ cm pieces and washed with water.
- Mineral solution was prepared with distilled water containing 5 g $(NH4)_2SO_4$;

- 1 g $K_2 HPO_4$; 1 g NaCl; 0.5 g $MgSO_4$ $7H_2O$ and 0.1 g $CaCl_2$ $2H_2O$ and adjusted to pH 4.5 with 0.5 M H_2SO_4.
- Sweet potato pieces (850 g, wet based) were mixed with 850 ml mineral solution and sterilized by autoclaving at 121°C for 20 min.
- After cooling, the free solution was decanted (about 650–700 ml).
- The sweet potato pieces were inoculated with the fungal spore suspension.
- An airtight steamer solid substrate fermenter was used with 3 trays of perforated aluminium,
- The fermenter was sterilized at 165°C for 3 h.
- The inoculated sweet potato solid substrate was mounted in the tray-fermenter and incubated under a constant supply of filtered and humidified air at a flow rate 0.8 l min^{-1}.
- The outside temperature of the fermenter was maintained at 26 ± 2°C. The mycelia biomass obtained was dried at 45°C.

2. Extraction of chitosan

Method I:

- Dried mycelia were treated with NaOH per at 40°C for 16 h.
- Under this condition the chitin present in the cell wall is not deacetylated.
- The alkaline insoluble material was washed with water and dried.
- The free cell wall chitosan was extracted from 1 g of AIM with 200 ml of 0.35 M acetic acid at 40°C for 16 h.

Method II:

- Dried mycelia were treated with NaOH containing sodium borohydride per g 45°C for 13 h.
- AIM was washed with water until neutral pH and dried.
- Dried AIM (1 g) was treated with 200 ml of 0.35 M acetic acid at 95°C for 5 h. After that AIM suspended solution was adjusted to pH 4.5 with 1MNaOH, treated with 4% (v/v) α-amylase and incubated in a shaking water-bath at 65°C, 200 rpm for 3 h.
- Enzyme treated suspensions were centrifuged at 1600 g for 15 min to obtain a clear chitosan solution.

After extraction by method I or II, chitosan was precipitated by adjusting the pH of the supernatant to 8–9 with NaOH solution. The chitosan precipitate was washed and freezed-dried.

The yield of chitosan derived from a fungal biomass largely depends on several factors, such as: the strain of fungi used, cultivation methods, such as submerged shaken culture, batch culture, continuous culture, solid state culture and surface culture and process parameters, such as pH, temperature, mixing rate, incubation time, particle size, and broth rheology.

3.6.1.4 Extraction of Chitin, Chitosan, from Shrimp by biological method

Several techniques to extract chitin from various crustaceans have been reported.

Conventionally, preparation of chitin from marine waste materials involves demineralization and deproteinization with the use of strong acids and bases. However, these reagents can cause a partial deacetylation of the chitin and hydrolysis of the polymer, resulting in final inconsistent physiological properties. Alternatively, some efforts have been directed towards the reduction of chemical treatments in a more eco-friendly processes such as bacterial fermentation and treatment by proteolytic enzymes which have been applied for the deproteinization of crustacean wastes [10].

Raw Material and Reagents

- The shrimp shells were washed thoroughly with distilled water. The shells were then stored at $-20°C$ until further analysis.

Enzyme Production

- Bacterial Strain, Bacillus cereus is used as the source of protease enzyme.
- The protease production was conducted in SWP medium consisting of: SWP40.0,
- Ammonium chloride 2.0, K_2HPO_4 0.5, KH_2PO_4 0.5, and $MgSO_4 \cdot 7H_2O$ 0.1 (pH 8.0).
- Cultivations were performed on a rotatory shaker at150 rpm for 72 h at 30°C, in 250-ml conical flasks with a working volume of 25 ml.

Deproteinization of Shrimp Wastes by Protease

- Shell wastes were mixed with water at a ratio of 1:2 (w/v), minced then cooked for 20 min at 90°C.
- The cooked sample was then homogenized for about 2 min. The pH of the mixture was adjusted to 8.0.

- Then, the shrimp waste proteins were digested with crude enzyme using different enzyme/substrate (E/S) ratio.
- After incubation for 3 h at 40°C, the reaction was stopped by heating the solution at 90°C during 20 min to inactivate the enzyme. The shrimp waste protein hydrolysates were then centrifuged at 5,000×g for 20 min to separate insoluble and soluble fractions.
- The solid phase was washed and then dried for 1 h at 60°C. The supernatant was used for analysis of protein concentration, lyophilized, and then used as shrimp waste protein hydrolysate (SWPH).

Deproteinization by Alkali

- Shrimp wastes were mixed with NaOH solution and were allowed to react at room temperature for 4 h. After filtration, the solid residue was washed with deionized water until a neutral pH, and then dried at 65°C in an oven.

Demineralization

- Demineralization was carried out in a dilute HCl solution. Solid fractions obtained after hydrolysis by crude protease were treated with HCl in 1:10 ratio for 6 h at room temperature (25°C) under constant stirring.
- The chitin product was filtered through four layers of gauze with the aid of a vacuum pump and washed to neutrality with deionized water and then freeze-dried (Figure 3.4).

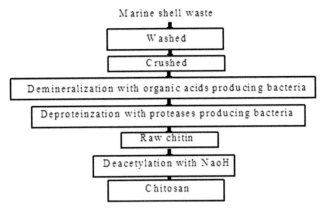

Figure 3.4 Procedure for extraction of chitosan by biological method.

3.7 Chitosan Derivatives

In order to improve or impart new properties to chitosan, chemical modification of the chitosan chains, generally by either grafting of small molecules or polymer chains onto the chitosan backbone or by quaternization of the amino groups, has been investigated. Chitosan chains possess three attractive reactive sites for chemical modification: two hydroxyl groups (primary or secondary) and one primary amine. The site of modification is dictated by the desired application of the final chitosan derivative. For example, the preservation of the primary amine is highly desirable for transfection application [11, 12].

3.7.1 Carboxymethylchitosan

Carboxymethyl-chitosan (CMCS), a new attractive biocompatible and biodegradable polymer, is obtained from the reaction of chloro acetic acid and chitosan in alkaline condition with easy building blocks for application in drug delivery, tissue engineering, and viscosupplementation.

Advantages: Because it contains cationic amine groups and anionic carboxyl groups in matrix, the swelling, drug permeation, and release properties could be controlled by the pH change.

The preparation of *O*-carboxymethylchitosan involves swelling the chitin in dimethylsulfoxide, treatment with concentrated sodium hydroxide solution, reaction with monochloroacetic acid and deacetylation in hot sodium hydroxide. Numerous studies have been conducted to obtain carboxyalkyl-chitosan. In these derivatives, the principle of transformations refers to a reaction by means of Schiff's reaction. The reaction between the carboxylic group present in organic acid and the amine group of chitosan leads to the formation of ceto-imine with low stability. This intermediate was then reduced with sodium cyanoborohydride or sodium borohydrate, in order to obtain a substitute chitosan which was stable and insoluble.

3.7.2 Mono-Carboxymethylated Chitosan

Mono-carboxymethylated chitosan (MCC) is a polyampholytic polymer, able to form visco-elastic gels in aqueous environments or with anionic macromolecules at neutral pH values. MCC appears to be less potent compared to the quaternized derivative of chitosan. Nevertheless, MCC was found to increase the permeation and absorption of low molecular weight heparin (LMWH; an anionic polysaccharide) across intestinal epithelia.

Chitosan derivative provokes damage of the cell membrane, and therefore they do not alter the viability of intestinal epithelial cells. The permeation is increased due to increase in pores formed in the cell membrane and there is no need to alter viability of epithelial cells.

3.7.3 *N*-Succinyl Chitosan

N-succinyl-chitosan was obtained by introduction of succinyl groups into chitosan terminal of the glucosamine units. Polyion complexes may be formed between the NH_3^+ and COO-groups in succinyl chitosan molecule.

Advantages:

- *N*-succinyl-chitosan displays good water soluble property at various pHs
- *N*-succinyl-chitosan is used as a wound dressing material,
- used as a cosmetic material.
- biocompatible, low toxicity and long term retention in the body.
- *N*-succinyl-chitosan is valuable as a drug carrier to readily prepare its conjugates with various drugs to avoid vexatious complications in cancer chemotherapy.

3.7.4 *N*-Acetylated Chitosan

The gastrointestinal tract poses a variety of morphological and physiological barriers to the expression of a target gene. *N*-acetylated chitosan is used as a gene delivery carrier to solve this problem. In particular, this result was most significant in the case of the duodenum, where the LacZ gene was expressed most effectively through the use of *N*-acetylated chitosan.

3.7.5 *N*-Trimethyl Chitosan

N-trimethyl chitosan (TMC) is one of the most commonly- studied chitosan derivatives. It was developed to improve the properties of chitosan and to overcome the main barrier in the use of chitosan in pharmaceutical applications, that is, its poor aqueous solubility at physiological pH. TMC has a fixed positive charge on the quaternary amino-group and the derivative is therefore highly soluble both in neutral and basic environments. The derivative has shown promising results when studied as a drug delivery system.

Advantages:

- Trimethyl chitosan (TMC) is a partially quarternized derivative of chitosan which is well soluble in a wide pH range (pH 1–9).
- TMC shows mucoadhesive properties.
- TMC has been proven to be a potent intestinal absorption enhancer of peptide and protein drugs, especially in neutral environments where chitosan is ineffective as an absorption enhancer.

The improvement in mucoadhesion and nanoparticles internalization capabilities makes trimethylchitosan-based nanosystems suitable carriers for the oral administration of macromolecules and, in particular, of peptides.

3.7.6 *N*-Trimethylchitosan Chloride

N-Trimethylchitosan chloride (NTMC), a chitosan derivative synthesized by the partial quaternization (trimethylation) of chitosan, is characterized by fixed positive charges irrespective of pH. This quaternized polymer forms complexes with anionic macromolecules.

Advantages:

- Trimethyl chitosan chloride (NTMC) is freely soluble over a wide pH range.
- NTMC has been proven to be a potent intestinal absorption enhancer for peptide drugs, even in neutral environments.
- This polymer is able to open tight junctions, which seal the paracellular pathways thereby facilitating the paracellular diffusion of peptide drugs. The process has the additional advantage of being reversible after removal of the polymer, leading to the resealing of the tight junctions.

3.7.7 Succinate and Chitosan Phthalate

Chitosan esters, such as Chitosan succinate and Chitosan phthalate have been used successfully as potential matrices for the colon-specific oral delivery of sodium diclofenac by converting the polymer from an amine to a succinate form, the solubility profile is changed significantly. The modified polymers were insoluble under acidic conditions and provided sustained release of the encapsulated agent under basic conditions. The same researchers also synthesized an iron cross-linked derivative of hydroxylated chitosan succinate, as a matrix for oral theophylline beads [13].

3.7.8 Amphiphilic Chitosan Derivatives

Aggregation of amphiphilic non-ionic polymers, block copolymers, and hydrophobically modified polyelectrolytes including block polyelectrolytes in organic solvents or aqueous solutions has been reported over the past two decades.

These polymeric amphiphilic self-assemblies have found widespread applications in the field of drug delivery systems because they are capable of forming aggregates in aqueous solutions. These aggregates are composed of a hydrophobic core and a hydrophilic shell. They are good vehicles for delivering hydrophobic drugs, as the drugs are protected from possible degradation by enzymes. The morphology of nanoparticles produced from amphiphilic polymers can be varied by changing the composition of hydrophobic and hydrophilic composites on the polymer chains. Various forms of morphologies have been reported, such as sphere, vesicles, rods, lamellas, tubes, large compound micelles, and large compound vesicles.

Amphiphilic chitosan derivatives can be prepared by grafting a hydrophilic polymer such as poly (ethylene glycol) (PEG) to the chitosan backbone. In that the chitosan-g-PEG forms aggregates in aqueous solution spontaneously by intermolecular hydrogen bonding. The other method is hydrophobical modification of chitosan, in which the modified chitosan self-aggregates can be used as a matrix for a plasmid DNA delivery system. The aggregation of hydrophobically modified chitosan can occur either within a single chain or among different chains.

3.7.9 Graft-Copolymerization of Chitosan

The graft-copolymer of Chitosan with 4-(6-methacryloxyhexyloxy)-4 nitro-biphenyl was synthesized by radical polymerization. Graft-copolymerization was carried out under homogeneous conditions with Azobisisobutyronitrile as initiator and 2% acetic acid as solvent. Homopolymerization was also carried out to obtain material for comparative analysis [14].

3.7.10 Thiolated Chitosan Conjugate

This modification of chitosan can be achieved by covalent attachment of isopropyl-S-acetylthioacetimidate to chitosan. Chitosan-TEA solutions of strength 0.5% and 1% showed the transition from sol to gel within 30 min. The polymer generated within this study represents a promising novel tool for various drug delivery systems, where in situ gelling properties are advantageous [15, 16].

3.7.11 Cyclodextrin (CD)-Chitosan Derivative

Cyclodextrins (CD) offer an alternative approach; these cyclic oligosaccharides have the ability to form non-covalent complexes with a number of drugs altering their physicochemical properties. β-cyclodextrins was successfully grafted onto a chitosan chain polymer with a cyclodextrin grafting yield of 7% and a CD-Chitosan yield of 85%. The synthesized Chitosan-CD polymer exhibits characteristics of a possible mucoadhesive drug delivery system with some inclusion properties from B-cyclodextrins. β-CD monoaldehyde was coupled to chitosan by a reductive amination reaction. Chitosan dissolved in aqueous acetic acid was reacted with β-CD monoaldehyde in the presence of sodium cyanoborohydride to provide CD-Chitosan with a grafting yield of cyclodextrin of approximately 7%. The cyclodextrin-Chitosan was precipitated with aqueous sodium hydroxide. The white precipitate was successfully washed with water and ethanol and dried to give β-Cyclodextrin-Chitosan [17].

A few other chitosan derivatives and their potential applications are given in Table 3.1.

3.8 Applications of Chitosan as Nanobiomaterial

Chitosan and its derivatives have been widely used with several modifications due to the advantages they possess as a polymer and the flexibility it poses to the researcher in modifying it according to their requirement. Chitosan is used in food industry, cleaning water, aquaculture and many other industries in the form of films, membranes, particulate systems, encapsulations etc. The use of chitosan as a delivery system can be clearly described based on the property for which it is being used rather than the system developed, since chitosan in nanofabrication is mostly used as particulate system than any other system like liposomes, emulsions etc.

3.8.1 Mucoadhesive Property

The mucoadhesive property of chitosan is one of the most exploited areas considering research in chitosan based particulate delivery systems. The adhesive property of chitosan is considered to be due to the ionic interactions as mentioned above but Sogias et al. in 2008 performed various studies on chitosan and detailed its adhesion property which was not widely understood. Their finding showed that when the number of amine groups were reduced in

Table 3.1   Chitosan derivatives and their applications

Derivatives	Potential Uses
N-Acyl chitosans	Textiles, membranes and medical aids
N-Carboxyalkyl(aryl) Chitosan	Chromatographic media and metal ion collection
O-Carboxyalkyl Chitosan	Molecular sieves, Viscosity builders, Metal ion collection
Natural polysaccharides Complexes,	Flocculation and metal ion chelation Desalting filtration, dialysis Enzymology, Dialysis Enzyme Immobilization Food additive, Anticholesterolemic
Chitosan Hydrochloride (Ch-HCl) and N Carboxymethylchitosan (N-CMCh)	Significantly enhanced intraocular drug penetration with respect to an iso viscous drug solution containing poly (vinyl alcohol) and to commercial ofloxacin eye drops.
Chitosan Aspartate, Glutamate, Lactate and Hydrochloride	Enhances the rate of drug release due to enzyme catalyzed hydrolysis by glucosidase at pH 7.0 in the treatment of inflammatory bowel disease.
Chitosan/Cellulose Acetate	Enhanced loading efficiency of drug into microspheres
Chitosan–Glutathione (GSH) Conjugate	Improved mucoadhesive and permeation enhancing properties.
Galactosylated, Chitosan-Graft-Poly (Vinyl Pyrrolidone)	Potential hepatocyte-targeting gene carrier
Poly(Ethylene Glycol)-Cross-linked N-Methylene Phosphonic Chitosan	Exhibits increased water swelling and hygroscopicity. The film forming capacity and swelling properties made this cross linked derivative suitable as medical material item

the chitosan by performing half acetylation the solubility increased but adhesion capacity decreased. They performed experiments using sodium chloride, urea and ethanol solutions in different concentrations to prove that it was not just electrostatic forces of attraction responsible for mucoadhesion but also hydrogen bonding and hydrophobic effects. The amine and carboxylic group of the molecules form hydrogen bonds with mucus leading to adhesion. The opposite charges of mucin and chitosan lead to attraction and retention of drug for continuous and improved bioavailability of the drug. The neutral and acidic pH strengthens adhesion property of chitosan. The higher molecular weight and degree of acetylation, the stronger the adhesion [18].

The cohesive properties of chitosan were improved as follows:

i. chitosan – multivalent anion complexes – The formation of complexes with multivalent anionic drug treatments, multivalent nonionic polymeric excipients, and also multivalent inorganic anions. This strategy is effective to a very limited extent, as the cationic substructures of chitosan being accountable for mucoadhesion via ionic interactions while using the mucus are blocked in such cases.

ii. Enhancing the cationic character – More cationic character of the polymer is provided by the trimethylation of the primary amino group of chitosan. It was found that when trimethylated chitosan is added to PEGylated, its mucoadhesive properties were improved up to 3.4-fold.

iii. Increasing the degree of deacetylation – This mucoadhesion is directly related to the degree of deacetylation of chitosan: actually, if chitosan degree of deacetylation increases, the number of positive charges also increases, which leads to improved mucoadhesive properties.

Of all the derivatives of chitosan, thiolated chitosan was most widely studied for mucoadhesion property. The immobilization of thiol groups on the chitosan increases the bioavailability and residence time of the drug apart from 100 fold increase in the adhesion. Apart from thiols PLGA showed 3.3 fold increase in the adhesion of drugs [19]. One such study focused on bioadhesive nature of chitosan and thiolated chitosan in an *ex-vivo* study. The study showed that higher concentration of thiolated groups on the surface of nanoparticles increased adhesion due to covalent bonds formed [20].

The use of chitosan in PEGylated form is also quite common, this is due to the interpenetration of PEG chains into mucus which increases the binding. Trimethyl chitosan and its PEGylated nanocomplexes were used in an experiment to increase the absorption of insulin by retaining it in the mucus. This study clearly showed the synergy between chitosan derivatives and PEG and it also showed that a particular number of PEG molecules were required to show maximum effect [21]. Dudhani and Kosaraju, prepared chitosan nanoparticles with catechin to observe 60% release of encapsulated catechin in presence of enzymes, thus increasing the bioavailability of catechin [18].

Morillonite, a type of clay was formed as nanocomposites with chitosan which gave rise to a hybrid substance with properties of wound healing and adhesion apart from several other properties. It has become a product with wide applicability in biomedical and tissue engineering [22]. Another preactivated chitosan was derived by treatment of chitosan with thioglycolic acid and thipyrazole giving a conjugate with protected sulfyhydryl moieties from oxidation and thus stabilizing the polymer matrix. It showed good mucoadhesive, permeation and antimicrobial properties [23].

Although the effect of chitosan and its derivatives on the mucoadhesion was studied widely little effort was made to study the effect of vehicles used in formulation on the mucoadhesion. Sakloetsakun et al., 2010 reported a first case study of vehicle effect on the residence time of chitosan nanoparticles. The vehicles used in the study were PEG 300, PEG 6000, carprylic triglyceride, cremophore EL and miglycol 840. It was found that PEG 300 and miglycol 840 could enhance the mucoadhesion and also the residence time in stomach. It was believed that the enhanced mucoadhesion of these particles was due to an effect similar to that when particles were PEGylated.

3.8.2 Permeation Enhancement

The positive charge on chitosan is responsible for the penetration enhancement of chitosan, which interact with cell membrane and cause structural reorganization of the junctions and its associated proteins. The intracellular tight junctions open causing the entry of macromolecular structures into the cell through paracellular transport. It was proved that cationic character is needed for enhancement. Just as high molecular weight and degree of acetylation promote mucoadhesive property they also enhance permeation property, they are used in combination with other polymers like acrylates to further enhance the property [24]. Penetration enhancement of chitosan is most commonly used with the other properties of chitosan like mucoadhesion, antimicrobial etc. It is used in ocular, nasal, pulmonary, vaginal, transdermal systems.

The buccal penetration capacity of chitosan and trimethyl chitosan in solution or particulate form was experimented to find that, methylation did not change the mechanism of penetration and also the nanoparticles showed higher penetration than the chitosan solution due to their mucoadhesive property [25]. After it was observed that chitosan and trimethyl chitosan have the same mucoadhesive property it was further studied due to the solubility advantage it possessed over chitosan. Chitosan is insoluble in the pH of GIT and this derivative of chitosan has much higher solubitility at neutal and acidic pH. It was observed that as the quaternization degree increases the penetration increased in trimethyl chitosan derivatives [26]. The alkylation of chitosan when studied with methyl, propyl, butyl or hexyl quaternization, they followed an increase in penetration and disruption of junctions with increase in lipohillicity of substituent and decreasing order for viability [27].

Citrate salts of chitosan have been used in enhancing permeation it is believed that citrates bind to calcium ions which regulate the gaps of the tight junctions. A study involved *ex-vivo* penetration experiments of acyclovir and ciprofloxacin HCl with chitosan citrate and chitosan HCl. Although the penetration capacity of both the polymers was found to be same, the protease inhibition capacity of citrate salt is much higher [28]. Carvediol is given as transdermal patch for sustained release, with soyabean extract and chitosan which showed maximum *in-vitro* penetration when compared to chitosan. These patches were also able to reduce hypertension in induced rats which showed their efficient usage as patches [29].

Chitosan was used for oral delivery of proteins like insulin and was found to be a safe permeation enhancer. It showed that after although the junctions are small for aggregates of chitosan in mucus to enter them they are wide enough to allow for the entry of soluble insulin though the junction [30].

3.8.3 Wound Healing

Restoration of skin after damage is a natural phenomenon but the process may be affected by infection, rate of healing, fluid loss or any other such complications which delay the healing process. Although most of the wounds heal without complications, it is the chronic non-healing wounds which are a challenge to wound care product researchers. These wounds are generally as a result of compromised wound physiology and occur with diabetes, venous stasis or prolonged local pressure. The wound healing process comprises of four steps namely [31],

- Haemostasis – formation of a temporary plug by platelets present in the exposed blood.
- Inflammation – phagocytes clear the matter and destroy the ingested matter.
- Proliferation – new vessels are formed to carry oxygenated blood.
- Maturation/remodelling – collagen is produced around the new vasculature and is remodelled which takes several years.

The initial classification of wound dressings was passive, interactive and bioactive products. Bioactive dressings deliver a substance which is active in wound healing which can be due to the dressing itself or due to a compound added to it. They include proteoglycans, collagen, non-collagen proteins, alginates or chitosan. Chitin and its derivatives are capable of accelerating wound healing at molecular, cellular and systemic levels [32]. Chitosan is used in wound healing for as a hemostat, which clots blood and blocks nerve ending reducing pain. Apart from this it has effects on biological activities, macrophage function and antimicrobial effect [32]. The use of chitin was first noted when shark cartilage accelerated wound healing. It was then noticed that *N*-acetyl-D-glucosamine was a wound accelerator [33]. Well it was found that even chitosan converts gradually to *N*-acetyl-D-glucosamine which increases growth of fibroblasts, deposition of collagen and increases hyaluronic acid synthesis at wound site [31]. It also promotes polymorphonuclear activation, fibroblast activation, giant cell migration and cytokine production [32]. Hence, all these properties as shown in Figure 3.5 make chitosan a useful healing accelerator. Apart from this its antibacterial activity prevents infection of open cuts during their repair thus reducing the time required to heal than when no chitosan is present in the product. Several studies were conducted using chitosan and its derivatives in various forms to prepare hydrogels, beads, nano/microparticles, nanofibers, scaffolds, sponges etc. Several dressings of chitin and is derivatives are now available in market and their order of tissue regeneration is as follows non-wovens [34–36], nanofibrils, composites, films [37], scaffolds and sponges [32].

Most of the work for use of chitosan in wound healing was done at micro level and in the form of hydrogels which is not the scope of present chapter [37–41]. Preparation of hydrogels was common for wound healing and specially polycross-linkable chitosan solution formed a hydrogel within

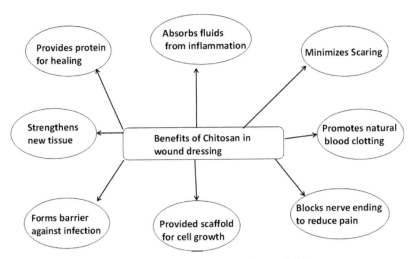

Figure 3.5   Wound healing mechanisms of chitosan.

60s and stopped bleeding from cut tail of mouse in 30s [42]. Jayakumar et al, in 2011 have reviewed on the studies conducted in wound healing properties of chitosan and its derivatives, with emphasis on hydrogels. Polyelectrolyte (PEC) systems have been commonly used with chitosan and its derivatives in the preparation of sponges, microcapsules, microparticles or dry coated tablets. Wang et al. (2002) prepared a novel chitosan-alginate PEC membrane which after sterilization with γ-radiation was stable and promoted cell growth with no cytotoxic effect. The prepared membrane was found to have comparable efficacy with a commercial Opsite® dressing [42]. PEC systems chitosan were also prepared with glutamic acid, hyaluronic acid, oligonucleotides etc. [32].

Complexes of chitosan have been prepared with a substance which shows antimicrobial activity to make them better wound healing accelerators. Nano-TiO_2 with chitosan was prepared as artificial skin whose activity was found to be similar with Duoderm, a marketed product [43]. Just like nano-TiO_2, nano-ZnO was also prepared as bandage to be used as an effective wound healing accelerator [44]. Chitosan due to its antiulcer property along with other properties was found to be effective with rhodium complex in reducing ulceration as well as healing open wounds [45]. Peptides like heparin formed complexes with chitosan to form hydrogels or films [33]. Apart from films peptides were immobilized on chitosan beads to prepare scaffolds for wound healing, such attempts have been made with various peptides to give enhanced wound healing [46].

When the use of chitin in nano-fabrication for biomedical application comes it is clear that nanofibres are the most efficient due to their large surface area to volume ratio and high porosity. The most common method of preparation of nanofibres is using electro-spinning technique. The preparation of nanofibres of chitosan as such is a hurdle due to the technique used in its preparation. Chitosan has polycationic nature in acidic aqueous environment and due to large amine groups present in its structure, it possess high surface tension requiring high energy to be used during electrospinning. It is also commonly noted that during formation of fibres particles are formed while using chitosan. Apart from these issues chitosan readily swells when in contact with water and is unstable. Hence, it is generally used in combination with other polymers, used as a coating over polymers thus delivering the same effects [47] or may form composites of nanofibres in films.

When it comes to activity of chitosan and its derivatives in comparison to chitin was observed that the collagenase increasing activity of chitosan was superior to chitin and higher tensile strength was observed with higher degree of acetylation in chitosan [48]. Chitosan film alone has poor tensile strength whatever be the degree of deacetylation when compared to composites of other materials used in dressing. In one study chitin nanofibres were incorporated in chitosan by solution casting method so that they form nanofillings which provide the required tensile strength to film without change in water-permeability [49].

Polyvinyl alcohol (PVA) nanofibers are prepared by electrospinning and heated to stabilise, these were then coated with chitosan to develop a superior dressing with high tensile strength when compared to either of the fibres alone due to high interfibre bonding of chitosan and the mechanical properties of PVA fibres [47].

A very special attempt was made in the recent past for topical delivery of chitosan with fatty acids. Ionic self-assembling micelles were formulated based on ionic interaction between amine groups of chitosan and carboxylic groups of oleic or linolenic acids which impart amphihillic nature to chitosan with the fatty acid part being the hydrophobic edge. When these chitosan-oleate micelles are topically delivered they get looser and form individual components on application and hence can act in wound healing by delivering poorly soluble drugs and keeping them intact due to their mucoadhesive property [50]. Thus it can be stated that a wide use of chitosan as accelerator for wound healing is happening with certain products under clinical trials and in market.

3.8.4 Gene Delivery

Gene delivery refers to treatment of diseases by inserting genetic material either to prevent the expression of certain protein or to cause the expression. As the DNA is negatively charged the cell membrane which is also negatively charged prevents its entry into the cell. Even if the naked DNA enters the cell, it is digested by nucleases in cytoplasm [51]. Hence, arises a need for delivery system to transfer the genes.

In the past viral vectors like adenovirus or retrovirus were commonly used for the delivery of genes into the system [52]. Although these viral vectors have high transfection and easy entry into the cell, they have many disadvantages like causing inflammation, immunogenic reactions, low target specificity, mutation and poor incorporating capacity. This gave rise to a need of using non-viral vectors and chitosan due to its versatile capacity was most sought carrier for gene therapy.

Chitosan is the only available positively charged biocompatible, biodegradable polymer with strong interaction for negatively charged DNA and also protecting them from nucleases [53]. Chitosan in its natural form has limitations due to which it is utilised mostly in the grafted copolymerized form in gene delivery. It has been observed that physicochemical character-istics of chitosan particles and complexation efficiency play a crucial role in chitosan acting as gene delivery and overcoming the cell membrane barrier for entry into the cell. For the first time chitosan was used *in-vitro* drug delivering in 2006.

The formulation technique involved in preparation of chitosan – nucleic acid particles plays a significant role in gene splicing. Electronic interactions is the most common, simple and straight forward method of preparation of chitosan nanoparticles, where the cationic polymer and anionic plasmid DNA and siRNA interact. There are two preparation processes in this which affect the particle size: ionic gelation technique which gives particles over a wide range of distribution (20–700 nm) and complex coacervation which gives a narrow distribution of particles in 100–250 nm [54]. Chitosan for gene therapy was used for cancer, antigen and targeted therapy mostly and is given orally, nasal or mucosal delivery.

The chitosan nanoparticles enter the cell either by interacting with the negatively charged membrane or by endocytosis in either of which they end up in an endosome. Cellular uptake isn't a rate limiting step when comes to the entry of particles less than 200 nm with positive surface charge. The pDNA is delivered into the nucleus and sRNA in the cytoplasm to elicit action of expression or suppression of gene. Chitosan releases the RNA into the cell by degradation and increasing the osmolarity in lysozyme but what is less understood is that how pDNA enters the nucleus [54]. The DNA binding capacity of chitosan is pH dependent the lower the pH more protonated the backbone is and higher binding of DNA [55]. Since a size of less than 200 nm is necessary for efficient carrying of the nucleic acid into the cell the factors influencing it are molecular weight and degree of deacetylation of chitosan used, ratio of polymer to nucleic acid and concentration of nucleic acid effect the size of particles [54]. It was shown in a study of A549 that decrease in molecular weight and degree of acetylation decreased the luciferase expression. It was also found that there was huge impact on the presence and absence of serum in the transfection process but inclusion of alginate increased transfection. The amine and hydroxyl groups of chitosan have been chemically modi-fied to form derivatives. They were modified with hydrophobic cholesterol groups to form nano-self assembling structures in aqueous media. It was observed that experiments were performed using deoxycholic acid, alkylated chitosan, trimethylated chitosan to induce quaternization and *N*-acetyl histamine conjugated chitosan to increase the transfection. Chitosan oligomers and lactose based chitosan is more commonly used to target liver cells for therapy [52].

Yang et al. in 2010 demonstrated the stability of chitosan-epidermal growth factor protein nanoparticles in murine albumin and also showed short term transgene expression in lung and intestine. Tumour targeting was done by modifications with folic acid [56], thiols, albumin [57], etc. to target specific tumour cells. For administering sRNA it was observed that they are modified with PLGA, poly-arginine etc. for better loading and transfection. Finally after many studies it was showed that 15–21 monomers of chitosan are the most effective as vectors due to its low viscosity and high solubility in water [51]. Kedjarune-Leggat & Leggat in 2011 summarized on chitosan in gene delivery investigations and targeting. They also suggested that chitosan was limited in use as non-viral delivery of genes as its pharmacokinetic was complex and they were removed from the blood to be deposited in organs.

3.8.5 Vaccine Delivery

Chitosan has well established mucoadhesive property as discussed above. When this property was used to administer antigens to the body it was called vaccine delivery. Since chitosan has good mucoadhesive property and enhances the permeation of drug with it, nasal vaccine of chitosan is a very widely exploited area. It was also assumed that since only when the antigens are exposed to the mucus antibody secretion occurs, it is a way of effective immunization against pathogens which enter through mucosal layer. It was also given as oral vaccines but nano-formulation was not very necessary to prepare in nasal delivery since it wasn't exposed to very low pH and also high concentration of enzymes. Chitosan nanoparticles were prepared using coacervation/precipitation technique and were incorporated with bovine serum albumin or tetanus toxoid or any other protein [58]. Clinical trials are being carried out on chitosan-based vaccines for influenza, pertussis, and diphtheria are in different stages of clinical trials [59]. Apart from preparing nanoparticles, nanofibers of chitosan were prepared with shigella subunits and their protection was tested. These nanofibres were intranasally administered and immunised guinea pigs on exposure to the bacteria they were found to be protectant [60]. *S. aureus* being one of the most common problems now is due to the resistance developed against various antibiotics. An amidase encapsulated *O*-carboxymethyl chitosan nanoparticles were developed for sustained action of immunity development against the bacteria [61]. Chitosan was used in thermosensitive form as coating over silica nanoparticles for vaccine delivery with antigens [62]. Oral vaccines have been developed with chitosan in liposomal formulations. Liposomes always have a threat to degenerate due the presence of enzymes, hence techniques have been studied to sandwich chitosan in the liposomal membrane and stabilizing it with agents like polyvinyl alcohol [63]. Chitosan and its derivatives were used to prepare emulsions also for nasal delivery to develop immune response with ovalbumin and cholera toxin in a study only to find that both nanoparticles and emulsions were successful in inducing immunity [64]. *In-vitro* studies of tetanus toxoid in low molecular weight chitosan nanoparticles showed high and long lasting immune response than the fluid vaccine at low concentrations of chitosan [63]. A part from human investigations, aquaculture being a fast growing industry was struck with vibriosis of fish and a chitosan based DNA vaccine was developed which showed protection against this disease [65].

3.9 Conclusion

Chitosan and its derivatives are used in a variety of industry like weight loss, water purification, dying etc. apart from pharmaceutical industry where it is used in drug delivery. The derivatives are being used in variety of systems like nasal, pulmonary, GIT, ocular, transdermal to deliver drugs in nanoform as complexes or particles or fibres. Thus such widely studied polysaccharide must be explored as nanobiomaterial for further variations in future.

References

[1] Chin-San, W. (2005). A comparison of the structure, thermal properties, and biodegradability of polycaprolactone/chitosan and acrylic acid grafted polycaprolactone/chitosan. *Polymer* 46, 147–155.

[2] Florence and Christine (2013). Chitosan-based biomaterials for tissue engineering. *Eur. Polym. J.* 49, 780–792.

[3] Vipin, B., Pramod, K. S., Nitin, S., Om, P. P., and Rishabha, M. (2011). Applications of chitosan and chitosan derivatives in drug delivery. *Adv. Biol. Res.* 5, 28–37.

[4] Badawy, M. E. I., and Entsar, I. R. (2011). A biopolymer chitosan and its derivatives as promising antimicrobial agents against plant pathogens and their applications in crop protection. *Int. J. Carbohydr. Chem.* 2011:460381.

[5] Elgadir, M., Salim, U., Sahena, F., Aishah, A., Ahmed, J. K. C., Zaidul, I. S., et al. (2011). Impact of chitosan composites and chitosan nanoparticle composites on various drug deliverysystems: a review. *J. Food Drug Anal.* 23, 619–629.

[6] Jag, P., Hari, O. V., Vijay, K. M., Satyendra, K. M., Deepayan, R., and Jitendra, K. (2014). Biological method of chitin extraction from shrimp waste an eco-friendly low cost technology and its advanced application. *Int. J. Fish. Aquat. Stud.* 1, 104–107.

[7] Kamala, K., Sivaperumal, P., and Rajaram, R. (2013). Extraction and characterization of water soluble chitosan from parapeneopsisstylifera shrimp shell waste and its antibacterial activity. *Int. J. Sci. Res. Publ.* 3, 1–8.

[8] Shaofang, L., Jie, S., Lina, Y., Chushu, Z., Jie, B., Feng, Z., et al. (2012). Extraction and characterization of chitin from the beetle *Holotrichia parallela* motschulsky. *Molecules* 17, 4604–4611.

[9] Gurpreet, S. D., Surinder, K., Satinder, K. B., and Mausam, V. (2012). Green synthesis approach: extraction of chitosan from fungus mycelia. *Crit. Rev. Biotechnol.* 33, 379–403.

[10] Laila, M., Olfa, G. B., Kemel, J. Islem, Y., and Moncef, N. (2010). Extraction and characterization of chitin, chitosan, and protein hydrolysates prepared from shrimp waste by treatment with crude protease from *Bacillus cereus* SV1. *Appl. Biochem. Biotechnol.* 162, 345–357.

[11] Raphae, R., Ragelle, H., Anne, R., Nicolas, D., Christine, J., and Veronique, P. (2011). Chitosan and chitosan derivatives in drug delivery and tissue engineering. *Adv. Polym. Sci.* 244, 19–44.

[12] Hemant, K. S. Y., Gunjan, B. J., Mangla, N. S., and Hoskote, G. S. (2011). Naturally occurring chitosan and chitosan derivatives: a review. *Curr. Drug Ther.* 6, 2–11.

[13] Nunthanid. J., and Huanbutta, K. (2008). Development of time-, pH and enzymecontrolled colonic drug delivery usingspray-dried chitosan acetate and hydroxypropyl methylcellulose. *Eur. J. Pharm. Biopharm.* 68, 253–259.

[14] Pozzo, D. A., Fagnoni, M., and Guerrini, M. (2000). Preparation and characterization of poly (ethylene glycol)-crosslinked re acetylated chitosans. *Carbohydr. Polym.* 42, 201–2166.

[15] Ramos, V. M., Rodriguez, N. M., and Henning, I. (2006). Poly (ethylene glycol)-crosslinked N-methylene phosphonic chitosan. Preparation and characterization. *Carbohydr. Polym.* 64, 328–336.

[16] Makhlof, A., Werle, M., Tozuka, Y., and Takeuchi, H, (2010). Nanoparticles of glycol chitosan and its thiolated derivative significantly improved the pulmonary delivery of calcitonin. *Int. J. Pharm.* 397, 92–95.

[17] Krauland, H., Alexander, H. H. M., and Bernkop-Schnurch, A. (2005). Viscoelastic properties of a new in situ gelling thiolated chitosan conjugate. *Drug Dev. Ind. Pharm.* 3, 885–893.

[18] Wang, J. J. et al. (2011). Recent advances of chitosan nanoparticles as drug carriers. *Int. J. Nanomed.* 6, 765–74.

[19] Sakloetsakun, D., Perera, G. and Hombach, J. (2010). The impact of vehicles on the mucoadhesive properties of orally administrated nanoparticles: a case study with chitosan-4-thiobutylamidine conjugate. *AapsPharmscitech* 11, 1185–1192.

[20] Bravo-Osuna, I., Vauthier, C. and Farabollini, A. (2007). Mucoadhesion mechanism of chitosan and thiolated chitosan-poly (isobutyl cyanoacrylate) core-shell nanoparticles. *Biomaterials* 28, 2233–2243.

[21] Jintapattanakit, A. (2009). The role of mucoadhesion of trimethyl chitosan and PEGylated trimethyl chitosan nanocomplexes in insulin uptake. *J. Pharm. Sci.* 98, 4818–4830.

[22] Salcedo, I. et al. (2012). In vitro biocompatibility and mucoadhesion of montmorillonite chitosan nanocomposite: a new drug delivery. *Appl. Clay Sci.* 55, 131–137.

[23] Mller, C. et al. (2013). Thiopyrazole preactivated chitosan: combining mucoadhesion and drug delivery. *Actabiomaterialia* 9, 6585–6593.

[24] Elgadir, M. A. et al. (2014). Impact of chitosan composites and chitosan nanoparticle composites on various drug delivery systems: a review. *J. Food Drug Anal.* 23, 619–629.

[25] Sandri, G. et al. (2006). Histological evaluation of buccal penetration enhancement properties of chitosan and trimethyl chitosan. *J. Pharm. Pharmacol.* 58, 1327–1336.

[26] Jonker, C., Hamman, J. and Kotze, A. (2002). Intestinal paracellular permeation enhancement with quaternised chitosan: in situ and in vitro evaluation. *Int. J. Pharm.* 238, 205–213.

[27] Benediktsdttir, B. E. et al. (2014). N-alkylation of highly quaternized chitosan derivatives affects the paracellular permeation enhancement in bronchial epithelia in vitro. *Eur. J. Pharm. Biopharm.* 86, 55–63.

[28] Bonferoni, M., Sandri, G. and Rossi, S. (2008). Chitosan citrate as multifunctional polymer for vaginal delivery: evaluation of penetration enhancement and peptidase inhibition properties. *Eur. J.* 33, 166–76.

[29] Sapra, B., Jain, S. and Tiwary, A. K. (2009). Transdermal delivery of carvedilol in rats: probing the percutaneous permeation enhancement mechanism of soybean extract-chitosan mixture. *Drug Dev. Ind. Pharm.* 35, 1230–1241.

[30] Sonaje, K., Chuang, E. and Lin, K. (2012). Opening of epithelial tight junctions and enhancement of paracellular permeation by chitosan: microscopic, ultrastructural, and computed-tomographic observations. *Mol. Pharm.* 9, 1271–1279.

[31] Paul, W., and Sharma, C.P. (2004). Chitosan and alginate wound dressings: a short review. *Trends Biomater. Artif. Organs* 18, 18–23.

[32] Jayakumar, R., Prabaharan, M., and Kumar, P., (2011). *Novel Chitin and Chitosan Materials in Wound Dressing. Biomedical Engineering, Trends in Materials Science.* Available at: http://www.intechopen.com/source/pdfs/12794/InTech-Novel_chitin_and_chitosan_materials_in_wound_dressing.pdf

[33] Ueno, H., Mori, T., and Fujinaga, T. 2001. (2001). Topical formulations and wound healing applications of chitosan. *Adv. Drug Deliv. Rev.* 52, 105–115.

[34] Wang, C.-C., Su, C.-H., and Chen, C.-C. 2008. (2008). Water absorbing and antibacterial properties of N-isopropyl acrylamide grafted and collagen/chitosan immobilized polypropylene nonwoven fabric and its application on wound healing enhancement. *J. Biomed. Mater. Res. A,* 84, 1006–17.

[35] Wang, C.-C., et al., (2009). An enhancement on healing effect of wound dressing: Acrylic acid grafted and gamma-polyglutamic acid/chitosan immobilized polypropylene non-woven. *Mater. Sci. Eng. C* 29, 1715–1724.

[36] Murakami, K., et al., (2010). Enhanced healing of mitomycin C-treated healing-impaired wounds in rats with hydrosheets composed of chitin/chitosan, fucoidan, and alginate as wound dressings. *Wound Repair Regen.* 18, 478–85.

[37] Sezer, A. et al., (2007). Chitosan film containing fucoidan as a wound dressing for dermal burn healing: preparation and in vitro/in vivo evaluation. *AAPS Pharm. Sci. Tech.* 8:39.

[38] Azad, A. (2004). Chitosan membrane as a wound-healing dressing: Characterization and clinical application. *J. Biomed. Mater. Res. B Appl. Biomater.* 69, 216–222.

[39] Altiok, D., Altiok, E., and Tihminlioglu, F., (2010). Physical, antibacterial and antioxidant properties of chitosan films incorporated with thyme oil for potential wound healing applications. *J. Mater. Sci. Mater. Med.* 21, 2227–2236.

[40] Yang, X., et al. (2010). Cytotoxicity and wound healing properties of PVA/ws-chitosan/glycerol hydrogels made by irradiation followed by freeze–thawing. *Radiat. Phys. Chem.* 79, 606–611.

[41] Sung, J. H., et al. (2010). Gel characterisation and in vivo evaluation of minocycline-loaded wound dressing with enhanced wound healing using polyvinyl alcohol and chitosan. *Int. J. Pharm.* 392, 232–240.

[42] Wang, L., et al. (2002). Chitosan-alginate PEC membrane as a wound dressing: Assessment of incisional wound healing. *J. Biomed. Mater. Res.* 23, 833–840.

[43] Peng, C., Yang, M., and Chiu, W., (2008). Composite nano-titanium oxide–chitosan artificial skin exhibits strong wound healing effect—an approach with anti-inflammatory and bactericidal kinetics. *Macromol. Biosci.* 8, 316–327.

[44] Kumar, P. S. (2013). *Flexible and Micro Porous Chitinous Nanocomposite Bandages for Wound Dressing*. Available at: http://ir.inflibnet.ac.in:8080/jspui/handle/10603/13057

[45] Afzaletdinova, N., and Murinov, Y., (2000). Synthesis and wound-healing and antiulcer activity of a chitosan-rhodium (iii) complex. *Pharm. Chem. J.* 34, 248–249.

[46] Bae, J. W., et al. (2009). EPDIM peptide-immobilized porous chitosan beads for enhanced wound healing: Preparation, characterizations and in vitro evaluation. *Mater. Sci. Eng. C* 29, 697–701.

[47] Kang, Y., Yoon, I., and Lee, S., (2010). Chitosan-coated poly(vinyl alcohol) nanofibers for wound dressings. *J. Biomed. Mater. Res. B Appl. Biomater.* 92, 568–576.

[48] Minagawa, T., Okamura, Y., and Shigemasa, Y. (2007). Effects of molecular weight and deacetylation degree of chitin/chitosan on wound healing. *Carbohydr. Polym.* 67, 640–644.

[49] Shelma, R., Paul, W., and Sharma, C. (2008). Chitin nanofibre reinforced thin chitosan films for Wound healing application. *Trends Biomater. Artif. Organs* 22, 111–115.

[50] Bonferoni, M. C., et al. (2014). Ionic polymeric micelles based on chitosan and fatty acids and intended for wound healing. Comparison of linoleic and oleic acid. *Eur. J. Pharm. Biopharm.* 87, 101–106.

[51] Saranya, N., et al. (2011). Chitosan and its derivatives for gene delivery. *Int. J. Biol. Macromol.* 48, 234–238.

[52] Lee, K. Y. (2011). Chitosan and its derivatives for gene delivery. *Macromol. Res.* 15, 195–201.

[53] Zheng, F., et al. (2007). Chitosan nanoparticle as gene therapy vector via gastrointestinal mucosa administration: results of an in vitro and in vivo study. *Life Sci.* 80, 388–396.

[54] Raftery, R., O'Brien, F. J., and Cryan, S.-A., (2013). Chitosan for gene delivery and orthopedic tissue engineering applications. *Molecules* 18, 5611–5647.

[55] Liu, W., et al. (2005). An investigation on the physicochemical properties of chitosan/DNA polyelectrolyte complexes. *Biomaterials* 26, 2705–2711.

[56] Shi, B., et al. (2013). Development of a novel folic acid chitosan supported imidazole Schiff base for tumor targeted gene delivery system and drug therapy. *J. Control. Release* 172:e98.

[57] Karimi, M., Avci, P., and Mobasseri, R., (2013). The novel albumin–chitosan core–shell nanoparticles for gene delivery: preparation, optimization and cell uptake investigation. *J. Nanopart. Res.* 15:1651.

[58] Van der Lubben, I. M., et al., (2001). Chitosan and its derivatives in mucosal drug and vaccine delivery. *Eur. J. Pharm. Sci.* 14:201–207.

[59] Illum, L., Jabbal-Gill, I., and Hinchcliffe, M. (2001). Chitosan as a novel nasal delivery system for vaccines. *Adv. Drug Deliv. Rev.* 51:81–96.

[60] Jahantigh, D., et al. (2014). Novel intranasal vaccine delivery system by chitosan nanofibrous membrane containing N-terminal region of ipad antigen as a nasal shigellosis vaccine, studies in guinea pigs. *J. Drug Deliv. Sci. Technol.* 24, 33–39.

[61] Smitha, K. T., et al. (2014). Amidase encapsulated O-carboxymethyl chitosan nanoparticles for vaccine delivery. *Int. J. Biol. Macromol.* 63, 154–157.

[62] Gordon, S., et al. (2010). In vitro and in vivo investigation of thermosensitive chitosan hydrogels containing silica nanoparticles for vaccine delivery. *Eur. J. Pharm. Sci.* 41, 360–368.

[63] Rescia, V. C., et al. (2011). Dressing liposomal particles with chitosan and poly(vinylic alcohol) for oral vaccine delivery. *J. Liposome Res.* 21, 38–45.

[64] Nagamoto, T., et al. (2004). novel chitosan particles and chitosan-coated emulsions inducing immune response via intranasal vaccine delivery. *Pharm. Res.* 21, 671–674.

[65] Li, L., et al. (2013). Potential use of chitosan nanoparticles for oral delivery of DNA vaccine in black seabream *Acanthopagrus schlegelii* Bleeker to protect from *Vibrio parahaemolyticus*. *J. Fish Dis.* 36, 987–995.

4

Design and Characterization of Lipid Mediated Nanoparticles Containing an Anti-Psychotic Drug for Enhanced Bio-Availability

**Jawahar Natarajan, Gowtham Reddy Naredla
and Veera Venkata Satyanarayana Reddy Karri**

Department of Pharmaceutics, JSS College of Pharmacy, Ootacamund,
Jagadguru Sri Shivarathreeswara University, Mysuru, India

Abstract

This study deals with the investigation carried out on the formulation and evaluation of solid lipid nanoparticles containing olanzapine so as to improve its oral bioavailability. to decrease dosing frequency, maintain prolonged therapeutic levels of the drug following dosing and to reduce the dose to achieve same pharmacological effect. Based on the literature studies various excipients viz. Lipids (glyceryl tripalmitate) and surfactants (soy lecithin, Pluronic F-68) were selected for the preparation of solid lipid nanoparticles. In preformulation studies, saturation solubility studies were carried out in different solvents and the solubility was found to be highest in 0.1 N HCl. Compatibility of the selected lipids, surfactants and olanzapine were carried by FT-IR peak matching method. Partitioning the behavior of olanzapine was determined in the selected lipid to determine the affinity of the drug towards selected lipid. In formulation development SLN were prepared by microemulsion method and the effect of certain process and formulation variables such as lipid concentration, stirring speed, stirring time and surfactant concentration on particle size was studied. SLNs were evaluated for zeta potential, entrapment efficiency drug loading, Polydispersity and *in vitro* release studies. In order to elucidate mode and mechanism of drug release, the *in vitro* data was transformed and interpreted at graphical interface constructed using various kinetic models. Oral bioavailability studies were carried out in albino Wistar rats where two different formulations viz. GTP-SLN and olanzapine suspension were administered orally and pharmacokinetic parameters revealed that GTP-SLN high bioavailability than drug suspension.

4.1 Introduction

Olanzapine is a psychotropic agent that belongs to the thioenobenzodiazepine class and is indicated for acute and maintenance treatment of schizophrenia [1]. The current formulations of olanzapine suffer from the low bioavailability (40%) and wide variety of side effects are associated with current dosage forms of the drug. It is eliminated extensively by hepatic first-pass metabolism [2]. To overcome the hepatic first-pass metabolism and to enhance bioavailability, intestinal lymphatic transport of drugs can be exploited [3, 4]. Lipid-based nanocarriers have been investigated for delivery of various drugs acting on central nervous system [5, 6]. Solid lipid nanoparticles (SLNs) are one of carrier systems having more advantages than other colloidal delivery systems with regard to biocompatibility and scale up [7]. The particle matrix of SLNs is composed of solid lipids and includes triglycerides (e.g. Tristearin), partial glycerides (e.g. Glyceryl monostearate), fatty acids (e.g. Stearic acid), steroids (e.g. Cholestrol) and waxes (e.g. Cetyl palmitate) [8]. A clear advantage of SLNs is the fact that the lipid matrix is made from physiological lipids, which decreases the danger of acute and chronic toxicity compared to polymeric nanoparticles [9]. The unique properties of lipids viz., physicochemical diversity and the ability to enhance oral bioavailability of poor water soluble drugs have made them very attractive carriers for oral formulations [10]. Hence, the aim of the present work was to study the improvement of bioavailability by incorporating olanzapine, a poorly bioavailable drug in SLNs.

4.2 Experimental Part

4.2.1 Preformulation Studies

Preformulation may be described as a phase of the research and development process where the formulation scientist characterizes the physical, chemical and mechanical properties of a new drug substance, in order to develop stable, safe and effective dosage forms. Ideally, the Preformulation phase begins early in the discovery process such that appropriate physical, chemical data is available to aid in the selection of new chemical entities that enter the development process. During this evaluation possible interaction with various inert ingredients intended for use in final dosage form are also considered.

4.2.1.1 Solubility studies

The solubility of the Olanzapine was determined in various solvents by adding an excess amount of drug to 10 ml of solvents in conical flasks. The flasks were kept at $25\pm 0.5°C$ in isothermal shaker for 72 h to reach equilibrium. The equilibrated samples were removed from the shaker and centrifuged at 4000 rpm for 15 min. The supernatant was taken and filtered through 0.45 μm membrane filter. The concentration of Olanzapine was determined in the supernatant after suitable dilution by using UV-Visible spectrophotometer at 258 nm.

4.2.1.2 Compatibility study

I.R spectroscopy can be used to investigate and predict any physiochemical interaction between different components in a formulation and therefore it can be applied to the selection of suitable chemical compatible excipients while selecting the ingredients, we would chose, those which are stable, compatible, cosmetically and therapeutically acceptable.

Infrared spectra matching approach was used for detection of any possible chemical interaction between the drug, lipid and surfactants. A physical mixture of drug, lipid and surfactants was prepared and mixed with suitable quantity of potassium bromide. This mixture was compressed to form a transparent pellet using a hydraulic press at 15 tons pressure. It was scanned from 4000 to 400 cm^{-1} in a FTIR spectrophotometer (FTIR 8400 S, Shimadzu). The IR spectrum of the physical mixture was compared with those of pure drug, lipid and surfactants and peak matching was done to detect any appearance or disappearance of peaks.

4.2.1.3 Development of calibration curve

A stock solution of Olanzapine was prepared by dissolving 100 mg of drug in 10 mL of 0.1 N HCl and phosphate buffer pH 6.8 made up to 100 ml (with different buffers viz., 0.1 N HCl and pH 6.8) to give a stock solution of concentration 1 mg/mL. From this stock solution, 5–25 μg/mL dilutions were prepared. The λ max of the drug was determined by scanning one of the dilutions between 400 and 200 nm using a UV-visible spectrophotometer. At this wavelength, the absorbance of all the other solutions was measured against a blank. Standard curve between concentration and absorbance was plotted.

4.2.1.4 Partition coefficient studies

Partitioning behaviour of olanzapine was determined with glyceryl tripalmitate (GTP) as lipid. 10 mg of Olanzapine (OL) was dispersed in a mixture of melted lipid (1 g) and 1 mL of hot phosphate buffer pH 6.8 (PB) and shaken for 30 min in an water bath shaker maintained at 10°C above the melting point of lipid. The aqueous phase of the above mixture was separated from the lipid by centrifugation at 10000 rpm for 20 min. The clear supernatant obtained was suitably diluted with 0.1 N HCl and OL content was determined in UV-Visible spectrophotometer at 258 nm against solvent blank. The partition coefficient was calculated as

$$PC = \frac{(COLI - COLA)}{COLA}$$

Where,
   COLI = the initial amount of OL added (10 mg)
   COLA = the concentration of OL in pH 6.8 PB

4.2.2 Preparation of Solid Lipid Nanoparticles (SLN) by Microemulsion Technique

This potential method was established for production of SLNs with understanding of two phase systems composed of an inner and outer phase (e.g. o/w microemulsion). Microemulsion is clear or slightly coloured solution, thermodynamically stable, micro-heterogeneous dispersion. They are composed of lipophilic phase (lipid), surfactant, co-surfactant, and water. Precisely addition of microemulsion to water, leads to precipitation of lipid phase forming fine particles. This method requires equipment to grind the lipid dispersion to the required size. The microemulsion needs to be produced at a temperature above the melting point of the lipid. The lipids (fatty acids and/or glycosides) are melted; drug is incorporated in molten lipid. A mixture of water, co-surfactant(s) and the surfactant is heated to the same temperature as the lipids and added under mild stirring to the lipid melt. A transparent, thermodynamically stable system is formed when the compounds are mixed in the correct ratios for microemulsion. Thus the microemulsion is the basis for the formation of nanoparticles of a requisite size. This microemulsion is then dispersed in a cold aqueous medium under mild mechanical mixing in the ratio of hot microemulsion to water (1:25–1:50). This leads to rapid recrystallization of the oil droplets on dispersion in cold aqueous medium. Thus SLNs are formed due to precipitation.

4.2.2.1 Optimization of lipid quantity

Solid lipid nanoparticles were prepared by varying the amount of the lipid (50, 65 and100 mg). The amount of soy lecithin (1% w/w w.r.t. lipid), Stearyl amine (1% w/w w.r.t. lipid) and Pluronic F-68 (1% w/v) were kept constant. The stirring time and stirring speed were kept constant at 4 h and 2500 rpm respectively.

4.2.2.2 Study on the effect of formulation process variables

The effect of formulation/process variables such as stirring time, stirring speed, surfactant concentration on the particle size was studied. To investigate the effect of formulation/process variables, each time one parameter was varied, keeping the others constant. From the results obtained, optimum level of those variables was selected and kept constant in the subsequent evaluations.

Effect of Stirring time Five different batches of solid lipid nanoparticles were prepared corresponding to 1, 2, 3, 4 h stirring time keeping the following parameters constant.

Lipid concentration	: 50 mg
Soyalecithin	: 1% w/w (w.r.t. lipid)
Stearylamine	: 1% w/w (w.r.t. lipid)
Stirring speed	: 2000 rpm
Surfactant concentration (Pluronic F-68)	: 1.2% w/v

Effect of Stirring speed Four different batches of solid lipid nanoparticles were prepared corresponding to 1000, 1500, 2000 and 2500 rpm stirring speed keeping the following parameters constant.

Lipid concentration	: 50 mg
Soyalecithin	: 1% w/w (w.r.t. lipid)
Stearylamine	: 1% w/w (w.r.t. lipid)
Stirring time	: 4 h
Surfactant concentration (Pluronic F-68)	: 1.2% w/v

Effect of Surfactant concentration Four different batches of solid lipid nanoparticles were prepared corresponding to 1.5% and 2% w/v surfactant (Pluronic F-68) concentration keeping the following parameters constant.

Lipid concentration	: 50 mg
Soyalecithin	: 1% w/w (w.r.t. lipid)
Stearylamine	: 1% w/w (w.r.t. lipid)
Stirring speed	: 2000 rpm
Stirring time	: 4 h

4.2.2.3 Preparation of drug loaded batches

Drug loaded solid lipid nanoparticles were prepared by microemulsion method using glyceryl tripalmitate (GT) as lipid, olanzapine as drug, soyalecithin was used as lipophilic surfactant, Pluronic F-68 was used as hydrophilic surfactant, Stearyl amine was used as positive charge inducer.

4.2.3 Evaluation of Solid Lipid Nanoparticles

4.2.3.1 Particle size, zeta potential and polydispersity index

Particle size and zeta potential of the solid lipid nanoparticles were measured by photon correlation spectroscopy using a Malvern Zetasizer Nano ZS90 (Malvern Instruments, Worcestershire, UK), which works on the Mie theory. All size and zeta potential measurements were carried out at 25°C using disposable polystyrene cells and disposable plain folded capillary zeta cells, respectively, after appropriate dilution with original dispersion preparation medium. In order to investigate the effect of Stearyl amine on zeta potential two batches were prepared with and without Stearyl amine and their zeta potential were measured.

4.2.3.2 Entrapment efficiency and drug loading

Entrapment efficiency and drug loading of freeze dried Solid lipid nanoparticles (SLNs) was determined according to the procedure described previously. Weighed quantities of SLNs (5 mg) were dissolved in 0.1 N HCl under water bath at 70°C for 30 min and then cooled to room temperature to preferentially precipitate the lipid. Drug content in the supernatant after centrifugation (4000 rpm for 15 min) was determined by UV visible spectroscopy at 258 nm using 0.1 N HCl as blank.

$Drug\ entrapment\ efficiency\ (\%)$
$$= \frac{Analyzed\ weight\ of\ drug\ in\ SLNs}{Theoretical\ weight\ of\ drug\ loaded\ in\ SLNs} \times 100$$

$$\text{Drug loading } (\%) = \frac{\text{Analyzed weight of drug in SLNs}}{\text{Analyzed weight of SLNs}} \times 100$$

4.2.3.3 Differential scanning calorimetry

Differential scanning calorimetry DSC Q200 V24.4 Build 116 was used. The instrument was calibrated with indium for melting point and heat of infusion. A heating rate of 20°C/min was employed throughout the analysis in the 25–200°C. Standard aluminium sample pans were used for all the samples; an empty pan was used as reference. The thermal behaviour was studied under a nitrogen purge; triplicate run were carried out on each sample to check reproducibility.

4.2.4 *In vitro* Release Studies

The release of Olanzapine from the SLNs was studied under sink conditions. Glyceryl Tripalmitate (GTP-SLN) which showed higher drug content and entrapment efficiency (NF-2 and NF-4) were evaluated for *in vitro* release. SLNs equivalent to 0.41 mg were suspended in 1 ml of dissolution media (0.1 N HCl and Phosphate buffer pH 6.8) and put in dialysis bags (MWCO 12000, Hi-Media). The dialysis bags were placed in 50 mL of dissolution medium and stirred under magnetic stirring at 37°C. Aliquots of the dissolution medium were withdrawn at each time interval and the same volume of fresh dissolution medium was added to maintain a constant volume. Samples withdrawn from 0.1 N HCl and pH 6.8 phosphate buffer were analyzed for Olanzapine content spectrophotometrically at 258 nm against solvent blank. Followed by *in vitro* release studies the data was fitted into various drug release models to examine the pattern of drug release.

4.2.5 *In vivo* Oral Bioavailability Studies

Albino Wistar rats (Male and Female) weighing 200±20 g were used for oral bioavailability studies. All animal experiments were approved by Institutional Animal Ethical Committee, J.S.S. College of Pharmacy, Ooty (Proposal no. JSSCP/IAEC/M.PHARM/PH.CEUTICS/02/2012-13). All the rats were fasted for 12 h before the experiments but had free access to water.

Animals were grouped in to 3 groups and 6 animals were taken in each group (Male: Female, ratio1:1) Group 1 received control, Group 2 received drug suspension and group 3 received GMS Solid lipid nanoparticles. Drug suspension (0.3% w/v CMC) and solid lipid nanoparticles were administered orally by oral feeding tube at dose of 0.18 mg (calculated based on human dose of 10 mg, conversion factor = 0.018). Blood (0.5 mL) was collected by cyano-orbital puncture at 0, 0.25, 0.50, 1, 2, 3, 4, 6, 8, 12 and 24 h after administration separately. Blood samples were placed into eppendorf tubes containing 0.3 mL of anticoagulant (citrate) solution and centrifuged immediately. After centrifugation, the plasma obtained was stored at –20°C until further analysis.

4.2.6 Bioanalytical Method Development and Analysis

Reverse phase HPLC method is the most popular mode for analytical and preparative separations of the compounds in chemical, biological, pharma-ceutical and food samples. In reversed phase mode, the stationary phase is non polar and the mobile phase is polar. The polar compounds gets eluted first in this mode and non polar compounds are retained for longer time. In present study, methods for the estimation of the olanzapine present in the blood plasma samples were developed and validated. For the estimation of olanzapine in blood plasma, the chromatographic variables, namely pH, solvent strength, solvent ratio, flow rate, addition of peak modifiers in mobile phase, nature of the stationary phase, detection wavelength and internal stan-dard were studied and optimized for the separation and retention of the drug. The following are the optimized chromatographic conditions, preparation of standard and sample solutions and the methods used for the estimation of olanzapine in plasma.

4.2.6.1 Chromatographic conditions

Shimadzu gradient HPLC system was used with following configurations: LC-20 AD 230V Solvent delivery system (Pump), Manual Injector 25 μl (Rheodyne), SPD-M20A 230V Photo diode array detector, LC solutions data station, Stationary phase: Phenomenex Gemini C18 (250 × 4.6 mm i.d., 5 μ), Mobile phase: Acetonitrile: 25 mM Potassium Dihydrogen Orthophosphate (pH-6.5), Mobile phase ratio: 50:50, Flow rate: 1.0 ml/min, Sample volume: 20 μl, Detection: 254 nm, Data station: LC Solutions. The mobile phase was filtered through 0.22 μ membrane and degassed using ultra sonicator. All the experiments were carried out at room temperature.

4.2.6.2 Preparation of olanzapine standard stock solution

10 mg of olanzapine was transferred into a 10 mL volumetric flask and the volume was made up to the mark with mobile phase to give 1 mg/mL (1000 µg/mL) solution. From this stock solution, 10 mL of 100 µg/mL solution was prepared.

4.2.6.2.1 *Standard stock solution of IS (Internal standard)*

10 mg of Atorvastatin working standard was accurately weighed and transferred into a 10 ml volumetric flask and dissolved in ACN and made up to the volume with the same solvent to produce a 1 mg/ml of Atorvastatin. The stock solution was stored in refrigerator at -20 ± 20 C until analysis.

4.2.6.3 Preparation of analytical calibration curve solutions

From the standard stock solution 11–220 ng/ml standard solutions were prepared and stored below 8°C until further analysis.

4.2.6.4 Preparation of blank plasma

Blank plasma (0.5 ml) was transferred into 2.0 ml Eppendorf tube and 0.1 ml of mobile phase was added. The resulting solution was vortexed for 5 min. The plasma was extracted using SPE cartridge with ACN-WATER (50:50) and analyzed.

4.2.6.5 Preparation of bio-analytical calibration curve samples

0.4 mL of 11, 15, 20, 30, 110, 130, 180, 220 ng/mL of olanzapine solutions were transferred to 2.0 mL centrifuge tube respectively, to this 0.5 ml of Internal Standard was added. The resulting solution was vortexed for 5 min. The plasma was extracted using SPE cartridge with ACN-WATER (50:50) and analyzed.

4.2.6.6 Preparation of plasma samples

Plasma samples (0.5 ml) obtained from study subjects was transferred into 2.0 ml Eppendorf tube and 0.5 ml of Internal Standard was added. The resulting solution was vortexed for 5 min. The plasma was extracted using SPE cartridge with ACN-WATER (50:50) and analyzed.

4.2.6.7 Method of analysis

The bio-analytical calibration curve samples and plasma sample solutions were injected with above chromatographic conditions and the chromatograms were recorded. The quantification of the chromatogram was performed using peak area.

4.3 Results and Discussion

4.3.1 Preformulation Studies

4.3.1.1 Solubility studies

The saturation solubility of olanzapine was determined in different solvents viz. 0.1 N HCl, Phosphate buffer pH 6.8, pH 7.4 and distilled water. The results are given in Table 4.1. Olanzapine was found to be highly soluble in 0.1 N HCl. The solubility in phosphate buffer pH 6.8, pH 7.4 and distilled water was found to be 197.7 µg/mL, 185.4 µg/mL and 45.46 µg/mL respectively.

4.3.1.2 Compatibility Studies

The spectra obtained from IR studies at wavelength from 4000 cm^{-1} to 400 cm^{-1} are shown in Figure 4.1. After interpretation of the above spectra it was confirmed that there was no major shifting, loss or appearance of functional peaks between the spectra of drug, lipid, physical mixture of drug and lipid (3236.66, 3051.49, 1587, 1419, 758.05 cm^{-1}). From the spectra it was concluded that the drug was entrapped into the polymer matrix without

Table 4.1   Solubility profile of olanzapine in different media

Solvent	Solubility
0.1 N HCl	241.3 ± 6.1 mg/mL
Phosphate buffer PH 6.8	197.7 ± 2.8 µg/mL
Phosphate buffer PH 7.4	185.4 ± 4.1 µg/mL
Distilled water	45.46 ± 0.92 µg/mL

Table 4.2   Functional groups present in the IR spectrum

Sl. No	Functional Group	Wave No. in Std Drug Spectra	Wave No. in Mixture Spectra	Observation
1	(N–H) S	3236.66	3236.66	No shifting
2	Thio (C–H) S	3051.49	3051.49	No shifting
3	Ar (N–H)	1587	1589.40	No shifting
4	Ar (C–N)	1419	1419	No shifting
5	Ar (C–H)	758.05	758.05	No shifting

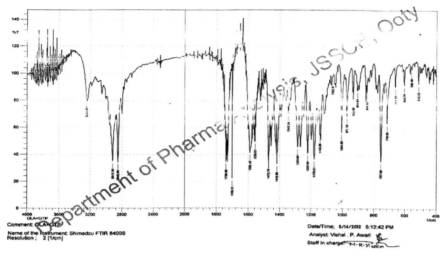

Figure 4.1   IR Spectra of olanzapine and glyceryl tripalmitate.

any chemical interaction. From the IR study it was concluded that, the selected lipid glyceryl tripalmitate is found to be compatible in entrapping the selected drug olanzapine.

4.3.1.3 Development of calibration curve

Calibration curve of the drug was developed to found out the linearity between concentration of drug in the solution and its optical density. It was concluded that the perfect linearity between the concentration and absorbance was observed when the concentration range was from 5 μg/mL to 25 μg/mL. Figures 4.2 and 4.3 shows the calibration of olanzapine using 0.1 N HCl and phosphate buffer pH 6.8. The "Slope (K)" and "Intercept (β)" value was found to be 0.074 and 0.025 for 0.1 N HCl, and 0.062 for phosphate buffer pH 6.8.

4.3.1.4 Partition coefficient studies

Olanzapine is a hydrophobic drug with a log P value of 2.199. The partition coefficient was found in order of Glyceryl tripalmitate (TP) (282.5). GTP is more lipophilic so it has higher affinity for olanzapine. It is suspected having more imperfections in its lipid matrix to accommodate the drug. Hence, an initial study of the partitioning nature of the drug between the melted lipid and aqueous media can provide some clues about the entrapment in SLN formulation.

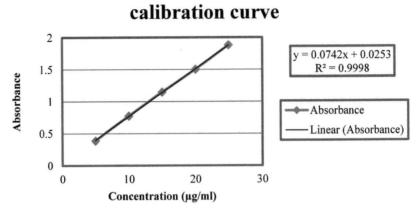

Figure 4.2   Calibration curve of Olanzapine in 0.1 N HCl (λ_{max}= 258 nm).

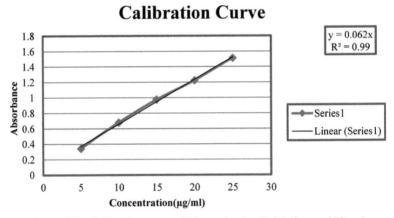

Figure 4.3   Calibration curve of Olanzapine in pH 6.8 (λ_{max}= 258 nm).

4.3.2 Effect of Formulation Process Variables

Solid lipid nanoparticles were prepared by micro-emulsion technique by varying the amount of the lipid. Eight different batches were prepared separately with Glyceryl tripalmitate (GTP) as lipid. It is showed in Table 4.3 and it was observed that at higher concentration of the lipid (65 mg) there was an increase in the particle size as observed. The possible reason might be that amount of lipid was high compared to surfactants used. Surfactant concentration was not sufficient or enough to effectively cover the lipid nanoparticles hence, larger particles were observed. Thus, due to lower particle size in the run 2 with 277.5 nm so this batch was considered as the optimized batch.

Table 4.3 Composition of drug-loaded batches

Formulation Code	Drug	Lipid Concentration	Soyalecitin (W.R.L)	Pluronic F-68	Stearylamine (W.R.L)
NF-1	10 mg	65 mg	1%	1.5%	1%
NF-2	10 mg	50 mg	1%	2%	1%
NF-3	10 mg	65 mg	1%	1.5%	1%
NF-4	10 mg	50 mg	1%	1.5%	1%
NF-5	5 mg	50 mg	1%	1.5%	1%
NF-6	5 mg	65 mg	1%	2%	1%
NF-7	5 mg	50 mg	1%	2%	1%
NF-8	5 mg	65 mg	1%	1.5%	1%

The effect of stirring time and stirring speed on particle size is given in Table 4.4 and 4.5 it was observed that particle size was in range of nm in Glyceryl tripalmitate (GTP). It is obvious that with increase in stirring time and stirring speed there was decrease in particle size up to 4 h in stirring time and 2500 rpm in stirring speed.

4.3.3 Evaluation of Solid Lipid Nanoparticles

4.3.3.1 Zeta potential

Zeta potential is the potential difference between the dispersion medium and the stationary layer of fluid attached to the dispersed particle. The significance

Table 4.4 Influence of stirring time on particle size

Stirring Time (H)	Particle Size (nm)	Polydispersity
1	1195	0.490
2	1068	0.470
3	328.3	0.42
4	294.2	0.357

Table 4.5 Influence of stirring speed on particle size

Stirring Speed (rpm)	Particle Size (nm)	Polydispersity
1000	944.4	0.589
1500	750.9	0.570
2000	310.6	0.377
2500	277.5	0.449

of zeta potential is that its value can be related to the stability of colloidal dispersions. The zeta potential indicates the degree of repulsion between adjacent, similarly charged particles in dispersion. For molecules and particles that are small enough, a high zeta potential will confer stability, i.e. the solution or dispersion will resist aggregation. When the potential is low, attraction exceeds repulsion and the dispersion will break and flocculate. It has been reported that positively charged solid lipid nanoparticles have a better uptake by intestinal lymphatics than neutral or negatively-charged particles. Thus, two batches were prepared with and without Stearyl amine(positive charge inducer) keeping other formulation variables as constant. It was found that batch prepared without Stearyl amine had zeta potential of –1.87 mV, which might be due to negative charge distributed at surface of SLNs due to Glyceryl tripalmitate, while batch prepared with Stearyl amine had zeta potential of 35.5 mV shown in Table 4.6 Stearylamine contains lipophilic hydrocarbon chain (18 carbons), which is accommodated in the lipid core projecting the amine group into the aqueous phase, and induces positive charge.

4.3.3.2 Polydispersity

The Polydispersity index (PI) is the measure of size distribution of the nanoparticles formulation. PI was measured using Malvern zetasizer. PI values range from 0.000 to 1.000 i.e. monodisperse to very broad particle size distribution. PI values of all the formulations indicate that particle size distribution was uni-modal. The optimized batch having least particle size (277.5 nm) had a PI of 0.449 was showed in Figure 4.4.

4.3.3.3 Entrapment efficiency and drug loading

The lipid core was found to affect the extent of entrapment efficiency (Table 4.7). As observed with GTP-SLN the maximum entrapment efficiency was 87.2 ± 0.54% (NF-2) and 68.32 ± 1.32% (NF-4) respectively (Figure 4.5). The entrapment efficiency was higher when amount of drug taken for preparation of SLN was more (10 mg), higher entrapment

Table 4.6 Influence of Stearyl amine on zeta potential

Amount of Lipid (mg)	Amount of Stearyl Amine (% w/w w.r.t. Lipid)	Zeta Potential (mV)
50	–	–1.87
50	1%	35.5

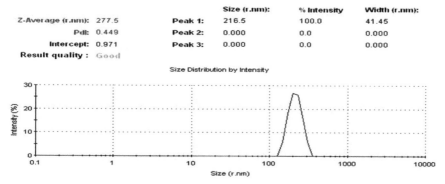

Figure 4.4 Particle size distribution for NF-2 formulation.

Table 4.7 Entrapment efficiency for different batches

Formulation Code	Entrapment Efficiency	Drug Loading
NF-2	87.29%	3.23
NF-4	68.32%	3.15

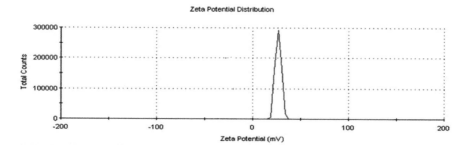

Figure 4.5 Zeta potential of NF-2 formulation.

was obtained. GT-SLN showed higher entrapment efficiency because GTP is more lipophilic and can accommodate more drug in the lipid matrix. These results correlate well with partition coefficient studies.

4.3.3.4 Differential scanning colorimetry

The DSC curve of Olanzapine (OL) shown in Figure 4.6 showed a melting endotherm at 194.92°C. The peak intensity corresponding to the melting of olanzapine decreased in thermograms of OL-loaded TP (167.24°C). These results indicate that only a small fraction of the drug substance existed in the crystalline state. Reduction in the melting point and the enthalpy of the melting endotherm was observed when the lipid was formulated as SLNs. Incorporation of OL inside the lipid matrix results in an increase in number of defects in the lipid crystal lattice, and hence causes a decrease in the melting point of the lipid in the final SLNs formulations. Freitas and Muller also observed the crystalline behaviour of Compritol SLNs differed distinctly from that of the bulk lipid. Small particle size of SLNs leads to high surface energy, which creates an energetically suboptimal state causing a decrease in the melting point.

4.3.3.5 *In vitro* release studies

In vitro dissolution studies were carried out in 0.1 N HCl and Phosphate buffer pH 6.8. The release profiles indicate that SLN formulations showed a retarded release of the drug from the lipid matrix when compared with plain olanzapine solution (OL-SOL). The *in vitro* release data and graph of

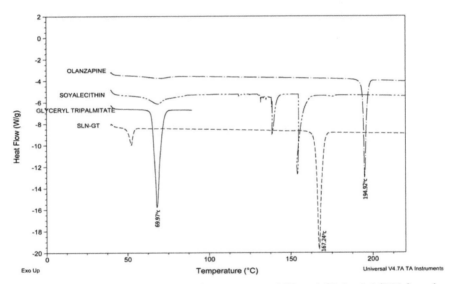

Figure 4.6   Differential scanning calorimeter curves of OL and OL-loaded SLN formulations. OL indicates Olanzapine; GTP – Glyceryl Tripalmitate.

SLN formulations and OL-SOL in 0.1 N HCl is shown in Figure 4.7. It was observed that OL-SOL showed 99.12% release in 4 hrs compared to 56.24% release for GTP-SLN at the end of same time. Cumulative percent of drug released at end of 24 h was 97.69%. This is due to fact that there is no barrier for diffusion at dialysis membrane interface for olanzapine molecules. Hence, higher release was observed in case of OL-SOL. But, GTP-SLN released the drug in a sustained manner. It is evident that there is an inverse relationship between the percent drug release and the partition coefficient of olanzapine. Similar results were obtained when dissolution studies were carried out in phosphate buffer pH 6.8 (pH 6.8) (Figure 4.8). The *in vitro* release data and graph of SLN formulations and OL-SOL in phosphate buffer pH 6.8 is shown in Figure 4.8. It was observed that OL-SOL showed 90.24% release in 3 hrs compared to 30.41% release for GTP-SLN at the end of same time. GTP-SLN showed sustained release of the drug up to period of 24 h Cumulative percent of drug released at end of 24 h was 90.45%. Both SLN formulations showed slower release in 0.1 N HCl compared to PB pH 6.8, which is due to higher solubility of olanzapine in 0.1 N HCl. Olanzapine is freely soluble in 0.1 N HCl and poorly soluble in PB PH 6.8.

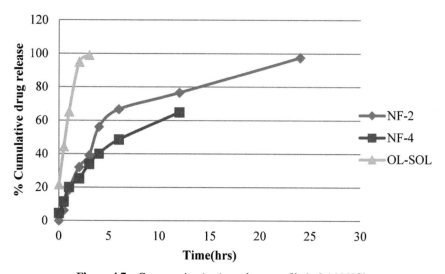

Figure 4.7   Comparative *in vitro* release profile in 0.1 N HCl.

Comparative In vitro release profile in Phosphate buffer pH 6.8

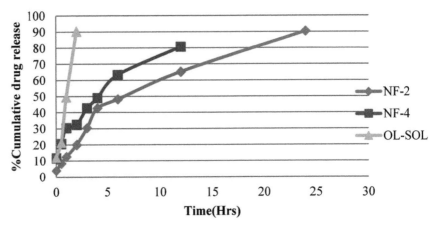

Figure 4.8   Comparative *in vitro* release profile in phosphate buffer 6.8.

4.3.3.6 Release kinetics

In order to elucidate mode and mechanism of drug release, the *in vitro* data was transformed and interpreted at graphical interface constructed using various kinetic models. The *in vitro* release data obtained for SLN formulations, in phosphate buffer pH 7.4, was fitted into various kinetic models. The results are shown in the best linearity was obtained in higuchi plot for SLN formulation indicating the release from matrix as a square root of time dependent process. The Higuchi's plot for NF-2 is shown in Figure 4.9.

Release mechanism: By incorporating the release data in Korsmeyer-Peppa's equation, the mechanism of the drug release can be indicated according the value of release exponent 'n'. Peppa's plot for GTP-SLN is given in Figure 4.10.

The release exponent values 'n' for GTP-SLN was found to be 0.64. Since, the release exponent 'n' values were between 0.5–1, it indicates that the SLN formulations undergo anomalous diffusion.

4.3.4 Bioanalytical Method Development and Analysis

Bio-analytical calibration curve of olanzapine was prepared in mobile phase and response factor was calculated. The results are given in Figure 4.11.

Higuchi plot for NF-2

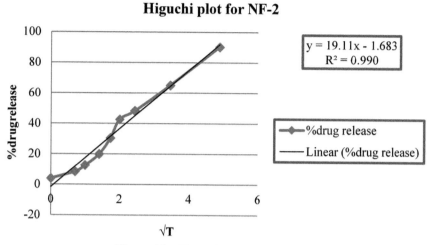

y = 19.11x - 1.683
R² = 0.990

Figure 4.9 Higuchi's plot for GTP-SLN.

Korsmeyer Peppa's plot for NF-2

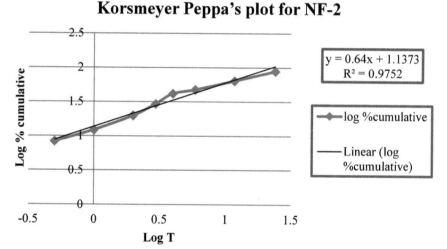

y = 0.64x + 1.1373
R² = 0.9752

Figure 4.10 Korsmeyer Peppa's plot for GTP-SLN.

The chromatograms of olanzapine showed stable baseline. The regression equation in the range of 11–220 ng/ml was as follows: $y = 0.006x + 0.012$, $R^2 = 0.995$.

The concentration of olanzapine was determined in plasma samples separated at different time intervals by HPLC analysis. The concentration of plasma samples was determined from the area of the chromatographic peak

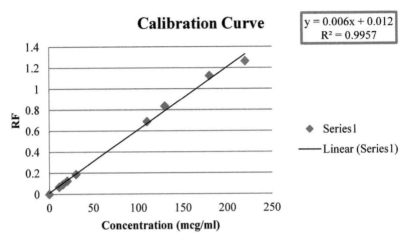

Figure 4.11   Bio-analytical calibration curve of Olanzapine.

using the calibration graph. Chromatogram of blank plasma, pure drug + IS, and Plasma + formulation (NF-2) are shown in Figures 4.12–4.14.

Pharmacokinetic parameters of olanzapine after oral administration are shown in Table 4.8. Peak concentration (C_{max}) and time of peak concentration (T_{max}) were obtained directly from the individual plasma-concentration

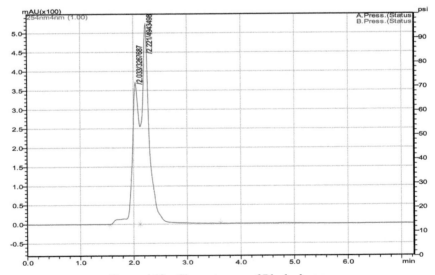

Figure 4.12   Chromatogram of Blank plasma.

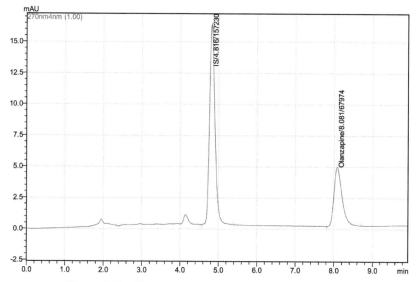

Figure 4.13 Chromatograph of pure drug and IS in plasma.

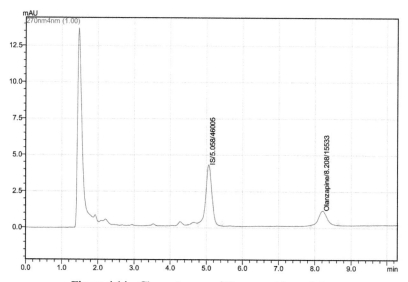

Figure 4.14 Chromatogram of Plasma and formulation.

time profiles. The area under the concentration-time curve from time zero to time t ($AUC_{0\rightarrow t}$) was calculated using the trapezoidal method. The area under the curve (AUC) determines the bioavailability of the drug for the

Table 4.8 Pharmacokinetic parameters of olanzapine after oral administration (mean $\pm$ S.D.) (*p < 0.05)

Parameter	OL-SUSP	GTP-SLN
C_{max} (μg/mL)	0.199 ± 0.056	0.450 ± 0.125*
T_{max} (h)	4 ± 0.083	6 ± 0.166*
$t_{1/2}$ (h)	4.07 ± 0.04	13.86 ± 0.069*
$AUC_{0 \rightarrow t}$ (μg h/mL)	2.695 ± 0.374	6.354 ± 0.905*
$AUC_{0 \rightarrow \infty}$ (μg h/mL)	5.17 ± 0.7128	13.62 ± 0.797*
K_{eli} (1/h)	0.17 ± 0.01	0.05 ± 0.01
$AUMC_{0 \rightarrow \infty}$ (μg h/mL)	59.69 ± 2.93	299.4 ± 6.93*
MRT (h)	5.323 ± 0.25	9.513 ± 0.16*
F_r	1	4.338 ± 0.473

given the same dose in the formulation. The area under the total plasma concentration–time curve from time zero to infinity was calculated by:

$$AUC_{0 \rightarrow \infty} = AUC_{0 \rightarrow t} + C_t / K_e$$

Where, C_t is the olanzapine concentration observed at last time, and K_e is the apparent elimination rate constant obtained from the terminal slope of the individual plasma concentration–time curves after logarithmic transformation of the plasma concentration values and application of linear regression. The relative bioavailability (F_r) at the same dose was calculated as: $F_r = AUC_{SLN, 0 \rightarrow t}/AUC_{SOL, 0 \rightarrow t}$

The mean residence time (MRT) was estimated from MRT = $AUMC_{0 \rightarrow \infty}/AUC_{0 \rightarrow \infty}$

Figure 4.15 shows the concentration–time curve of Glyceryl tripalmitate solid lipid nanoparticles (GTP-SLN) and Olanzapine suspension (OL-SUSP). It is evident that there was increased absorption of olanzapine. The C_{max} value of olanzapine in SLNs (0.450 ± 0.125 μg/mL) was significant ($p < 0.05$) than that observed with OL-SUSP (0.199 ± 0.056 μg/mL). Twenty-four h after oral administration the olanzapine plasma concentration was still more than 0.5 μg/mL. The $AUC_{0 \rightarrow t}$ values of olanzapine after oral administration of GTP-SLN were 3.96-fold higher than those obtained with OL-SUSP. Possible explanation for significantly higher ($p < 0.05$) AUC, $t_{1/2}$ and MRT for GMS-SLN than OL-SUSP is due to slower release of olanzapine from SLNs compared to OL-SUSP which lead to lower clearance.

The enhanced bioavailability by the SLNs formulation might be attributed to direct uptake of nanoparticles through the GI tract, increased permeability by surfactants, and decreased degradation and clearance. Firstly, the uptake

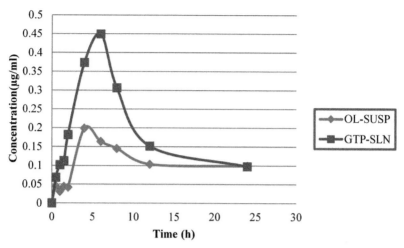

Figure 4.15 Concentration time profile after oral administration of Glyceryl tripalmitate solid Lipid nanoparticles (GTP-SLN) and Olanzapine suspension (OL-SUSP).

of olanzapine in the SLNs-encapsulated form could be up taken through the GI tract, where the particle size played a dominant role in absorption rate [11]. The mechanisms of such uptake include diffusion of particles through mucus and accessibility to enterocyte surface, epithelial interaction and cellular trafficking, and exocytosis and systemic dissemination. The size of SLNs in the range of 20–500 nm allows the efficient uptake in intestine, particularly in the lymphoid sections of this tissue; therefore bypass the liver first-pass metabolism [12]. Secondly, the surfactants, such as Pluronic F-68 and soy lecithin, have contributed to an increase in the permeability of the intestinal membrane or improved the affinity between lipid particles and the intestinal membrane, and also may exhibit bio-adhesion to the GI tract wall [13, 14]. Thirdly, by incorporation into nanoparticles, olanzapine can be embedded into a solid lipid matrix thus not only reducing its exposure to bacterium as well as enzymatic degradation during absorption process, but also offering a long time contact with the wall of intestine *in vivo* due to the nice adhesiveness of SLNs to the mucosal surface of intestine. Also, positively charged particles are better taken up by intestinal lymphatics than neutral or negatively charged particles [13]. Apical potential of epithelial cells of gastrointestinal tract as well as other cells in the body possess a negative charge on their surface due to the presence of negatively charged proteins on the outer membrane of the cells there by better permeability and uptake would occur for positively charged colloidal particles due to electrostatic attraction between oppositely charged surfaces. In addition, Olanzapine-SLNs could provide olanzapine with a long circulation effect *in vivo* with sustained-release property, which prolonged the drug residence time in systematic circulation and resulted in better bioavailability.

4.4 Conclusion

In conclusion, microemulsion technique was suitable for producing solid lipid nanoparticles. Lipophilic drugs like olanzapine can be successfully incorporated into the lipid (Glyceryl tripalmitate). The formulated solid lipid nanoparticles showed a significant increase in oral bioavailability compared to pure drug suspension. Higher relative bioavailability would be due to avoidance of first-pass hepatic metabolism by intestinal lymphatic transport, which circumvents the liver. The dose of the olanzapine SLN needs to be corrected in accordance with increased bioavailability, to minimize its dose related adverse effects. SLNs provided sustained release of the drugs, and these systems are the preferred drug carriers for lipophilic drugs to overcome these drugs oral bioavailability problems.

References

[1] Bhana, N., Perry, C. M. (2001). Olanzapine: a review of its use in the treatment of bipolar I disorder. *CNS Drugs* 15, 871–904.

[2] Mattiuz, E., Franklin, R., Gillespie, T., Murphy, A., Bernstein, J., Chiu, A., et al. (1997). Disposition and metabolism of olanzapine in mice, dogs, and rhesus monkeys. *Drug Metab. Dispos.* 25, 573–583.

[3] Suresh, G., Manjunath, K., Venkateswarlu, V., and Satyanarayana, V. (2007). Preparation, characterization, and *In vitro* and *In vivo* evaluation of lovastatin solid lipid nanoparticles. *AAPS Pharm. Sci. Tech.* 8, E1–E9.

[4] Paliwal, R., Rai, S., Vaidya, B., Khatri, K., Goyal, A. K., Mishra, N., et al. (2009). Effect of lipid core material on characteristics of solid lipid nanoparticles designed for oral lymphatic delivery. *Nanomedicine* 5, 184–191.

[5] Alam, M. I., Baboota, S., Ahuja, A., Ali, M., Ali, J., and Sahni, J. K. (2011). Nanostructured lipid carrier containing CNS acting drug: formulation, optimization and evaluation. *Curr. Nanosci.*, 7, 1014–1027.

[6] Mustafa G., Baboota, S., Ahuja, A., and Ali, J. (2012). Formulation development of chitosan coated intra nasal ropinirole nanoemulsion for

better management option of parkinson: an *In vitro Ex Vivo* evaluation. *Curr. Nanosci.* 8, 348–360.

[7] Date, A. A., Joshi, M. D., Patravale, V. B. (2007). Parasitic diseases: liposomes and polymeric nanoparticles versus lipid nanoparticles. *Adv. Drug Deliv. Rev.*, 59, 505–521.

[8] Mehnert, W., Mäder, K. (2001). Solid lipid nanoparticles production, characterization and applications. *Adv. Drug Deliv. Rev.* 47, 165–196.

[9] Blasi, P., Giovagnoli, S., Schoubben, A., Ricci, M., and Rossi, C. (2007). Solid lipid nanoparticles for targeted brain drug delivery. *Adv. Drug Deliv. Rev.* 59, 454–477.

[10] Chakraborty, S., Shukla, D., Mishra, B., and Singh, S. (2009). Lipid- An emerging platform for oral delivery of drugs with poor bioavailability. *Eur. J. Pharm. Biopharm.* 73, 1–15.

[11] Hussain, N., Jaitley, V., Florence, A. T. (2001). Recent advances in the understanding of uptake of microparticulates across the gastrointestinal lymphatics. *Adv. Drug Deliv. Rev.* 50, 107–142.

[12] Huang G., Zhang, N., Bi, X., and Dou, M. (2008). Solid lipid nanoparticles of temozolomide: potential reduction of cardiac and nephric toxicity. *Int. J. Pharm.* 355, 314–320.

[13] Manjunath, K., and Venkateswarlu, V. (2005). Pharmacokinetics, tissue distribution and bioavailability of clozapine solid lipid nanoparticles after intravenous and intraduodenal administration. *J. Control Rel.* 107, 215–228.

[14] Duchêne, D., and Ponchel, G. (1997). Bioadhesion of solid oral dosage forms, why and how? *Eur. J. Pharm. Biopharm.* 44, 15–23.

5

Nanogels: The Emerging Carrier in Drug Delivery System

Prashant Sahu[1], Samaresh Sau[2], Arun K. Iyer[2] and Sushil K. Kashaw[1,*]

[1]Department of Pharmaceutical Sciences, Dr. Hari Singh Gour University, Sagar [MP], India
[2]Use-inspired Biomaterials & Integrated Nano Delivery (U-BiND) Systems Laboratory, Department of Pharmaceutical Sciences, Wayne State University, Detroit, Michigan, USA
*Corresponding Author

Abstract

Nanogels are engorged nanosized system composed of hydrophilic, or amphiphilic polymer unit established as carriers for drug delivery, planned for impulsively include bioactive molecules by formation of ionic bonds, hydrogen bonds, or hydrophobic conjugation. Nanogels are the nano-sized version of their parent hydro gels, engulfing high water uptake ability, swelling capability, degradability and *pH*-sensitivity styles them appropriate for responsive nanocarrier shows promising and advanced drug delivery system that plays an important role by demonstrating the problems related to chronic and contemporary therapeutics such as non-specific possessions and poor stability aggravate release kinetics approach from slow release while circulating to quick release at the targets will be beneficial as potent medicated drug carriers. The nano-structured hydrogel particles pooled with bacterial enzymes have revealed to activate antibiotic release by debasing the polymeric core. The targeted nanogel specially delivers drugs to either macrophages or on to the attacking microbes leading to drug accumulation at bacterial infection sites, consequently providing lesion site-responsive drug release action, which improves bacterial growth inhibition. The antigen-specific immune retorts induced by novel Nanogel vaccine have successfully possessed protected animal against the pathogen such as *S. pneumonia* making ideal for tuneable degradabi- lity, broadcasting fine blood compati- bility, cyto-compatibility and cellular localization when examined on various cell lines. Thus, the pursuit of this chapter is to concisely describe the recent development of nanogel drug delivery system in terms of its efficacy in various diseases and treatments.

5.1 Introduction

Nanogel was first familiarized to express cross-linked dual-functional net- works of a poly-ions and a non-ionic polymer for carriage of polynucleotide. Nanogels consist of nano-sized particles synthesized by physically or chem- ically cross-linked polymer systems that surges in an optimum solvent [1] (Figure 5.1).

Novel expansion in the arena of nanotechnology have instigate the neces- sity for emerging nanogel structures which established their impending to transport drugs or bioactive in measured, and controlled manner. Expansion of nanogel as the promising field of polymer sciences it has now become predetermined to formulate insolent nano-systems which can be capable for treatment, analysing as well as clinical trials development [2].

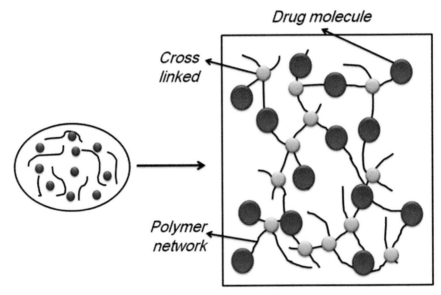

Figure 5.1   A simple nanogel system.

5.2 Properties of Nanogels

Nanogel possesses useful properties which create its optimum candidature for drug delivery, drug carriage, sustained release features, etc. (Figure 5.2). The chief properties of nanogels are as follows:

5.2.1 Good Drug Loading Capacity

High drug loading capacity of nanogels is mainly contingent upon the functional group existing in the polymeric division. The functional groups plays dynamic part in drug delivery and drug release possessions and specific functional groups displays enhance prospective of conjugation with drugs/peptides/proteins/bioactive in targeting solicitations. The vital formation of hydrogen bonding, van der Waals forces in nanogel system exist due to these functional groups which grades in drug resounding bulk, on site drug release characteristics etc. Good drug packing features also take place between the interfaces of drug protein particles due to existence of these dangling functional groups [3].

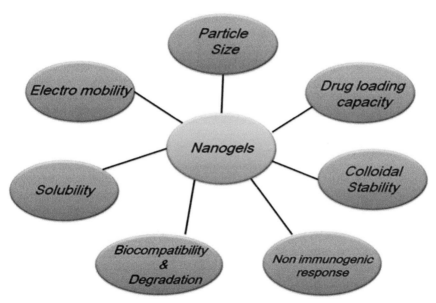

Figure 5.2   Properties of nanogel.

5.2.2 Solubility

Nanogels can proficiently solubilize the hydrophobic drugs and biomolecules in their gel structure. They are also capable to solubilize investigating agents in their gel system resourcefully [4].

5.2.3 Colloidal Stability

Polymeric nanogels systems (polymeric micelles) illustrate enhance stability equated to surfactant micelles and shows low critical micellar concentrations, slow rate of separation and prolonged preservation of entrapped drugs or bioactives [5–8].

5.2.4 Particle Size

The magnitude (size) of the nanogels generally ranges between 20–200 nm in diameter [9]. This actual size range plays vital role in evading the quick renal separation but are small tolerable to avoid the uptake by the reticuloendothelial system. Due to their nano-sized feature they can effortlessly cross blood brain barrier (BBB) and demonstrate budding permeation abilities [10].

5.2.5 Biocompatibility and Degradability

Nanogel-constructed drug delivery system is extremely biocompatible and biodegradable, due to this features it is a very favourable drug delivery ground in the present situation. The most beneficial feature of nanogels is their quick swelling and de-swelling property [11].

5.2.6 Electro Mobility

The chief feature of nanogel system is that they can easily fabricate and be synthesised without much expenditure of energy and evade consuming the more mechanical arrangements and machineries. Simple sonication and homogenization are sufficient for the effective frame-up and encapsulation of drug/bioactives/proteins/peptides in nanogel system [13].

5.2.7 Non-Immunologic Response

Nanogel drug delivery systems generally eliminate any type of immunological responses or effects related to drug delivery, treatment and diagnostic phenomenon [14].

5.2.8 Others

Nanogel can summarise and transport both hydrophilic and hydrophobic drugs/bioactives which is administered and exaggerated by *pH*, temperature, functional group (hydrophilic/hydrophobic groups) in the polymeric system, cross-linking compactness of the gels, surfactant concentration, and the type of cross-linker extant in the nanogel network [15].

5.3 Classification of Nanogels

Nanogel generally falls in two chief categories, they are as follows (Figure 5.3).

5.3.1 Responsive Type

It is further divided into following types:

5.3.1.1 Non-responsive nanogels
They basically swell in optimum solvent like water due to simple absorption phenomenon.

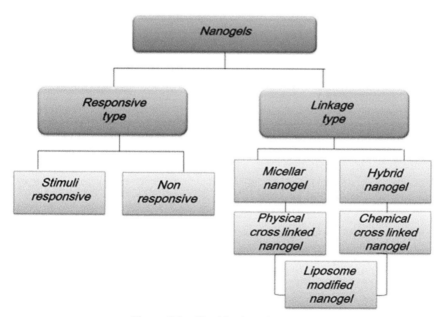

Figure 5.3 Classification of nanogel.

5.3.1.2 Stimuli-responsive nanogels

They swell or de-swell on contact to environmental variations like *pH*, temperature, humidity, magnetic field, ionic strength etc. [16]. Multi-responsive nanogels are responsive to more than one environmental provocation [17].

5.3.2 Linkage Type

The second group is constructed on the type of linkage in the nanogel network. This classification further subdivided into the following sub categories:

5.3.2.1 Physical cross-linked gels

Physically cross-linked gels or pseudo gels are synthesized by the weak association forces like Van der Waal's forces, hydrophobic interactions, electrostatic interactions or hydrogen binding etc. [18]. Varieties of simple approaches are offered for the preparation of physically cross-linked gels. These classifications are very profound and the responsive performance greatly subjective by polymer concentration, temperature, *pH*, ionic strength of medium, cross-linkers etc. [19]. The linkage of amphiphilic block copolymers and complexation of oppositely charged polymeric chains outcomes in the formation of micro- and nanogels in only few minutes. Physical gels can also be designed by the combination and/or self-assembly of polymeric chains [20].

5.3.2.2 Liposomes modified nanogels

Liposome altered nanogels are developing as decent drug carriers in the extensive variety of disease ailments. These nanogels display liposomes in their network matrix and release of liposomes governed by the various factors like, *pH*, temperatures and ionic charge, for example liposome constituting succinylated poly (glycerol) expressively transport calcein to the cytoplasm at *pH* 5.5 [21]. Also liposome-aided nanogel fabricated by the poly (N isopropylacrylamide) polymeric moieties reveals thermo responsive triggered phenomenon of drug release when given transdermally.

5.3.2.3 Micellar nanogels

The micellar nanogels are prepared synthesized by the supra-molecular self-assembly of amphiphilic blocks or graft co-polymers in aqueous media. Micellar nanogel exhibits pendant core casing organization [22]. The core is fabricated of hydrophobic block unit which is enclosed by the hydrophilic polymer section. The hydrophilic unit plays a vital role in the development of hydrogen bond which results in the steady core shell assembly around the micelles. These core shell micellar nanogel are stimuli receptive in nature which reveals significant stability. The core shell nanogel shows large surface area to deliver fine accommodation of drugs or bioactives prepared by physical entrapment method [23].

Thus, the drug or bioactive falls in hydrophobic core are restricted from the enzymatic deprivation and hydrolysis. Scientific fraternity reconnoitring the development of Y-shaped micelles organized by biodegradable, non-toxic and *pH* responsive polymers such as PLGA, chitosan, poly (oleic acid-N-isopropylacrylamide), dextran etc. for the effective drug delivery.

5.3.2.4 Hybrid nanogel

Hybrid nanogels composed of nano-sized particles detached in the organic or inorganic media. Hybrid nanogels are very reactive due to their structural organisation. Scientist have evaluated that the hybrid nanogel synthesized in an aqueous media by self-assembly of polymers are very effective in quick drug release to the targeted site. Many polymers such as pullulan- PNIPAM, polysaccharides etc. are employed for the preparation of these nanogels [24]. Cholesterol-pullulan hybrid nanogel has been widely explored and initiates to be a promising drug carrier in the exceptional delivery of DNAs, RNAs, proteins, peptides, oligosaccharides and other bioactive molecules. Hybrid nanogels are also well effectual in the onsite and on demand delivery of insulin and anticancerous agents. These polymeric nano-carriers are very precise in functions and responsive towards the stimuli for example, the delivery of CHP through Pullulan deportment cholesterol nanogel produced by self-assembled or combined method are well-accomplished of delivering CHP on the targeted site by virtue of *pH* and thermo-sensitive triggered technique [25].

5.3.2.5 Chemically cross-linked nanogels

Chemically cross-linked nanogels are developed by the covalent bonding method. The feature of chemically cross-linked gel is greatly influenced the chemical bonding and the type of functional unit [26] or group present in the gel network. Wide variety of nanogel has been magnificently prepared by this technique. The procedure of polymerization between hydrophilic polymer and hydrophilic-hydrophobic co-polymer exist in the presence of multifunctional cross-linkers by the support of vinyl monomers. These cross-linkers play a crucial role in the physiochemical properties of nanogel and managed the permanency of nanogel. The size, shape, texture and morphology of nanogel predisposed greatly by the property of cross-linkers for example, biodegradable PLGA-PEG di block nanogel shows the size range in between 10–200 nm in the presence of central thiol group attained by the disulphide cross-linking method. The whole polymerization method acquired by virtue of 'Green chemistry' combination mechanism [27].

5.4 Method of Preparation of Nanogel

5.4.1 Photolithographic Technique

Photolithographic technique is employed to achieve the 3-D nanogel structure. Photolithography shows reversible cast gels with decent surface area that lodge large volume of drugs or bioactive. This technique is categorised in to five main steps, they are as follows:

1. Discharge of UV cross-linked polymer which acts as a substrate containing larger surface area on the previously dried photo resist coated water.
2. Beading of polymer into the wafer like assembly by pressing the quartz prototype onto the polymer. This moulded template then visible to UV radiations.
3. Quartz template was detached by detection the particles displaying thin enduring interrelating film layer.
4. Lastly, the residual thin layer was detached by plasma comprising oxygen molecules.
5. Collection of fabricated particles is done by fine dissolution of substrate in water comprising buffers [28, 29].

Photolithography technique subsidizes PRINT method (Particle repetitions in non-wetting templates). PRINT shows creation of fine nano-sized, controlled moulded steady nanogel system. Variety of drugs, DNAs, RNAs, proteins, peptides, oligosaccharides etc. are encapsulated in the exceptional targeted delivery via nanogel system developed by using PRINT mechanism.

5.4.2 Micro-Moulding Method

Micro-moulding method is very comparable to the photolithography; only difference is the low cost production of micro-moulding technology compare to photolithography. Micro-moulds originate in wide variety of forms like square-prism, disks, strings etc. Micro-moulding technique displays cell suspension in a hydrogel pioneer which covers photo-initiator in aqueous media such as PEGDA, MeHA (methaacrylated hyaluronic acid) etc. [30]. The obtained solution was accommodated in plasma cleaned hydrophilic PDMS section and additionally it was photo cross-linked by acquaintance to UV light. The dumped cell-loads microgels particles were isolated, washed and finally collected.

5.4.3 Bi-Polymers Synthesis Technique

Naturally occurring carbohydrate based bi-polymers like chitosan, albumin, hyaluronic acid; dextran etc. creates to formulate nanogels system. Wide varieties of techniques was used in the effective preparation of nanogels, microgels and hydrogels of these bi-polymers like self-assembly mechanism, swelling methods etc. they are chiefly classified into following 4 categories [31].

1. Spray drying method
2. Chemical cross-linking
3. W/O (water in oil) heterogeneous emulsion
4. Aqueous homogenous gelation

5.4.4 Water in Oil (W/O) Heterogeneous Emulsion Method

W/O heterogeneous emulsion method unlocks two key methods, they are as follows:

1. Emulsification of aqueous particles of hydrophilic bi-polymers in continues lipophilic phase containing oil soluble surfactants.
2. Process of cross-linking of bi-polymers with hydrophilic cross-linkers [32].

5.4.5 Inverse Mini Emulsion Method

Inverse mini emulsion technique is composed of mixture of hydrophilic bi-polymer and continuous hydrophobic phase with continuous high speed homogenizer or stirrer. Hydrophobic surfactants like span-80, Aerosol OT are employed in this technique for the improved permanency of nanogel network which results in attainment of mini emulsion. The hydrophilic particle in mini emulsion network includes varieties of DNAs, RNAs, drugs etc. These hydrophilic particles further cross-linked with the cross-linkers agents such as gluteraldehyde. The obtaining microgels particles were then precipitated, centrifuges at high rate and washed many times with suitable organic solvents like iso- propanol and finally proceeds to lyophilisation. Generally, gelatine aided nanogels are synthesized by mini-emulsion mechanism. Mini-emulsion technique consists of 2 keys element i.e. fusion and fission technique. The mini-emulsion of one polymer like gelatine is attached with another polymer solution such as poly (butylenes-co-ethylene) block poly (ethylene oxide) cross-linked by means of gluteralde-hyde; the mini-emulsion was then ultra-sonicated resulting in the fabrication of stable gel particles [33]. This gel particle shows the exceptional size of 100–400 nm. The mini-emulsion process needs large amount of surfactant for the development of stable gel particles discloses the particles size in nano and micro-meter.

5.4.6 Reverse Micellar Method

Reverse micellar technique includes W/O (water in oil) scattering mechanism, equivalent to the inverse mini-emulsion method. The reverse micellar technique uses large quantity of hydrophilic surfactant moieties for the thermodynamically stable manufacture of micellar solution comprising aqueous particles dispersed in continuous oil phase. The gel particles shows decent nano-sized particles vacillating about 10–150 nm. For the generation of consistence nano-sized particles the cross-linkers play a vital part, and suitable concentration of oil soluble surfactant is very important for the thermodynamically stable nano-sized gel particles [34].

5.4.7 Membrane Emulsification Method

Membrane emulsification technique is a new method for the fabrication of nanogel particles having fine spherical size with homogeneous size circulation. This method comprises the unvarying size membrane filtration phenomenon, usually Shirasu porous glass (SPG) membrane with fine homogenous pore size ranging 0.1–20 mm [35]. The emulsion permitted to pass through the membrane at suitable pressure producing the uniform size particles. The resulting particles shows homogenous size distribution arrangement which is dispersed in the continuous phase leads to the development of W/O or O/W emulsions. This method permits the generation of multiple emulsion system like W/O/W, O/W/O and solid/O/W dispersion system and the composed particles demonstrates wide varieties of shapes like spherical, hollow spherical, micro-spherical, core shell micro-particles, micro capsules etc.

Membrane emulsification technique recently merged with the stage wise cross-linking method for the development of nanogel showing uniform and thermodynamically steady particles. Conjoining with ion gelation method the generation of fine nanoparticles achieved which further filter through the membrane and distributed in ionic solidifier followed by chemical solidifier where cross-linking methodology occurs and the synthesized nanogel particles were washed and collected [36].

5.4.8 Heterogeneous Free Radical Polymerization

Heterogeneous polymerization methodology of water soluble monomers involves with an assistance of cross-linkers (multifunctional, di functional and mono-functional) for the fabrication of stable organic micro-gels or nanogels. Heterogeneous free radical polymerization technique for the synthesis of micro-gel and nanogels comprises various techniques like Inverse micro emulsion, Inverse mini-emulsion, Precipitation process, Dispersion polymerization and heterogeneous controlled/living radical polymerization [37].

5.4.8.1 Inverse micro emulsion

It gives thermodynamically stable microemulsion system on addition of emulsifier above the critical concentration. Inverse micro emulsion method includes stable dispersion of aqueous particles having large amount of hydrophobic surfactants in continuous organic medium. Polymerization method develops within the aqueous particles, which gives thermodynamically hydrophilic nanoparticles revealing size about 10–100 nm. Inverse microemulsion produces fine, stable and uniform nanogel system utilization noteworthy polymer variety like dextran, poly-caprolactone, PLGA, Chitosan etc.

5.4.8.2 Inverse mini-emulsion polymerization

This method involves water in oil type (W/O) polymerization which comprises aqueous particles of hydrophilic monomers consistently distributed in a continuous phase by virtue of hydrophobic surfactants. Mechanical stirring of mini emulsion system offers the stable and uniform dispersion of system which is further sonicated for the accomplishment of fine nano-sized particles. The polymerization befalls within the aqueous droplets on accumulation of polymerization initiator radicals [38].

5.4.8.3 Precipitation polymerization

Precipitation polymerization includes the development of uniform mixture at initial stage and the polymerization start in homogenous solution. This resulting polymer solution is un-swellable but possesses decent solubility in the aqueous medium. The cross-linker at this stage plays crucial part for the uniform cohort of nano-sized particles. Introduction of cross-linker results in the generation of nano-sized particles, but display uneven morphology and high polydispersity. The shape and size of the particles can be controlled by the concentration of supplementary monomer in aqueous solution. Concentration of cross-linker also effect the gel particles possessions for example increasing the concentration of cross-linkers decreases the equilibrium swelling of nano gel particles. Development of nanogel by precipitation polymerization shows diverse nanogel delivery for the ailment of various diseases through variety of routes for example nanogel synthesised by PMAA-PEG nanogel particles by precipitation polymerization shows brilliant oral delivery of proteins.

5.4.8.4 Dispersion polymerization

This method comprises solubilization of elements like monomers, stabilizer, polymerization initiators and surfactants in a continuous phase. Polymerization initiates due to presence of initiator which gives uniform reaction mixture. The obtained polymer solution is insoluble in the continuous medium which results in the formation of stable uniform nano-sized particles by virtue of colloidal stabilizers. Many polymeric nanogel systems are synthesised by using polymer dispersion method for example; PHEMA particles were fabricated by PEO-b-poly (1, 1, 2, 2-tetrahydroperfluorodecyl acrylate) di block co-polymer as a stabilizer in supercritical carbon dioxide and methacryloyl – terminated PMMA in a mixture of 2-butanol/toluene (ratio 55:45 wt/wt). These nanogel systems obtained by dispersion polymerization are expressively employed in the effective delivery of DNAs, RNAs, peptides and proteins [39].

5.4.8.5 Heterogeneous controlled/living radical polymerization

This is an innovative technique of fabrication of nanogel particles with the major advantage of polymer protein/peptides conjugates preparations. Controlled/living radical polymerization or CRP explored as favourable carrier for the delivery of prodrugs, protein peptides conjugation at targeted site avoiding any toxicities. It engulfs many methodologies for CRP such as ATRP (atom transfer polymerization), SFRP consistence free radical polymerization, RAFT (Reversible addition fragmentation chain transfer). Evaluating with other techniques RAFT method is the most advantageous for the CRP.

This method is very effective for the generation of gel particles possessing size about 10–150 nm for example it contains single step for the fabrication of PEGylated nanogel of polymer such as poly (*N*, *N*-dimethylaminomethyl methacrylate with an aid of amphiphilic RAFT agent trithio-carbonate having hydrophobic dodecyl chain commencing polymerization [40].

5.4.9 Conversion of Macrogels to Nanogels

This technique is used for the conversion of bulk gel network (macroscopic nanogel having walled to wall bonding) to the nano-sized gels. This technique offers easy preparation of nanogel eliminating the control of synthetic factors. The nanogels are easily manufactured by the method of bulk polymerizations results in generation of solid gel network with macro-porous block. Further these blocks are compacted, grounded and sieved to obtain gels of nano size. The main disadvantage of this method is that, it is very tedious and result in significant loss of materials.

5.4.10 Chemical Cross-Linking Method

Chemical cross-linking technique utilizes synthetic cross-linkers for attaining nano-sized particles. For example polymeric nanogel made from biodegradable dextran consumes the cross-linker carbodiimide and results in the carbodiimide coupling. For the coupling process free radical polymerization method takes place which is termed as Michael's addition reaction. This technique is engaged in many nanogel system fabrications like pullulan bearing cholesterol nanogel, PEG pullulan cholesterol nanogel preparation etc. [41].

5.5 Characterization of Nanogel

Nanogel should be evaluated fundamentally and effectively for obtaining uniform product. Nanogels demonstrate variety of attractive physical properties that shows certain precise quality and features. For example, nanogels possess decent scattering of visible lights which results in the white appearance. The process of multiple scattering of light is due to the refraction mechanism ascending from the nanogel particles showing the higher refractive index. The small bundles of the photon which are probing the nanogels particles scatter numerous times by micro-scale assemblies before they leave the nanogel due to the shortage of optical absorption. Nanogel shows optically transparent organization of the particles even at the elevated phase volume and at high divergence in refractive index. Nanogel demonstrates larger surface area to volume ratio, hence bears greater elastic module [42]. The greater surface area shows greater stability of particles. There are numerous other character-ization parameters possess the several discriminate features of nanogel drug delivery system (Figure 5.4), they are as follows:

5.5.1 Morphological Analysis

Improved resolution and technical adaptabilities of electron microscope have offered this methodology a useful element in the evaluation of the actual morphology of a nanogel system. The main electron microscope utilizes for the characterization purpose of nanogel is as follows:

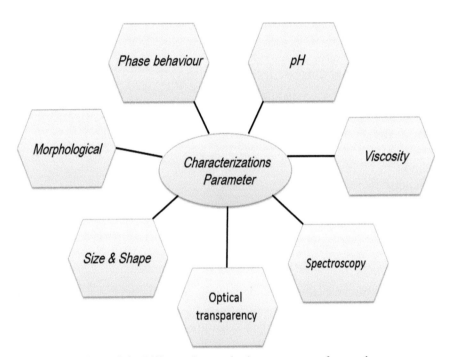

Figure 5.4   Different characterization parameters of nanogel.

5.5.1.1 Scanning Electron Microscopy (SEM)

SEM gives the three dimensional picture of the nanogel particles. The sample of nanogel is observed under the accelerating voltage ranging 20 kV to 50 kV which provides the comprehensive surface morphological image of nanogel. The occurred image further examined by the high performance softwares such as Leica imaging system to obtain automatic investigated results of the surface morphology and the shape of the nanogel particles.

5.5.1.2 Transmission Electron Microscopy (TEM)

TEM is very delicate method for the assessing and evaluation of the morphology of nanogel. In the TEM phenomenon the nanogel sample is stained with the negatively charged 1% aqueous solution of phosphotungstic acid or by the dropping 2% acetate solution in to the 200 μm mesh size pioloform coated copper grid or a microscope carbon coated grid employing a micropipette and the sample examined under the transmission electron microscope at 50 kV–80 kV. TEM offers qualitative analysed digital image and supply the morphological particles view [43].

5.5.2 Size and Shape

The dimension of the nanogel particles plays vital role in therapeutic activity of formulation. Polydispersity, surface charge, size and shape of the particles governed effectively in the nanogel remedial property. The particle size can be evaluated by the TEM, Zeta-size analyzer, SEM and the atomic force microscopy (AFM). The surface charge can be examined by the Zeta-size analyzer or Zeta potential analyzer. The polydispersity index assessed the homogeneity of the nanogel formulations. It can be effectively analysed by the dynamic scanning colorimetry (DSC) via photon correlation spectroscopy. Polydispersity index can also be assessed by the Abbe-refractometer. Laser diffraction method can also be engaged for the assessment of particle size distribution design. Laser diffraction is a volume based methodology that displays particle size pattern in appearance of volume of identical particles and partial mean of the volume distribution [44]. Photon correlation or mini electrode technique can also be used for the effective determination of the surface charge of nanogel formulation.

5.5.3 Viscosity

Viscosity or thickness of the nanogel demonstrates the aqueous, oleaginous and surfactant constituent and their reliability, which can be evaluated at diverse shear rate at different temperatures on the basis of Brookfield rotator viscometer. Viscosity of nanogel governs the stability and drug release properties. The viscosity of the nanogel can be augmented by decreasing the surfactant and co-surfactant concentration. The viscosity of the nanogel can be effectively decreases with increase in the concentration of aqueous phase of the nanogel formulation [45].

5.5.4 Phase Behaviour

Phase behaviour and the phase inversion distribution can be effectively evaluated by the assessment of electrical conductance. Hydrophilic nanogel system shows high electrical conductance which results in the percolative performance of particles of nanogels. Determination of the dielectric constant measured the dynamic and the morphological features of nanogel formulation. The interfacial tension play vital role in the phase behaviour arrangement of nanogel. It can be obtained from the assessment of morphology of nanogel having low phase density when rotating in the cylindrical packed high density spinning particle apparatus.

5.5.5 Optical Transparency

The nanogel particles shows "Rayleigh scattering phenomenon" which is uniform parallel to the polarizable molecules stirring from atmospheres. Nanogel particles are fine transparent molecules and possess very low dispersion but decent refractive index. Nanogel is clear and transparent in nature but reveals a slight shade which is light blue in colour due to the scattering of low wavelength light [46]. But they shows reddish tint colour when they exposed to the white light due to the scattering away of blue light. These inclusive optical features make nanogel a suitable candidate for the micro-rheology evaluation.

5.5.6 Spectroscopic Analysis

The nanogel formulations composed of useful polymers including natural and synthetic, which displays their own physio-chemical properties. For the good assessment of nanogel the structural description is very significant. Nanogel composed of polymeric back bone gives varieties of functional groups, these functional groups absorb wavelength of different type and of varied range. For the advance evaluation like conjugation, prodrug examination, tri-block or di-block evaluations etc. the nanogel formulations are subjected to numerous spectroscopic evaluations such as FTIR Spectroscopy, NMR Spectroscopy, DSC analysis, HPLC analysis etc. [47].

5.5.7 *pH*

pH is the determination of acidity or basicity of compound or material. The *pH* of nanogel plays a crucial role in the delivery of bioactives to the targeted site. The *pH* responsive polymer like chitosan displays *pH* range of 5–6 which significantly affects its release and delivery of drug load at the precise targeted area. Therefore, assessment of *pH* factors influence prominently in expansion of nanogel formulations for the effective release, targeting and mode of delivery of nanogel system.

5.6 Routes of Administration of Nanogel

Routes of administration greatly influence the therapeutic efficiency of nanogel formulations. Nanogel acts as capable carrier in varieties of delivery of drugs, bioactives, natural extracts, protein, peptides etc. via diverse routes of administration [48]. Nanogel can be delivered in the body by numerous routes for example oral, transdermal, ocular, parenteral, rectal or vaginal and pulmonary (Figure 5.5).

The nanogel possesses numerous discriminate features by various routes of delivery such as, parenteral delivery preferred controlled release, nutrient

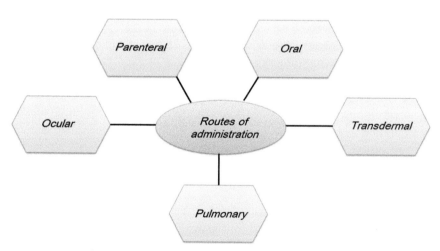

Figure 5.5   Different routes of administration of nanogel.

carrier etc, transdermal delivery epitome for sustained release, *pH* triggered phenomenon of drug release and many more. The nanogel drug delivery system is enormously complimentary for quick absorption with significant therapeutic efficiency. In oral drug delivery, the absorption of drug in gastrointestinal tract (GIT) is directly proportional to the nanosized particles of nanogel. Nanogel possesses noteworthy drug delivery in ophthalmic field, they have been extensively envisioned for their functionality in treatment of cataract, lens damage disorder, retinal damage etc. with extreme advantage of sustained release of drug at targeted site evading contact to normal cells [49]. Nasal or pulmonary delivery of nanogel has been widely visualized for the diverse benefits of this nano-formulation. The broad benefits of various routes of delivery of nanogel are broadly elucidated as follows:

5.6.1 Parenteral Drug Delivery System

Parenteral drug delivery of hydrophobic drug is the major hindrance in absorption at targeted site due to low concentration. The nano fine property of nanogel system raises parenteral delivery at new height. The parenteral delivery of nanogels shows good residing time and significant clearance period. Additional remarkable advantage of parenteral nanogel administration is controlled of low dose which results in fringe toxicity due to presence of surfactants.

In medical and pharmaceutical research, the usage of C-3 and C-4 alcohols as surfactants over PEG-400 or PEG-600 has evolved as good option to overcome the difficulties of toxicity and to preserve the extended stability of system. Nanogel comprising thalidomide shows good therapeutic effective concentration at a very low dose of about 25 mg. Chlorambucil nanogel synthesized by the high energy ultrasonication process reveals noteworthy chemotherapic efficiency against the ovarian cancer. The nanogel formulation containing Chlorambucil displays significant tumour suppression rate in adeno-carcinoma when assessed *in-vivo* compare to the free drug therapy. Hydrophobic drug such as carbamazepine when encapsulated in nanogel delivery system reveals significant release kinetics profile.

5.6.2 Oral Drug Delivery System

Oral route is the most superior route of administration of varieties of drugs and bioactives. Oral route is exhibits patent compliance feature and effective route of achieving therapeutic target along with comfort of personal medication. Despite of suitable route, oral drug delivery shows noteworthy disadvantages also such as GIT degradation of drug, low therapeutic efficiency, difficulties to geriatric patients and nuisance to paediatrics patients. Hydrophobic drugs reveal poor absorption at the targeted site through oral delivery due to the related stability matters. Intestinal absorption over oral drug delivery system is very contracted of peptides molecules due to the quick hydrolysis and enzymatic deprivation. The chief advantage of the nanogel systems is the nano-sized particles attained by ultrasonication, centrifugation and stirring possessing decent penetrating ability into the GIT membrane compare to the other drug delivery system. The oral nanogel delivery system needs good optimization and evaluation methods for the good and stable product [50].

A di-terpenoid pseudo alkaloid "Paclitaxel" is a effective antineoplastic agent and used varieties of tumour conditions. Polymeric nanogel fabricated by the PLGA coated with chitosan shows cationic charge on the surface of nanogel, introduction of oleophilic solvent such as pine nut oils and egg lecithin as emulsifiers possess brilliant stability when ultrasonicated at high rate. Paclitaxel loaded PLGA-Chitosan nanogel shows significant bioavailability and sustained release kinetics. The nanogel loaded absorbed drug possess uniform distribution in liver, kidneys and lungs offering noteworthy targeting. The paclitaxel nanogel reveals good intestinal permeability when administered orally.

The self-emulsifying nanogel drug delivery carriers shows elevated transport rate and fine absorption compared to the free drug delivery. The transport rate and absorption is directly proportional to the composition. Self-emulsifying nanogel system expected as a good drug delivery of proteins and peptides. For example, antihypertensive agent ramipril when given via nanogel system shows 3 times greater absorption of ramiprilat. Ramipril nanogel system is very suitable for the geriatric and paediatric patients.

In gastrointestinal tract (GIT) the lipids molecules are easily absorbed via lipid absorption mechanism. But the absorption of proteins and enzymes decreases severely due to enzymatic deprivation. Thus, nanogel matrix system is one of the best alternatives for the efficient delivery of proteins and enzymatic molecules showing decent increase in the proteins and peptides. Utilization of nanogel system in oral drug delivery system produce promising therapeutic activity which results in the enhance absorption of drug, distribution of molecules, good targeting efficacy, oral bioavailability, effective permeability, better cell or tissue targeting, imaging, and other therapeutic action [51].

5.6.3 Transdermal Drug Delivery

Transdermal route of drug delivery is another most desirable route after the oral drug administration. Delivery of drug into the systemic circulation via dermal route is widely visualized due to variety of benefits like sustained release kinetics, controlled delivery of drug, ease of self-administration, bypass first pass metabolism, elimination of GIT deprivations, bowel distress etc.

Transdermal nanogel drug delivery system has been synthesized for the ailment of a variety of ailments like eczemas, cardiovascular disorders, Parkinsonism, Alzheimer's disease, depression and cancer. The major problem in this route is the intact skin barriers which greatly governed the pharmacokinetic aspects and targeting of drugs. The cutaneous blood vessels and lymph vessels should be optimized for fine targeting and absorption of drugs. Nanogel system can easily penetrate the intact skin barriers via epidermal pores and deliver drug moieties directly to the targeted site or systemic circulations showing fine therapeutic action. The efficient skin penetration of the can also be attained by the use of penetrator enhancer like oleic acid, malic acid, terpenes, esters, controlling the optimization parameters and vehicle aspects [52].

Magnetic nanogels are evolved as a favourable drug delivery carrier in contending the cancers through photodynamic chemotherapy. The chief phenomenon behind the photodynamic chemotherapy is the delivery of photo sensitizer such as Foscan into the subterranean tissue layers outside the epidermal area of the skin, commencing hyperthermia doe resultant free radical generation. Transdermal delivery of NSAIDs like aceclofenac, diclofenac and piroxicam possess good residing time at targeted site compares to the free delivery via conventional dosage form, for examples aceclofenac shows 3 times better therapeutic bioavailability when compared to aceclofenac delivered in tablet form. Nanogel as transdermal drug delivery carrier concerns as an effective tool having plasma concentration profiles and bioavailability avoiding the side effects and toxicities.

5.6.4 Ocular Drug Delivery System

The ocular drug delivery concerns as the delicate drug delivery system. Due to the particular environment and very sensitive pharmaco-kinetic profile, wide variety of challenge occurs in ocular drug delivery system. Nanogel system shows a great opportunity in ocular drug delivery stream with positive benefits like decent ocular bioavailability, decline in dose interval and good patient compliance for example Hydrocortisone nanogel delivery not only exhibits decent bioavailability but decline in dose administration can also be attained nanogel system [53]. Low viscosity and nano-fine particles size of nanogel produces fine absorption and quick entry in the ocular environment which results in better bioavailability like Pilocarpine nanogel constituting lecithin, propylene glycol and PEG-200 as co-surfactant with IPM (Isopropyl myristate) as oleaginous phase shows low refractive index with marginal viscosity leading to decent ophthalmic delivery. Polymeric nanogel-contact lens evolved as a impending option for the cataract and other eye dys-functioning conditions. The mineral oil nanogel contact lens gives decent drug reservoir carrier in cataract conditions, possessing good bioavailability of drug at ocular circulation, steady physio-chemical properties and reducing frequent dosing task. The main disadvantage of nanogel in the ocular delivery is low residing time in eye socket and faster drifting of particles from ocular environment.

5.6.5 Pulmonary or Intranasal Drug Delivery System

Intra nasal or pulmonary drug delivery system is evolved as a decent ther-apeutically carrier for the management and delivery of wide variety of medications showing fine and quick release of drug load at desired site. Pulmonary drug delivery system is not a new element, it owning its signif-icant efficacies from ancient time of Ayurvedic medicinal system offering stable formulation with elevated bioavailability at low dose. Intranasal drug delivery of nanogel shows non-invasive, patient friendly and sympathetic route eliminating the first pass metabolism and GIT enzymatic deprivation of drugs [54].

The chief task in the drug delivery system is to cross BBB (Blood Brain Barriers) exclusively for hydrophilic drugs due to the impervi-ous characteristics of endothelium of cerebral capillaries. The olfactory region of nasal mucosa offers a straight link between the nose and brain and by the application of nanogel system loaded with drugs shows good therapeutic action in disease like migraine, depression, schizophre-nia, Alzheimer's disease, Parkinson's disease and meningitis. Nanogel also offers astonishing properties in pulmonary drug delivery such as low dose delivery, negligible toxicities, guarding of bioactives, enhanced interaction of drug with nasal mucosa, and targeted delivery of drug to lymphoid tissues.

Therefore, employment of nanogel in intranasal or pulmonary drug delivery discovers the novel possibility in targeting drugs and bioactives to the brain unswervingly and aids the new platform in variety of CNS (Central nervous system) malady, for example nanogel delivery displays significant therapeutic concentration in cerebral hemisphere related to the intravenous delivery of Risperidone.

5.7 Application of Nanogels

Nanogels in a current scenario of present medical world is evolved as a encouraging drug delivery system and produces diverse features like on site drug delivery system, sustained release formulation, high drug encapsulation properties, water solubility, biodegradability, low toxicity etc. (Figure 5.6). Owing to these inherent properties and characteristics nanogel developed widely in many drug deliver grounds. Combined with polymers, metals and other active fragments nanogel turned out as exceptional drug delivery system [55]. The nanogel system generally engaged in many streams like pharmaceuticals, medicine, diagnosing tool, food processing etc. which is expansively expanded in Table 5.1.

The nanogel drug delivery system is broadly utilized in following fields.

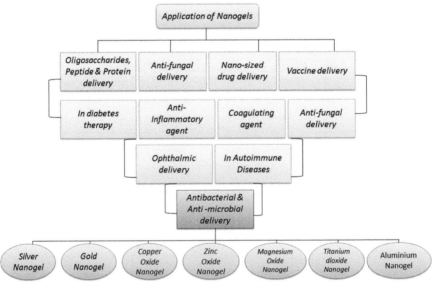

Figure 5.6   Applications of nanogel.

Table 5.1 Applications of nanogel as drug delivery system

Polymer	Nanogel System	Explanation
Heparin Pluronic gel	Self assembled nanogel	Fine RNAs enzyme delivery internalized in cells.
Polyethyleneimine & PEG	Polyplex nanogel	Enhance activity and reduced toxicity of fludarabine
Glycol chitosan grafted with 3-diethylaminopropyl group	pH responsive	Enhanced doxorubicin uptake
Pullulan/folate – pheophorbide	Self reduced polysaccharide based	Marginal toxicity of pheophorbide
Cross-linked polyethylene mine & PEG/Pluronic	Biodegradable nanogel	5-triphosphorylated ribavirin reduced
Acetylated chondroitin sulphate	Self organizing nanogel	Doxorubicin encapsulated
Polyacrylamide	Core magnetic nanogel	Radio pharmaceutical carrier for cancer radiotherapy
Cholesterol bearing pullulan nanogel	Sustained release nanogel	Recombinant marineinterleukine-12 sustained tumour immunotherapy
Poly9(N-isopropylacrylamide) and chitosan	Thermo sensitive magnetically active nanogel	Hyperthermia induced cancer treatment
PEo-b0PMA	Self organizing nanogel	Enhance cisplatin or doxorubicin delivery
Acetylated hyaluronic acid	Targeted nanogel	Doxorubicin loaded nanogel
Polyethyleneimine nanogel	Size dependent nanogel	Enhanced gene hTERT-CD-TK delivered for lung cancer
Acrylate group modified cholesterol bearing pullulan	Cross-linked assembled nanogel	Fine interleukin-12 encapsulation & plasma level
Hydroxypropylcellulose – poly(acrylic acid)	pH & thermo sensitive Cd(II) ion quantum dots	Optical pH sensing, cell imaging & significant drug loading
Pluronic polyethyleneimine/DNA complex	Temperature stimuli volume transition nanogel	Thermo responsive endosomal rupture by nanogel and efficient drug release.
Carboxymethyl chitosan-linolenic acid	Biodegradable nanogel	Adryamycin siRNA anti EGFR delivery

5.7.1 Nano-Sized Drug Delivery System

Nanogel composed of micro and nano molecules, drugs, and other bioactives expressing small molecular weight in their arrangement proficiently. This implausible feature contributes nanogel system a very encouraging drug carrier for controlled and sustained release delivery of drugs and other bioactives. Composite with polymers these systems widely used in the effective delivery of biomaterials and natural extracts. These nanogels are fabricated mostly with diverse ration of polymers accounting synthetic and natural polymers [56]. The different type of Nanogels preparation establishing small molecular weight demonstrated in Table 5.2.

5.7.2 Peptide and Protein Delivery

Nanogel can be proficiently delivered via diverse routes like oral, parenteral, topical etc. Oral route is the most favoured and the most suitable route of drug delivery but the main drawback of the oral route is the dilapidation of drug, poor bioavailability, low mucosal permeability, marginal stability, Gastrointestinal (GIT) irritation etc. In case of sensitive drugs like proteins and peptides, the oral drug delivery turns out to be a lurid element. The nanogel drug delivery system opens new avenue in difficulty associated with the oral drug delivery. Entrapment of drugs, peptides, proteins in nano-formulation not only enhances drug therapeutic efficacy but also offers on-time and on-site drug delivery. The main benefit of nanogel deceived protein or peptides delivery is that, it requires very little amount of concentration of drugs and delivers directly to the desired site which eventually grasped into the systematic circulation without GIT degradation.

Introduction of polymers like chitosan, dextrin, PLGA (Poly-lactic glycolic acid) augmented the nanogel activity like sustained and controlled release, enzymatic cleavage etc. Polymeric nanogel offer decent therapeutic delivery of proteins which are very important in the medicinal era, [57, 58] since some of the supplies at its pinnacle level, for example Cytokines appears as noteworthy bioactive for vaccine and immunotherapy [59]. The nanogel containing growth hormones, cellular peptides, insulin emerge as new therapeutic drug delivery candidate to contest many life disturbing diseases like cancer, arthritis, diabetes etc. PEGylation of polymeric nanogel is evolved as novel candidate in many diseases especially viral and bacterial diseases. For example, delivery of Bleomycin (an anticancerous antibiotic drug) via PEGylated chitosan nanogel is appeared as effective drug system in the ailment of squamous cell carcinoma (SCC) transdermally.

The oligosaccharides heparin (a sulphated natural glycosaminoglycan) also was effectively prepared as an effective anticoagulant nanogel injection preparation. The heparin entrapped nanogel formulation demonstrate fine stability at *pH* 7.4 in phosphate buffer for maximum 24 hrs and possess 10.5–10.8% un-fractioned release profile. In treatment of asthma the chitosan nanogel entailing hyaluronic acid as ligand has been assessed as effctive anti-asthmatic nano-formulation. Addition with intracellular delivery of Chitosan and PEG/heparin starts apoptosis in cancerous cell due to the stimulation of caspase proteins. Antihistamine activity of polymeric nanogel comprises of fluorescent heparin has been efficaciously estimated in rat mast cell. The *ex-vivo* trial proficiently established the inhibitory action of heparin nanogel in mast cell of rats [60].

5.7.3 Vaccine Delivery

Delivery of antigen or antibodies by mean of nano-formulation delivers the significant platform in the medical avenue. Conservative approach of drug delivery system containing the antigens or antibodies transact with the several tasks like integrity of antigen, deteriorations, stability, sterility, protection against infections, immune response etc. Nanogel system not only offers the promise against the conventional tasks but guaranteed the delivery of both type of antigen i.e., humoral and cellular with the chief assertion of keeping actual immunity after the single delivery of vaccine. Vaccine nanogel system reproduces microbes or pathogens usually aware, phagocytosed and preceded by the dedicated antigen presenting cells (APCs). Generally most of the organism and pathogens are perceived and removed in very short time by the protection process, which are not antigen definite and does not need any extended period of induction [61]. These contrivances are the imperious part of inborn immunity. Nanogel based cancer vaccines fetches the drug delivery system at its zenith innovative stature, articulating the improbable features like low adverse effects, stimuli specific, production of long term or life time immunity etc. Some of the utmost prominent polymeric nanogel vaccines are established in Table 5.3.

5.7.4 Gene Delivery

The appearance of Nucleic acids as biomedicines opens the mysteries of various fatal diseases treatment. Delivery of pDNA, oligonucleotides such as ODNs and si-RNA displays their strong impression on variety of undying diseases counting cancer and HIVs. Polymeric nanogel engilfed of active genes like si-RNA and DNAs displays imperious tools in the cancer chemo-therapy [62]. Cancer currently scattering like a mushrooms globally, not only distressing the under developed countries but upsetting the super powers also [63]. Angiogenesis plays crucial role in the expansion and the initiation of cancer and tumour cells. PEG conjugated VEGF-siRNA/EGFR-siRNA complexes with PEI (polyethylenimine) displays extraordinary inhibition of VEGF/EGF appearance at tumour site, potently limiting tumour growth without creating the systemic side effects like inflammation of cells in tumour mice model.

The gene delivery via polymeric nanogel is demonstrating as command-ing delivery carrier in competent anticipation of asthma, dropping the virus retention in lungs, prevention the inflammation in lungs and removing the hyper-infection related with asthma. One of the key landmark has been attained by the gene therapy in the ailment of pulmonary asthma is the topical delivery of 5% Imiquimod gel composed with the si-RNA targeting si-NPRA (natri-uretic peptide receptor-A). The administration of polymeric si-RNA fused with imiquimod cream effectively decline the airway hyper stimuli eosinophilia and pro-inflammatory cytokines IL-4 and IL-5 in lung homogenates when examined in asthmatic mice model [64].

5.7.5 Antiviral Nanogel Delivery

The drug delivery in the viral infection is always been the stimulating task. Research community discovering the trials related with the antiviral drug delivery system. After the efficaciously reconnoitring the antibacterial deliv-ery of nanogel system, wide-ranging research has make the nanogel as a real dug carrier against the virus. Viral diseases are most fatal and the life embel-lished diseases. Convention drug delivery against viruses comprises delivery of antiviral drug orally, parentally etc. which shows its own drawbacks like systemic toxicities, drug degradation, low therapeutic effects etc. With the appearance of nano-medicines the difficulties ascends due to the conventional drug delivery has been lessens to the great amount.

Polymeric silver nanogel has demonstrated the noteworthy antiviral activ-ity against HIV-1 without persuading the toxicity and at very little concen-tration. *In vitro* research established the improved antiviral activity of silver nanogel against the HIV-1 viral strain. The silver nanogel shows the antiviral activity at an early phase of viral replication by hindering the cell division and striking the replication and translation progression. They also limit the binding, conjugation, fusion and infectivity of CD4 dependent virions. They hold strong virus removing properties against the cell related and cell free virus strains. Ligand anchored gold nanogel possess decent antiviral activity on variety of viruses [65, 66]. Amphiphilic sulphate tailored ligand conjugated Gold nanogel confers to the HIV envelope glycoprotein and hinders the HIV contagion of T-cells nano-concentration when evaluated and analysed *in vitro* [67].

Ligand conjugated nanogel particles augment the local concentration of bioactive particles at the targeted site which results in the better receptor binding of nanogel particles resulting in the good therapeutic effects at targeted site. The polymeric gold nanogel are proficiently utilised against Herpes simplex virus Type-1 constituting MES showing strong antiviral activity targeting the virus through contrasting the binding to cellular heparin sulphate by its sulfonate end molecule. This blockage of the viral entrance into the cell results in removal of viral infection. The nanogel shows the ultra-fine action against the lethal Hepatitis-B virus. Research community professionally shows the silver nanogel delivery against the Hepatitis-B virus laterally with syncytial virus strain, HIV-1 virus and monkey-pox virus significantly [68].

5.7.6 Antifungal Nanogel Delivery

Fungal infection evolved as the big problem in present scenario of medicinal era subsidizing the higher risk and reason of impermanence and morbidity around the world. Inadequate medicines and studies against the fungal infections results in more distressing condition. Polymeric silver nanogel estimated as potent fungi-static and fungicidal agent when examined against pathogen like yeasts. Silver nanogel also shows good antifungal activity against *Candida spp.,* at very low concentration of about 1–2 µg/ml of silver (Ag). The silver nanogel displays outstanding antifungal property against the species of *Candida* and *Trichophyton mentagrophytes* showing the IC80 value at very low concentration of 1–10 µg/ml. The nanogel delivery suggestively exposes noteworthy action against the fungal pathogens as compare to the free drug like fluconazole or amphotericin B. The chief phenomenon behind the antifungal activity of nanogel is to target the cell membrane and distracting the membrane potential, which results in the pores development in the cell membrane. Pores formation hints the leakage of cellular contents, eventually cause the cell death. Transmission electron microscopy (TEM) examination strongly reveals the disruption of cell membrane and pores in *Candida albicans* when the silver nanogel particles interact with the cell membrane of *C. albicans* [69]. Antifungal metal nanogel also shows the subordinate actions like coating of particles, bio-stabilization of footwear materials, disinfectants of films against variety of fungal pathogens etc.

5.7.7 In Autoimmune Diseases

Nanogel demeanour liposomes entailed of MPA (mycophenolic acid) was effectively evaluated in treatment of autoimmune diseases. The MPA liposomal nanogel solubilised by addition of cyclodextrin, Irgacure 2959 was presented as photo-initiator and PEG (Polyethylene glycol) was concluded with an acrylate end group. When the contact of UV radiation takes place, photo-polymerization of PEG oligomer initiates [70]. The nanogel system produces numerous intrinsic properties such as lodging of the MPA (drug) load at targeted site, active penetrating to the cell membrane for the effectual binding to the receptor cell triggers the rapid therapeutic response. Polymeric nanogel delivery in the autoimmune conditions raises the therapeutic efficeincy to the targeted site and evades the frequent dosing of drug, which leads to eradicate the risk of renal damage, liver injury and local tissue damages.

5.7.8 Ophthalmic Delivery

pH sensitive nanogel delivery discloses the usefulness of this nano-medicine in the arena of ophthalmology. Polymeric *pH* activated nanogel composed of polyvinyl pyrrolidone-poly acrylic acid (PVP/PAAc) which is contrived by the polymerization of PAAc in aqueous solution of PVP starts by gamma radiation. Pilocarpine compressed PVP/PAAc nanogel reveals enhance ophthalmic drug delivery of *pH* sensitive nanogel at the targeted site with negligible toxicity [71].

5.7.9 Diabetes

Glucose receptive insulin entrapped nanogel injection has been effectively evaluated and measured. The charge interaction plays crucial role between the nanogel particles and the glucose molecules. Acidity of the nanogel particles rises by the accumulation of the dextran leads to fine interface of nanogel particles and evades the clearance from the body. Polymeric nanogel complexes with dextran entrapped insulin possess good antidiabetic activity when injected in the body with the benefit of less dose and better residing time. The elevated permeation of gel agreements the glucose molecules to enter and diffuse within the gel matrix. At elevated glucose level, large amount of glucose molecules diffuses through the gel and release of insulin befalls due to the ionic interaction between the gel and glucose molecules. Conversion of glucose to glucuronic acid take place displays the noteworthy contribution of the nanogel delivery system in the condition of diabetes [72].

5.7.10 Coagulating Agent

Polymeric nanogel explored as a strong carrier unruffled of proteins molecules to restrict bleeding. The biodegradable chitosan anchored protein nanogel shows very strong and important coagulation response against wounds, rashes, bites etc. [73].

5.7.11 Anti-Inflammatory Agent

Nanogels display very significant and effective anti-inflammatory action. Nanogel synthesized by PLGA/chitosan constitutes oleic acid as penetrator relish shows superfine anti-inflammatory activity. Combined delivery of ketoprofen and spantide II via PLGA/chitosan nanogel displays good residing time and noteworthy anti-inflammatory action assessed *in-vivo* [74]. The major benefit of PLGA/chitosan topical nanogel is the sustained release delivery with *pH* triggered release of ketoprofen and spantide II in allergic dermatitis and psoriatic plaque situations. The PLGA/chitosan topical nanogel elevates the percutaneous drug delivery of ketoprofen and spantide II and shows fine therapeutic action with low toxicities [75].

5.8 Disadvantages of Nanogel

1. It is very tedious and expensive process for the removal of solvent, surfactants molecule after the complete synthesis of nanogel.
2. Toxicities problem can ascend due to the residual of touches of surfactants or emulsifiers [76].

5.9 Conclusion

Nanogel delivery system assured promising drugs carrier which plays important role in delivery of numerous medicaments and therapeutics in bacterial and microbial diseases. They offer noteworthy features and evade problems related with formulation like stability, absorption and compatibility. Nanogels emerge as a promising candidate in multiple targeting like brain, lungs, colon, skin, GIT and heart with huge advantage of various routes of administration. In future the constraints like site specificity, selectivity, adverse effects and therapeutic efficiency can be avoids by controlling the precise factors related to formulation and delivery of nanogels which results in intensifying prospect of nanogel delivery system in drug delivery system. Additional and extensive *in-vivo* & clinical research should be conceded for the better economic fabrication of nanogel at commercial stages.

References

[1] Dreis, S., Rothweiler, F., Michaelis, M., Cinatl, J. J., Kreuter, J., and Langer, K. (2007) Preparation, characterization and maintenance of drug efficacy of doxorubicin-loaded human serum albumin (HSA) nanoparticles. *Int. J. Pharm.* 341, 207–214.

[2] Esmaeili, F., Ghahremani, M. H., Esmaeili, B., Khoshayand, M. R., Atyabi, F., and Dinarvand, R., (2008). PLGA nanoparticles of different surface properties: Preparation and evaluation of their body distribution. *Int. J. Pharm.* 349, 249–255.

[3] Garcia-Closas, R., Garcia-Closas, M., Kogevinas, M., Malats, N., Silverman, D., Serra, C., et al. (2007). Food, nutrient and heterocyclic amine intake and the risk of bladder cancer. *Eur. J. Cancer* 43, 1731–1740.

[4] Garcion, E., Lamprecht, A., Heurtault, B., Paillard, A., Aubert-Pouessel, A., Denizot, Philippe Menei, B., et al. (2006). A new generation of anticancer, drug-loaded, colloidal vectors reverses multidrug resistance in glioma and reduces tumor progression in rats. *Mol. Cancer Ther.* 5, 1710–1722.

[5] Gawde, K. A., et al. (2017). Synthesis and characterization of folate decorated albumin bio-conjugate nanoparticles loaded with a synthetic curcumin difluorinated analogue. *J. Colloid Interface Sci.* 496, 290–299. doi: 10.1016/j.jcis.2017.01.092

[6] Luong, D., et al. (2017). Folic acid conjugated polymeric micelles loaded with a curcumin difluorinated analog for targeting cervical and ovarian cancers. *Colloids Surf. B: Biointerfaces* 157, 490–502. doi: 10.1016/j.colsurfb.2017.06.025

[7] Luong, D., et al. (2017). Polyvalent folate-dendrimer-coated iron oxide theranostic nanoparticles for simultaneous magnetic resonance imaging

and precise cancer cell targeting. *Biomacromolecules* 18, 1197–1209. doi: 10.1021/acs.biomac.6b01885

[8] Bhise, K., et al. (2017). Nanomedicine for cancer diagnosis and therapy: advancement, success and structure–activity relationship. *Ther. Delivery* 8. Available at: https://doi.org/10.4155/tde-2017-0062

[9] Sahu, P., et al. (2017). pH responsive biodegradable nanogels for sustained release of bleomycin. *Bioorg. Med. Chem.* 25, 4595–4613. doi: 10.1016/j.bmc.2017.06.038

[10] Ghahremankhani, A. A., Dorkoosh, F., and Dinarvand, R., (2007). PLGA-PEG-PLGA tri-block copolymers as an in-situ gel forming system for calcitonin delivery. *Polym. Bull.* 59, 637–646.

[11] Gill, K. K., Nazzal, S., and Kaddoumi, A., (2011). Paclitaxel loaded PEG (5000)-DSPE micelles as pulmonary delivery platform, formulation characterization, tissue distribution, plasma pharmacokinetics, and toxicological evaluation. *Eur. J. Pharm. Biopharm.* 79, 276–284.

[12] Gupta, M., and Sharma, V., (2011). Targeted drug delivery system, A Review. *Res. Jour. Chem. Sci.* 1, 135–138.

[13] Gupta, P. C., Sinor, P. N., and Bhonsle, R. B., (1998). Oral sub-mucous fibrosis in India, A new epidemic? *Nat. Med J India*, 11, 113–116.

[14] Hana, X., Liua, J., Liua, M., Xiea, C., Zhana, C., Gua, B., et al. (2009). NC-loaded folate-conjugated polymer micelles as tumor targeted drug delivery system, Preparation and evaluation in vitro. *Int. J. Pharm.* 372, 125–131.

[15] Hao, Y. L., Deng, Y. J., Chen, Y., Wang, X. M., Jun, H., and Zhong, S. X. B. (2005). In vitro and in vivo studies of different liposomes containing topotecan. *Arch. Pharm. Res.* 28, 626–635.

[16] Sahu, P., et al. (2017). Assessment of penetration potential of pH responsive double walled biodegradable nanogels coated with eucalyptus oil for the controlled delivery of 5-fluorouracil: *In vitro* and *ex vivo* studies. *J. Controlled Release* 253, 122–136. doi: 10.1016/j.jconrel.2017.03.023

[17] Hasegawa, U., Sawada, S. I., Shimizu, T., Kishida, T., Otsuji, E., Mazda, O., et al. (2009). Raspberry-like assembly of crosslinking nanogels for protein delivery. *J. Control. Release* 140, 312–317.

[18] Hogg, N., (2007). Red meat and colon cancer, Heme proteins and nitrite in the gut. A commentary on diet-induced endogenous formation of nitroso compounds in the GI tract. *Free Radic. Biol. Med.* 43, 1037–1039.

[19] Sau, S., et al. (2017). Advances in antibody–drug conjugates: A new era of targeted cancer therapy. *Drug Discovery Today* 22, 1547–1556. doi: 10.1016/j.drudis.2017.05.011

[20] Houchin, M. L., and Topp, E. M., (2009). Physical properties of PLGA films during polymer degradation. *J. Appl. Polym. Sci.* 114, 2848–2854.

[21] Hureaux, J., Lagarce, F., Gagnadoux, F., Marie-Christine, R., Moal, V., Urban, T., et al. (2010). Toxicological study and efficacy of blank and paclitaxel-loaded lipid nanocapsules after i.v. administration in mice. *Pharm. Res.* 27, 421–430.

[22] Jain, R. A. (2000). The manufacturing techniques of various drug loaded biodegradable poly (lactide-co-glycolide) (PLGA) devices. *Biomaterials* 21, 2475–2490.

[23] Jain, A., Agarwal, A., Majumder, S., Lariya, N., Kharya, A., Himanshu, A., et al. (2010). Mannosylated solid lipid nanoparticles as vectors for site-specific delivery of an anti-cancer drug. *J. Control. Release* 148, 359–367.

[24] Jain, R. K., (2001). Delivery of molecular medicine to solid tumors, lessons from in vivo imaging of gene expression and function. *J. Cont., Rel.* 74, 7–25.

[25] Jaiswal, M. K., Banerjee, R., Pradhan, P., and Bahadur, D., (2010). Thermal behavior of magnetically modalized poly (N-isopropylacrylamide)-chitosan based nanohydrogel. *Colloids Surf. B* 81, 185–194.

[26] Jayakumar, R., Chennazhi, K. P., Nair, S. V., and Rejinold, N., (2011). The art, method, manner, process and system of preparation of alpha chitin nanogels for drug delivery and imaging applications. Indian Patent No. 357/CHE/2011 A.

[27] Jeong, B., Bae, Y. H., and Kim, S. W. (2000). In situ gelation of PEG-PLGA-PEG triblock copolymer aqueous solutions and degradation thereof. *J. Biomed. Mater. Res.* 50, 171–177.

[28] Kabanov, A. V., and Vinogradov, S., (2009). Nanogels as pharmaceutical carriers, finite networks of infinite capabilities. *Angew. Chem. Int. Ed.* 48, 5418–5429.

[29] Kayal, S., and Ramanujan, R. V. (2010). Doxorubicin loaded PVA coated iron oxide nanoparticles for targeted drug delivery. *Mater. Sci. Eng. C* 30, 484–490.

[30] Kievit, F. M., Wang, F. Y., Fang, C., Mok, H., Wang, K., Silber, J. R., et al. (2011). Doxorubicin loaded iron oxide nanoparticles overcome multidrug resistance in cancer in vitro. *J. Control. Release* 152, 76–83.

[31] Kim, S. C., Kim, D. W., Shim, Y. H, Bang, J. S., Oh, H. S., Kim S. W., et al. (2001). In vivo evaluation of polymeric micellar paclitaxel formulation, toxicity and efficacy. *J. Control. Rel.* 72, 191–202.

[32] Kim, S., Park, K. M., Ko, J. Y., Kwon, I. C., Cho, H. G., Kang, D., et al. (2008). Minimalism in fabrication of self-organized nanogels holding both anti-cancer drug and targeting moiety. *Colloids Surf. B* 63, 55–63.

[33] Kim, J. H., Bae, S. M., Na, M. H., Shin, H., Yang, Y. J., Min, K. H., et al. (2012). Facilitated intracellular delivery of peptide-guided nanoparticles in tumor tissues. *J. Control. Release* 157, 493–499.

[34] Kohli, E., Han, H. Y., Zeman, A. D., and Vinogradov, S. V. (2007). Formulation of biodegradable nanogel carriers with 5'-triphosphates of nucleoside analogs that display a reduced cytotoxicity and enhanced drug activity. *J. Control. Release* 121, 19–27.

[35] Lacoeuille, F., Hindre, F., Moal, F., Roux, J., Passirani, C., Couturier, O., et al. (2007). In vivo evaluation of lipid nanocapsules as a promising colloidal carrier for paclitaxel. *Int. J. Pharm.* 344, 143–149.

[36] Lee, J., and Yoo, H. S., (2008). Pluronic decorated nanogels with temperature responsive volume transitions, cytotoxicities and transfection efficiencies. *Euro. J. Pharm. Biopharm.* 70, 506–513.

[37] Lee, Y., Park, S. Y., Kim, C., and Park, T. G., (2009). Thermally triggered intracellular explosion of volume transition nanogels for necrotic cell death. *J. Control. Release* 135, 89–95.

[38] Lee, S., Yun, M. H., Jeong, S. W., Hoon, C., Kim, J. Y., Seo, M. H., et al. (2011). Development of docetaxel-loaded intravenous formulation, Nanogel-PM using polymer-based delivery system. *J. Control. Release* 155, 262–271.

[39] Li, N., Wang, J., Yang, X., and Li, L., (2011). Novel nanogels as drug delivery systems for poorly soluble anticancer drugs. *Colloids Surf. B* 83, 237–244.

[40] Li, Y., Pei, Y., Zhang, X., Gu, Z., Zhou, Z., Yuan, W., et al. (2001). PEGylated PLGA nanoparticles as protein carriers: Synthesis, preparation and biodistribution in rats. *J. Control. Release* 71, 203–211.

[41] Liu, X., Sun, J., Chen, X., Wang, S., Scot, H., Zhan, X., et al. (2012). Pharmacokinetics, tissue distribution and anti-tumour efficacy of paclitaxel delivered by polyvinylpyrrolidone solid dispersion. *J. Pharm. Pharmacol.* 64, 775–782.

[42] Lo, J. T., Chen, B. H., Lee, T. M., Han, J., and Li, J. L., (2009). Self emulsifying o/w formulations of paclitaxel prepared from mixed nonionic surfactants. *J. Pharm. Sci.* 99, 2320–2332.

[43] Maeda, H., Wu, J., and Sawa, T., (2000). Tumour vascular permeability and the EPR effect in macromolecular therapeutics, a review. *J. Control. Release* 65, 271–284.

[44] Mason, T. G., Wilking, J. N., Meleson, K., Chang, C. B., and Graves, S. M. (2006). Nanoemulsion, formation, structure and physical properties. *J. Phys. Cond. Matter.* 2, 56–60.

[45] Missirlis, D., Kawamura, R., Tirelli, N., and Hubbell, J. A., (2006). Doxorubicin encapsulation and diffusional release from stable, polymeric, hydrogel nanoparticle. *Euro. J. Pharm. Sci.* 29,120–129.

[46] Mohamed, F., and van der Walle, C. F. (2008). Engineering biodegradable polyester particles with specific drug targeting and drug release properties. *J. Pharm. Sci.* 97, 71–87.

[47] Mundargi, R., Babu, V., Rangaswamy, V., Patel, P., and Aminabhavi, T. (2008). Nano/micro technologies for delivering macromolecular therapeutics using poly (D, L-lactide-co-glycolide) and its derivatives. *J. Control. Release* 125, 193–209.

[48] Naik, S., Patel, D., Chuttani, K., Mishra, A. K., and Misra, A., (2012). In vitro mechanistic study of cell death and in vivo performance evaluation of RGD grafted PEGgylated docetaxel liposomes in breast cancer. *Nanomed., Nanotechnol.* 8, 951–962.

[49] Nair, L. S., and Laurencin, C. T., (2007). Biodegradable polymers as biomaterials. *Prog. Polym. Sci.* 32, 762–798.

[50] Oh, J. K., Bencherif, S. A., and Matyjaszewski, K. (2009). Atom transfer radical polymerization in inverse miniemulsion; A versatile route towards preparation and functionalization of microgel/nanogels for targeted drug delivery applications. *Polymers* 50, 4407–4423.

[51] Oh, N. M., Oh, K. T., Baik, H. J., Lee, B. R., and Lee, A., H. (2010). A self organized 3-diethylaminopropyl bearing glycol chitosan nanogel for tumor acidic pH targeting, in-vitro evaluation. *Colloids Surf. B.* 78, 120–126.

[52] Oishi, M., Miyagawa, N., Sakaru, T., and Nagasaki, Y. (2007). pH-responsive nanogel containing platinum nanoparticles, Applications to on-off regulation of catalytic activity for reactive oxygen species. *React. Funct. Polym.* 67, 662–668.

[53] Owens, D. E., and Peppas, N. A., (2006). Opsonization, biodistribution, and pharmacokinetics of polymeric nanoparticles. *Int. J. Pharm.* 307, 93–102.

[54] Pangburn, S. H., Trescony, P. V., and Heller, J. (1982). Lysozyme degradation of partially deacetylated chitin, its films and hydrogels. *Biomaterials* 3, 105–108.

[55] Park, J. H., Chi, S. C., Lee, W. S., Lee, W. M., Koo, Y. B., Yong, C. S., et al. (2009). Toxicity studies of Cremophor-free paclitaxel solid dispersion formulated by a supercritical anti solvent process. *Arch. Pharm. Res.* 32, 139–148.

[56] Passerini, N., and Craig, D. Q. M., (2001). An investigation into the effects of residual water on the glass transition temperature of polylactide microspheres using modulated temperature DSC. *J. Control. Release* 73, 111–115.

[57] Sau, S., et al. (2014). Cationic lipid-conjugated dexamethasone as a selective antitumor agent. *Eur. J. Med. Chem.* 83, 433–447. doi: 10.1016/j.ejmech.2014.06.051

[58] Almansour, A. I., et al. (2017). Design, synthesis and antiproliferative activity of decarbonyl luotonin analogues. *Eur. J. Med. Chem.* 138, 932–941. doi: 10.1016/j.ejmech.2017.07.027

[59] Alsaab, H. O., et al. (2017). PD-1 and PD-L1 Checkpoint signaling inhibition for cancer immunotherapy: Mechanism, combinations, and clinical outcome. *Front. Pharmacol.* 8, 561. doi: 10.3389/fphar.2017.00561

[60] Patnaik, S., Sharma, A. K., Garg, B. S., Gandhi, R. P., and Gupta, K. C., (2007). Photoregulation of drug release in azo-dextran nanogels. *Int. J. pharm.* 342, 184–93.

[61] Prakash, R., and Thiagarajan, P., (2011). Nanoemulsions for drug delivery through different routes. *Res. Biotech.* 2, 9–13.

[62] Tatiparti, K., et al. (2017). siRNA delivery strategies: A comprehensive review of recent developments. *Nanomaterials* 7, 77. doi: 10.3390/nano7040077

[63] Prasad, R. R., Singh, J. K., Mandal, M., Kumar, M., and Prasad, S. S., (2005). Profile of gall bladder cancer cases in Bihar. *Indian J. Med. Paediatr. Oncol.* 26, 31–35.

[64] Prasanth, V. V., Chakraborthy, A., Mathew,. S. T., and Mathapan, R., (2011). Microspheres – An overview. *Int. J. Res. Pharm. Biomed. Sci.* 2, 232–238.

[65] Sau, S., et al. (2017). Cancer cell-selective promoter recognition accompanies antitumor effect by glucocorticoid receptor-targeted gold nanoparticle. *Nanoscale* 6, 6745–6754. doi: 10.1039/C4NR00974F

[66] Mukherjee, S., et al. (2016). Green synthesis and characterization of monodispersed gold nanoparticles: Toxicity study, delivery of doxorubicin and its bio-distribution in mouse mode. *J. Biomed. Nanotechnology* 12, 165–181. doi: 10.1166/jbn.2016.2141

[67] Rejinold, N. S., Sreerekha, P. R., Chennazhi, K. P., Nair, S. V., and Jayakumar, R., (2011). Biocompatible, biodegradable and thermo-sensitive chitosan-g-poly (N-isopropylacrylamide) nanocarrier for curcumin drug delivery. *Int. J. Biol. Macromol.* 49, 161–172.

[68] Rejinold, N. S., Muthunarayanan, M., Chennazhi, K. P., and Jayakumar, R. (2011). Curcumin loaded fibrinogen nanoparticles for cancer drug delivery. *J. Biomed. Nanotechnol.* 7, 521–534.

[69] Sahu, M. K., and Ahmad, D. (2010). Development and optimization of fixed dose antihypertensive combination drugs using double layer sustained release microsphere technology. *Int. J. Pharm. Biomed. Res.* 1, 114–123.

[70] Salaun, F., and Vroman, I., (2009). Curcumin loaded nanocapsules, formulation and influence of the nanoencapsulation processes variables on the physico-chemical characteristics of the particles. *Int. J. Chem. React. Eng.* 7, A55–A61.

[71] Siegel, S. J., Kahn, J. B., Metzger, K., Winey, K. I., Werner, K., and Dan, N. (2006). Effect of drug type on the degradation rate of PLGA matrices. *Eur. J. Pharm. Biopharm.* 64, 287–293.

[72] Shah, P., Bhalodia, D., and Shelat, P. (2010). Nanoemulsion: a pharmaceutical review. *Syst. Rev. Pharm.* 3, 24–32.

[73] Sheihet, L., Garbuzenko, O. B., Bushman, J., Gounder, M. K., Minko, T., and Kohn, J., (2012). Paclitaxel in tyrosine-derived nanospheres as a potential anti-cancer agent, in vivo evaluation of toxicity and efficacy in comparison with paclitaxel in Cremophor. *Eur. J. Pharm. Sci.* 45, 320–329.

[74] Shigemasa, Y., Saito, K., Sashiwa, H., and Saimoto, H., (1994). Enzymatic degradation of chitins and partially deacetylated chitins. *Int. J. Biol. Macromol.* 16, 43–49.

[75] Shimizu, T., Kishida, T., Hasegawa, U., Ueda, Y., Imanishi, J., Yamagishi, H., et al. (2008). Nanogel DDS enables sustained release of IL-12 for tumor immunotherapy. *Biochem. Biophys. Res. Commun.* 367, 330–335.

[76] Shin, Y., Chang, J., H, Liu, J., Williford, R., Shin, Y. K., and Exarhos, G. J. (2001). Hybrid nanogels for sustainable positive thermosensitive thermosensitive drug release. *J. Control. Release.* 73, 1–6.

6

Fe, Co Based Bio-Magnetic Nanoparticles (BMNPs): Synthesis, Characterization, and Biomedical Application

Amirsadegh Rezazadeh Nochehdehi[1], Sabu Thomas[2], Minoo Sadri[3], S. M. Mehdi Hadavi[4], Yves Grohens[5], Nandakumar Kalarikkal[6] and Neerish Revaprasadu[7]

[1]Researcher, Biomaterials research group, Biomedical Engineering Department, Materials and Biomaterials Research Center (MBMRC), Tehran, Iran

[2]Professor, Polymer Science & Engineering, School of Chemical Sciences and Hon. Director of International and Inter University Centre for Nanoscience and Nanotechnology (IIUCNN), Mahatma Gandhi University (MGU), Kottayam, Kerala, India

[3]Department of Biochemistry and Biophysics, Education and Research Center of Science and Biotechnology, Malek Ashtar University of Technology, Tehran, Iran

[4]Professor, President of Materials and Biomaterials Research Center (MBMRC) and Iran National Institute of Materials and Energy (MERC), Tehran, Iran

[5]Université de Bretagne Sud, Laboratoire Ingénierie des Matériaux de Bretagne, BP 92116, 56321 Lorient Cedex, France

[6]Professor, School of Pure and Applied Physics, Mahatma Gandhi University, Kottayam 686 560, Kerala, India

[7]Professor, SARCHI Chair in Nanotechnology, Department of Chemistry, Faculty of Science and Agriculture, University of Zululand, KwaZulu-Natal, South Africa

Abstract

The nanotechnology has concentrated to study by scientists and researchers around the world to synthesize novelty multifunctional alloy nanoparticles which will be used in wide range of applications from industry to medicine.

In the recent years, one of the most common nanomaterial in the world is an alloy nanoparticles. In addition, Multifunctional alloy nanoparticles has been growing by scientists around the world. The various metallic, ceramic and polymeric compounds like Iron Oxides, Zinc Oxides, Iron-Cobalt, Nickel-Cobalt, Iron-Nickel, Titanium Dioxide, Ag doped gold, Copper alloys, PEG, PPA, PMMA, Chitosan, Hydroxyapatite and sort of that, will be produced by chemists, Physicists or materialist in the advanced laboratory. There are various method to synthesize alloy nanoparticles like precipitation, chemical and physical vapor deposition, thermal and plasma spray, laser deposition, mechanical alloying and so on. Because of an impressive and unique chemical, physical and an antimicrobial properties of nanoparticles along with their biocompatibility; makes these materials find specific applications in various industries. Thus, alloy nanoparticles have lots of applications in manufacturing, agriculture, environment, energy, electronics, and medicine. These use as an industrial coatings, lubricant oils, catalysts, gas sensors, magnetic separators, antioxidant, break down oil, breakdown volatile organic air pollutants, fuel cell electrodes, storage materials, lithium ion batteries, semiconductor (photovoltaic cells), solar steam device, storing and packaging of agricultural produces, nutrients absorber, food flavoring, perfumes, scratch resistance eyeglasses, fluorescent biological labels, contrast imaging, bone growth, drug and gene delivery, immunoassay, bio detection of pathogens, separation and purification of biological molecules and cells, cancer diagnosis and treatment, tumors destruction via heat therapy (hyperthermia), tissue engineering and etc. Developed and identifies of intelligent multifunctional alloy nanoparticles like Iron and Cobalt Based Magnetic Nanoparticles with biomedical application is purpose of this chapter. After studies done on previous research, it was found that Fe- and Co-based biomagnetic nanoparticles (BMNPs) have various biological applications including diagnosis and treatment, separation, immunoassay, drug delivery to eyes and brain, nano drug carriers, gene transmitter and magnetic resonance imaging, recently. Besides, the use of magnetic nanoparticles in cancer treatment via heat therapy (hyperthermia) method as an alternative to radiotherapy and chemotherapy has been considered and developed by many researchers. Nanoparticles have modern technologies and knowledge plus various useful applications in medicine.

6.1 Introduction

Nanotechnology has made remarkable progress in recent decades. In a short definition, nanotechnology is design, identification, production, schemes and systems by controlling the shape and size of materials on a nanoscale scale.

In fact, nanotechnology is a purely interdisciplinary knowledge which is related to materials science and engineering, medical science, pharmacy and drug design, veterinary medicine, biology, applied physics, molecular chemistry, and even mechanical engineering, electrical engineering, and chemical engineering [1–4]. Analysts believe that nanotechnology, biotechnology and information technology; are three realms which forms the third industrial revolution [4–9]. Generally, it can say that the nanotechnology products are most commonly used application from industry to medicine [10]. Besides, one of the most commonly used fields of nanotechnology which is attracted attention by many researchers around the world, recently; is medicine application of nanotechnology [11]. In fact, nanomedicine is a multifunctional applications of Nanotechnology products in medicine fields [12]. The purpose of medical nanotechnology to the prevention, diagnosis, care and treatment of diseases defined. In this regard, magnetic nanoparticles have been developed by many researchers around the world for various applications in the field of medicine. Recent years have witnessed an unprecedented growth in research in the area of Nanoscience. One of the most promising applications of nanoscience is in the field of medicine. Indeed, a whole new field of "**Nanomedicine**" is emerging. There is increasing optimism that nanotechnology applied to medicine will bring significant advances in the diagnosis treatment and prevention of disease. However, many challenges must be overcome if the application of nanomedicine is to realize the improved understanding of the pathophysiological basis of disease, bring more sophisticated diagnostic opportunities, and yield more effective therapies and preventive measures. The field of "nanomedicine" is the science and technology of diagnosing, treating, and preventing disease and traumatic injury; of relieving pain; and of preserving and improving human health, using molecular tools and molecular knowledge of the human body. It was perceived as embracing five main subdisciplines that in many ways are overlapping and underpinned by the following common technical issues: analytical tools (AT); nanoimaging (NI); nanomaterial and nanodevices (NM/ND); novel therapeutics and drug delivery systems (DDS); and clinical, regulatory and toxicological issues (RTI). Miniaturizations of devices, chip-based technologies and, on the other hand, ever more sophisticated novel nano-sized materials and chemical assemblies are already providing novel tools that are contributing to improved health care in the twenty-first century. Opportunities include superior diagnostics and biosensors, improved imaging techniques from molecules to man and not least, innovative therapeutics and technologies to enable tissue regeneration and repair. However, to realize nanomedicine's full potential, important challenges must be addressed. New regulatory authority guidelines must be developed quickly to ensure safe and reliable transfer of new advances in nanomedicine from laboratory to bedside. These aspects were viewed as complementary even though many of the technologies required are very different in being designed for ex vivo, cellular or in vivo/patient use [1]. Advances should begin with the optimization of existing technologies toward specific nanomedicine challenges. The development of new multifunctional, spatially ordered, architecturally varied systems for targeted drug delivery was seen as a priority. There is a pressing need to enhance expertise in scale-up manufacture and material characterization, and to ensure material reproducibility, effective quality control, and cost-effectiveness. These issues should be addressed urgently to enable rapid realization of clinical benefit (within five years). For realization to application within the next decade, new materials are needed for sensing multiple, complicated analyses *in vitro*, for applications in tissue engineering, regenerative medicine, and 3D display of multiple biomolecular signals. Telemetrically controlled, functional, mobile *in vivo* sensors and devices are required, including construction of multi-functional, spatially ordered, architecturally varied systems for diagnosis and combined drug delivery (theranostics). The advancement of bio-analytical methods for single-molecule analysis was seen as a priority. Nano-sized drug delivery systems have already entered routine clinical use in this field. The most pressing challenge is application of nanotechnology to design of multifunctional, structured materials able to target specific diseases or containing functionalities to allow transport across biological barriers. In addition, nanostructured scaffolds are urgently needed for tissue engineer-ing, stimuli-sensitive devices for drug delivery and tissue engineering, and physically targeted treatments for local administration of therapeutics (e.g., via the lung, eye, or skin). To realize the desired clinical benefits rapidly, the importance of focusing the design of technologies on specific target diseases was stressed: cancer, neurodegenerative, and cardiovascular diseases were identified as the first priority areas. Longer term priorities include the design of synthetic, bioresponsive systems for intracellular delivery of macromolecular therapeutics (synthetic vectors for gene therapy), and biore-sponsive or self-regulated delivery systems including smart nanostructures such as biosensors that are coupled to the therapeutic delivery systems [2]. There is an urgent need to improve the understanding of toxicological implications of nanomedicines in relation to the specific nanoscale properties currently being studied, in particular in relation to their proposed clinical use by susceptible patients. The nanoscale is the place where the properties of most common things are determined just above the scale of an atom. Nanoscale objects have at least one dimension (height, length, depth) that measures between 1 and 999 nanometers. In addition, due consideration should be given to the potential environmental impact and there should be a safety assessment of all manufacturing processes. Risk-benefit assessment is needed in respect of both acute and chronic effects of nanomedicines in potentially predisposed patients—especially in relation to target disease. A shift from risk assessment to proactive risk management is considered essential at the earliest stage of the discovery and then the development of new nanomedicines [3]. As the technologies are designed based on a clear understanding of a particular disease, disease-specific oriented focus is required for the development of novel pharmaceuticals. In addition, it will be important to establish a case-by-case approach to clinical and regulatory evaluation of each nanopharmaceutical. High priority should be given to enhancing communication and exchange of information among academia, industry, and regulatory agencies encompassing all facets of this multidisciplinary approach [4].

6.1.1 Magnetic Properties

The study of magnetic properties of materials at the nanoscale is an important area for the advancement of nanoscience and nanotechnology. It can be attributed to the fact that the nanoscale magnetic properties differ from their bulk counterparts. So, magnetic nanoparticles in the size range of 1–100 nm have attracted a great deal of attention due to their technological importance. The research has evolved to develop nanoparticles in applications such as magnetic resonance imaging (MRI) for medical diagnosis (MD), high-density magnetic recording (HDMR), magneto-optical switches (MOS), and controlled drug delivery (CDD). Basically, all materials can be divided into three categories according to their interaction with an external magnetic field: diamagnetism, paramagnetism, and cooperative magnetism [5–9]. The electron possesses a spin that is equivalent to the strength of the magnetic field (magnetic moment) of the electron itself, in an atom. Electrons are arranged in energy states of successive order, and for each energy state, there can only be two electrons of opposite spins as established by Paul's principle. The orbital motion of an unpaired electron around the nucleus and the spin of the electron about its own axis can generate magnetic moments. The magnetic moment of each electron pair in an energy level is opposed, and consequently, whenever an energy level is completely full, there is no net magnetic moment.

Based on this reasoning, we expect any atom of an element with an odd atomic number to have a net magnetic moment from the unpaired electron. In most of the elements, the unpaired electron is in the valence shell and can interact with other valence electrons leading to the cancelation of the net magnetic moment in the material. However, certain elements such as cobalt and nickel have an inner energy level that is not completely filled; that is, each atom in the metal has a permanent magnetic moment, equal in strength to the number of unpaired electrons [5]. Table 6.1 summarized the above-described behavior of materials when a magnetic field is applied [5, 10].

6.1.2 Magnetic Nanoparticles

Advances in nanotechnology and molecular biology have helped to translate multifunctional nanoparticles into biomedical applications by overcoming the shortcomings related to traditional disease diagnosis and therapy. Cancer is such a difficult disease to treat because of barriers in disease diagnosis and prognosis. The unique physical properties of magnetic nanoparticles (MNPs) enable them to serve as imaging probes for locating and diagnosing cancerous lesions and, simultaneously, as drug delivery vehicles that deliver

Table 6.1 The types of magnetism seen in materials: Blue arrows signify the direction of the applied field and blue arrows in the black circle signify the direction of the electron spin

Type of Magnetization	Description
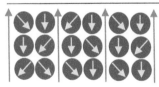 **Paramagnetism**	Atoms whose shells contain electrons with spins that are not compensated by another electron of an opposing spin will have a resultant magnetic moment; this is due to the unpaired electrons. These moments tend to positively with the applied field; however, they are kept from total alignment by thermal energy. This scenario is referred to as paramagnetism [5, 10].
Diamagnetism	Diamagnetism refers to a material that exhibits a negative magnetism. Even though the substance is composed of atoms that have no net magnetic moment (paired electrons), it reacts in a particular way to an applied field. Wilhelm Weber and Paul Lange in theorized that an applied field acts on a single electron orbit to reduce the effective current of the orbit, in turn producing a magnetic moment that opposes an applied field. Common diamagnetic materials are water, wood, most organic compounds and in the case of this work copper [5, 10].

Ferromagnetism

Ferromagnetism is the basic mechanism by which is associated with iron, cobalt, nickel, and some alloys or compounds containing one or more of these elements. In physics, several different types of magnetism are distinguished. Ferromagnetism (including ferrimagnetism) is the strongest type that such an atom itself is an elementary electromagnet produced by the motion of electrons about its nucleus and by the spin of its electrons on their own axes. The magnetism in ferromagnetic materials is caused by the alignment patterns of their constituent atoms, which act as elementary electromagnets. One requirement of a ferromagnetic material is that its atoms or ions have permanent magnetic moments. The magnetic moment of an atom comes from its electrons, since the nuclear contribution is negligible. Another requirement for ferromagnetism is some kind of interatomic force that keeps the magnetic moments of many atoms parallel to each other [5, 10].

Antiferromagnetism

If the atoms are in close enough contact with each other so that the electrons can be exchanged between neighboring atoms, a cooperative magnetization may occur which spontaneously aligns all atoms in a lattice and creates a synergistic and strong magnetic moment. When the spins between neighboring atoms are aligned parallel, the material is said to be ferromagnetic. In some cases, the spins between neighboring atoms are antiparallel and are referred to as antiferromagnetic. In antiferromagnetic materials, the resultant magnetization is small because the opposite spins cancel each other out.

Ferrimagnetism

Lastly, if two atoms have antiparallel magnetization of unequal magnitude, a resultant magnetization remains in the direction of the stronger magnetic moment and applied field. This is referred to as ferrimagnetism.

therapeutic agents preferentially to those lesions. The current efforts are being carried out to combine these two properties and to develop MNP-based nanotheranostics having imaging and therapeutic functionalities that will help toward the development of personalized medicine with scope for real-time monitoring of biological responses to the therapy. So, there is always a need to summarize the existing knowledge and current progress on engineering of different MNPs and their applications from the theranostic point of view. We will continue to investigate a variety of magnetic nanoparticles.

6.1.2.1 Iron and iron oxide nanoparticles

Iron is a ferromagnetic material with high magnetic moment density (about 220 emu/g) and is magnetically soft. Iron nanoparticles in the size range below 20 nm are super paramagnetic. Procedures leading to monodisperse Fe nanoparticles have been well documented [13]. Nevertheless, the preparation of nanoparticles consisting of pure iron is a complicated task, because they often contain oxides, carbides, and other impurities. A sample containing pure iron as nanoparticles (10.5 nm) can be obtained by evaporation of the metal in an argon atmosphere followed by deposition on a substrate [14]. When evaporation took place in a helium atmosphere, the particle size varied in the range of 10–20 nm [15]. Relatively, small (100–500 atoms) Fe nanoparticles are formed in the gas phase on laser vaporization of pure iron [16]. The common chemical methods used for the preparations include thermal decomposition of $FeCo_5$ (the particles so prepared are extremely reactive), reductive decomposition of some iron (II) salts, or reduction of iron (III) acetyl acetone; there is a chemical reduction with TOPO capping [17]. A sonochemical method for the synthesis of amorphous iron was developed in [18]. The method of reducing metal salts by NaBH has been widely used to synthesize iron-containing nanoparticles in organic solvents [19]. Normally, reductive synthesis of Fe nanoparticles in an aqueous solution with $NaBH_4$ yields a mixture including FeB [20, 21]. Well-dispersed colloidal iron is required for applications in biological systems such as MRI contrast enhancement and biomaterials separation. Nevertheless, the syntheses have as yet a difficulty in producing stable Fe nanoparticle dispersions, especially aqueous dispersions, for potential biomedical applications. The phase composition of the obtained nanoparticles was not always determined reliable. The range of specific methods was proposed to prepare nanoparticles of the defined phase composition. Thus, the α-Fe nanoparticles with a body-centered cubic (bcc) lattice and an average size of ~10 nm were prepared by grinding a high-purity (99.999%) Fe powder for 32 h [15, 22–23]. Iron oxides have received increasing attention due to their extensive applications, such as magnetic recording media, catalysts, pigments, gas sensors, optical devices, and electromagnetic devices. They exist in a rich variety of structures (polymorphs) and hydration states; therefore until recently, knowledge of the structural details, thermodynamics, and reactivity of iron oxides has been lacking. Furthermore, physical (magnetic) and chemical properties commonly change with particle size and degree of hydration. By definition, super paramagnetic iron oxide particles are generally classified with regard to their size into super paramagnetic iron oxide particles (SPIO), displaying hydrodynamic diameters larger than 30 nm, and ultra-small super paramagnetic iron oxide particles (USPIO), with hydrodynamic diameters smaller than 30 nm. USPIO particles are now efficient contrast agents used to enhance relaxation differences between healthy and pathological tissues, due to their high saturation magnetization, high magnetic susceptibility, and low toxicity. The biodistribution and resulting contrast of these particles are highly dependent on their synthetic route, shape, and size [24].

6.1.2.2 Cobalt-based nanoparticles

Cobalt nanoparticles depending upon the synthetic route are observed in at least three crystallographic phases: typical for bulk Co HCP, ε-Co cubic [25–27], or multiply twinned FCC-based icosahedral [28]. Conditions of synthesis reactions are influence on the final product structure; in rare cases of the determined phase, nanoparticles can be obtained. Often a size and phase selection was required to obtain Co nanocrystals with a specific size and even shape. Methods for the synthesis and magnetic properties of cobalt nanoparticles' different structures have been described in detail in a review. A popular approach is to synthesize colloidal particles by inversed micelle synthesis; the inverse micelles are defined as a micro reactor [29]. In order to obtain stable cobalt nanoparticles with a narrow size distribution, Co (AOT) reverse micelles are used; their reduction is obtained by using NaBHa as reducing agent. Such particles are stabilized by surfactants and are often monodispersed in size, but are also unstable unless kept in a solution. Nevertheless, the chemical surface treatment by auric acid highly improves the stability and cobalt nanoparticles could be stored without aggregation or oxidation for at least one week [29]. In many instances, it is possible to obtain Co nanoparticles coated by other ligands, which can be either dispersed in a solvent or deposited on a substrate; in the latter case, self-organized mono layer shaving a hexagonal structure can be obtained. In some instances of reduction with NaBH, it is possible to obtain Co–B nanoparticles. The size, composition, and structure of this kind of nanoparticles strongly depend on the concentration of the solution, pH, and the mixing procedure [30]. It is well known that the presence of oxides in magnetic materials, which form spontaneously when the metallic surface is in contact with oxygen, drastically changes the magnetic behavior of the particles. An enhanced magneto resistance, arising from the uniform Co resize and CoO shell thickness, has been reported [31]. This effect is caused by the strong exchange coupling between the ferromagnetic Co core and the antiferromagnetic CoO layer. However, up to now, the understanding of this effect has not been well understood. The ordered Co–Fe alloys are excellent soft magnetic materials with negligible magneto crystalline anisotropy [32]. The saturation magnetization of Fe–Co alloys reaches a maximum at Co content of 35 at.%; other magnetic characteristics of these metals also increase when they are mixed. Therefore, FeCo nanoparticles attract considerable attention. Thus, Fe, Co, and Fe–Co (20, 40, 60, 80 at.%) nanoparticles (40–51 nm) with a structure similar to the corresponding bulk phases have been prepared in a stream of hydrogen plasma [33]. The Fe–Co particles reach a maximum saturation magnetization at 40 at.% of Co, and a maximum 100-x coercive force is attained at 80 at.% of Co. Chemical reduction by NaBH also used for the preparation of FeCo nanoparticles [34]. X-ray data show that the ratio of Co to Fe is around 30:70 in the prepared nanoparticles.

6.2 Synthesis and Characterization of Magnetic Nanoparticles (MNPs)

6.2.1 Iron Oxide (Fe_3O_4) Nanoparticles (ION)

Magnetic nanoparticles possess some extraordinary physical and chemical properties and find applications in many industrial and biological fields. The experimental apparatus for preparing and coating Fe_3O_4 nanoparticles was description in various articles. In this regards [80], $FeCl_3$ (0.5 mol/L) and $FeSO_4$ (0.5 mol/L) solutions were prepared separately, and syringed into 250-mL three-neck round-bottom flask equipped with a mechanical stirrer. Immediately after adding the iron salt solutions, 5 mL $NH_3.H_2O$ (50%, v/v aqueous) was quickly syringed into the flask with stirring. A moment later, a black gelatinoid appeared, and $NH_3.H_2O$ was dropped into the mixture continuously and slowly until the pH value of the solution reached 9.0. Then, the reaction was continued for 30 min. The resulting black mixture was kept for 1 week and then aged at 70°C for 30 min. The precipitate was then collected by filtration under a magnetic field, rinsed with deionized water 3 times, and completely dried under vacuum at 60°C and pulverized. The Fe_3O_4 nanoparticles were then coated at 80°C for 1 h using a saturated sodium oleate solution. Finally, the precipitate was collected by filtration under a magnetic field and washed with ethanol and deionized water 3 times each, and then dried in vacuum at 60°C and pulverized. In other way, $FeCl_3.6H_2O$ and $FeSO_4.7H_2O$ were dissolved in deionized water according to a molar ratio of 3:2, and put into a flask. NaOH was slowly dropped into the solution with stirring. When the pH reached 6.5, a black gelatinoids appeared. NaOH was dropped into the mixture continuously until the pH increased to 13.0. The mixture was stirred vigorously for 30 min, and then aged for 30 min at 80°C. After the mixture was cooled to room temperature, the nanoparticles were separated under magnetic field and washed with distilled water until the pH decreased to 7.0. After keeping for a day, the Fe_3O_4 nanoparticles were coated [80]. Magnetic Fe_3O_4 nanoparticles less than 10 nm demonstrated super paramagnetism. In the absence of applied magnetics field, the magnetization vectors of the magnetic particles were order less because of thermal movement and did not exhibit any overall magnetism. However, the magnetization vectors of magnetic particles followed quickly an external magnetic field, making the particles highly magnetic. Table 6.2 lists the properties of magnetic Fe_3O_4 nanoparticles prepared by the original and improved methods.

IR spectra of surface-modified Fe_3O_4 nanoparticles could be attributed to the flex vibrations of $-CH_2$, indicating that surface modification of Fe_3O_4 nanoparticles was successful. In addition, repeated experiments showed that it was not easy to coat Fe_3O_4 nanoparticles prepared by co-precipitation using NaOH, and that Fe_3O_4 nanoparticles prepared either by the original or by the improved method had almost the same IR spectra. The TEM photographs of the surface-modified Fe_3O_4 nanoparticles prepared by the original method, indicating that the diameter of primary nanoparticles is about 10 nm. Besides, SEM photographs of surface-modified Fe_3O_4 samples prepared by the improved method, indicating that coating reduced the size as well as the agglomeration of the Fe_3O_4 nanoparticles. On the other hand, The XRD spectra of different kinds of Fe_3O_4 nanoparticles prepared by the original and the improved methods showed that, there are many weak but

Table 6.2 Saturation magnetization of samples synthesized using different methods

Preparation Methods Co-precipitation Using	Saturation Magnetization (emu/g)		
	Original Method, Uncoated	Improved Method, Uncoated	Improved Method, Coated with Sodium Oleate
$NH_3.H_2O$ in water	34.05	68.38	65.98
NaOH in water	41.26	42.97	40.98
$NH_3.H_2O$	50.61	52.09	45.02

obvious peaks showing that the Fe_3O_4 nanoparticles were oxidized due to lack of Ar protection. Meanwhile, in another process determined that, the characteristic peaks of Fe_3O_4 without interference from other phases of Fe_xO_y, to indicate the high purity of the Fe_3O_4 obtained. However, the saturation magnetization of the Fe_3O_4 nanoparticles prepared by co-precipitation using $NH_3.H_2O_4$ was higher than those prepared using NaOH. Well-shaped Fe_3O_4 nanoparticles can be obtained using any of the three improved methods. However, the Fe_3O_4 nanoparticles prepared by co-precipitation using NaOH cannot be coated successfully. The diameter of Fe_3O_4 nanoparticles prepared by co-precipitation using $NH_3.H_2O$ in alcohol was the smallest, but agglomeration was severe. Co-precipitation using NaOH can be easily carried out, requires the least time, and has the highest yield. Co-precipitation using $NH_3.H_2O$ in water needs the longest time. Co-precipitation using $NH_3.H_2O$ in alcohol calls for precise control of pH, consumes large quantity of alcohol, and has the least yield.

6.2.2 Cobalt-Based (FeCo) Nanoparticles (CBN)

The design, synthesis, characterization, and implementation of novel ferromagnetic nanoparticles and their alloys have been of significant interest over the past decade. FeCo-based alloys have specifically gained interest due to elevated magnetization along with a high Curie temperature ($\sim900°C$), high saturation magnetizations (MS = 2.4 T, for the bulk alloy at room temperature), the highest among binary alloys [83, 84] see Figure 6.1, high permeability, low magnetic losses at high frequencies, and being relatively strong and good mechanical strength [5, 85, 86]. This fact makes this binary alloy the best candidate as the soft component in making a rare earth-free exchange coupled permanent magnet. Since first report of fabrication of FeCo as soft magnetic material by Elmen in 1929 [87], there has been lots of research done to understand the physics of magnetism and improve their properties [81, 88–91]. These ferromagnetic alloys have been prepared several ways, including thermal decomposition, sonochemical reduction, arc-discharge, laser-pyrolysis, polyol process, and aqueous reduction by borohydride derivatives [82, 92–94]. These synthesis techniques have produced several shapes including spheres, cubes, dice, and wires. Aqueous reduction by borohydride has been used for producing monometallic nanoparticles of CoB and FeB [95–100]. Recently, Ekeirt et al. have shown that by using a capping agent, such as sodium citrate, they can eliminate the formation of FeB/Fe_2B nanoparticles to form elemental α-Fe. It's possible to design

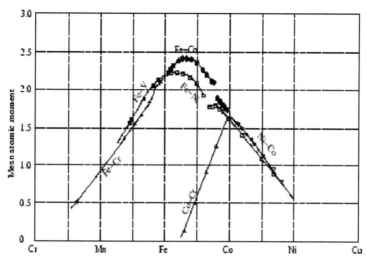

Figure 6.1 The Slater–Pauling curve showing the mean atomic moment for binary alloys of transition metals as a function of their composition [81].

very kind of nano-alloys particles with different specification and applications according to Slater–Pauling curve which is showing at Figure 6.1.

6.3 Synthesis and Characterization of Core/Shell Magnetic Nanoparticles (CS-MNPs)

6.3.1 Iron Oxide Core/Shell Nanoparticles (IOCSN)

6.3.1.1 Fe_3O_4@Ag core/shell nanoparticles

The study of magnetic iron oxide nanoparticles is interesting due to their properties that arise from a finite size, and surface effects that vigorously impacted magnetic properties relative to the bulk for showing remarkable phenomena, such as super paramagnetism [102], and their applications in a wide range, such as magnetic fluids hyperthermia (MFH) [103], magnetic resonance imaging (MRI) [104], and magneto-optics devices (MOD) [105]. The spherical-shaped magnetite nanoparticles (Fe_3O_4@Ag) have been prepared by using the Polyol method. $FeCl_3.6H_2O$ (1.35 g, 5 mmol) was dissolved in ethylene glycol (40 mL) in an ultrasonic bath to form a clear solution. The pH of the solution was acidic. Then, sodium acetate (3.7 g, 45 mmol) and polyethylene glycol (1.0 g, 0.05 mmol) were added drop wise to the solution while stirring for 30 min. Then, the acquired red precursor refluxed at 190°C for 16 h. After being cooled down to room temperature, the acquired black products were washed several times with ethanol and then dried at 80°C for 8 h, these black products are magnetite. To deposit silver on the magnetite particles, initially, 4 mM ethanolic silvering solution was prepared, and then, 3 mg Fe_3O_4 nanoparticles were dispersed in 30 mL of this solution in a polypropylene container by using an ultrasonic bath. The polypropylene container was used to avoid nonspecific silvering of the reaction vessel. At the end, silver coating was achieved by adding butylamine as a weak reductant of $AgNO_4$ in ethanol. The molar ratios of the ethanolic silvering solution to butylamine were changed as 1:0.5, 1:1, and 1:2, and nominated as S1, S2, and S3, respectively. The resultant solution was incubated for 45 min at 50°C in an ultrasonic bath. After being rinsed with ethanol, the obtained product was re-dispersed in ethanol under sonication for 5 min [101]. The X-ray diffraction pattern of as prepared magnetite showed reflection planes (111), (220), (311), (222), (400), (422), (511), and (440). As silver is deposited on the surface of magnetite, an additional four diffraction peaks are observed at 38.1, 44.3, 64.4, and 77.4, corresponding to (111), (200), (220), and (311) planes of silver, respectively, with face-centered cubic structure. The scanning electron micrograph shows clusters of fine particles clinging together, and the aggregates are spherical in shape. A statistical histogram shows that the mean diameter of the Fe_3O_4 particles is about 73 nm. The agglomerated clusters of particles are due to the magneto-static coupling between the particles. A statistical histogram of the fine particles on the surface of the powders shows that the primary particle size is about 16 nm in dimension. After reduction of silver ions on the surface of magnetite, the approximated silver layers were estimated using statistical histogram images, 3.5, 9, and 11 nm for these samples. Magnetic properties of nanoparticles were characterized using vibrating sample magnetometer (VSM with a maximal applied field of 10 KOe). It showed the magnetic hysteresis loops of the nanoparticles at room temperature. Besides, saturation magnetization (M_s) are detected around 91, 68, 53 and 32 emu.g-1 for Fe_3O_4, Fe_3O_4@Ag samples. The polyol reduction technique has been used to yield magnetite nanosphere with the saturation magnetization as high as a value of bulk (80 nm diameter) as well as low coercivity (50 Oe). This can be due to the nanoparticles being in the single-domain region agglomerate, and separated by polyethylene glycol, that enhanced the stability of its magnetic properties. Then, silver shell was coated on the magnetite nanoparticles by using butylamine as a weak reductant. The remaining magnetization was, nonetheless, strong enough, so that it could be expected to apply this composite to magnetic separation application, functional device assemblies, and MRI through further research and development [101].

6.3.1.2 Fe_3O_4@Chitosan core/shell nanoparticles

There are various ways to prepare Fe_3O_4 nanoparticles, which have been reported earlier, such as arc discharge [107], mechanical grinding [108], laser ablation [109], micro emulsions [110], and high temperature decomposition of organic precursors [111]. These methods may be able to prepare magnetite with controllable particle diameters. However, well-dispersed aqueous Fe_3O_4 nanoparticles have met with very limited success. Chemical co-precipitation method is a convenient and cheap method having the potential to meet the increasing demand for the direct preparation of well dispersed (water-base) Fe_3O_4 nanoparticles [112]. The sizes of nanoparticles can be well controlled by various surfactant in each processes [113, 114]. In this research Article [106], magnetite nanocrystals were prepared using $FeCl_2.4H_2O$. In brief, 2g $FeCl_2.4H_2O$ was dissolved in 30 mL HCl 1M. To this, 30 mL NaOH 3M was added drop wise with continuous stirring till black precipitate was formed. The precipitate was then washed several times with distilled water till neutral pH was obtained. The precipitate was then separated out using external magnetic field and dried at 100°C. The as-prepared nanoparticles were further used for the coating procedure. The obtained bare Fe_3O_4 nanoparticles were then coated with chitosan using the ultrasonication method. In brief, the obtained MNPs (1 g) were dispersed in 50 mL distilled water by ultrasonication for 20 min. This suspension was then added to 100 mL 1% chitosan (in 2% acetic acid) solution. Then, the mixture was ultrasonicated for 30 min. After that, it was washed three times to remove excess CH and allowed to settle down using an external magnetic field. The chitosan-coated MNPs were collected and dried at 50°C. The CH-coated nanoparticles (200 mg) were then dispersed in 25 mL of doubled distilled water and 10 mL of GLD was added to it and mechanically stirred for 8 h. The mechanical stirring was enough to cross-link CH. These resulting nanoparticles were washed with ethanol and water three times. The XRD patterns revealed that the desired phase formation occurs in thees samples. The main characteristic peaks were obtained with the (HKL) values of (220), (311), (400), (422), (511), and (440). The FTIR spectra of Fe_3O_4 MNPs were found to be a Fe_3O_4–CH/GLD MNPs along with CH over the range of 450–4000 cm^{-1}. The band observed at 565 cm^{-1} corresponds to the intrinsic stretching vibration (Fe_{tetra}–O) of metal–oxygen at tetrahedral site, whereas the band observed at 455 cm $tetra^{-1}$ corresponds to the stretching vibration (Fe_{Octa}–O) of metal–oxygen at octahedral site. The bands observed at 3376 and 1620 cm^{-1} correspond to surface-adsorbed water molecules on Fe_3O_4. The bands observed at 634, 692, and 1403 cm^{-1} in the spectrum of CH correspond to the bending vibration of C–H and the band observed at 1627 cm^{-1} corresponds to N–H bending vibration which showed efficient coating of CH on Fe_3O_4 MNPs. Thermo gravimetric (TG) curves of uncoated and CH/GLD-coated Fe_3O_4 nanoparticles measured by a thermo gravimetric analyzer. As TG was performed under N_2 atmosphere, the oxidation of MNPs was greatly reduced. It provides additional quantitative evidence on the structure of coating on surface of nanoparticles. For further confirmation, X-ray photoelectron spectroscopy (XPS) of coated MNPs was used. The SEM images of Fe_3O_4 and Fe–CH/GLD MNPs showed the formation of spherical nanoparticulates with sizes 20.573.8 nm and 22.273.4 nm, respectively which have good dispersibility (nonagglomerated) as compared to uncoated nanoparticles. The average particulate diameter is found to increase after functionalization with CH/GLD molecules. Improvement in dispersibility after coating with CH/GLD may be attributed to the presence of the nonmagnetic surface layer of CH/GLD which readily decreases the interparticle interaction, i.e., dipole–dipole interaction and thus enhances the dispersibility. To understand the effect of coating on the magnetization behavior of Fe_3O_4 system, they have carried out the M vs H measurements as a function of applied field and temperature. On the basis of SQUID measurements, it can be seen from the hysteresis curves for bare and functionalized Fe_3O_4 MNPs at 100 and 300 K that almost negligible coercivity or remanence existed, indicating the super paramagnetic behavior of Fe_3O_4 MNPs before and after coating. This is because magnetization is proportional to the amount of weight for the same magnetic material. Organic coating (CH/GLD) layers on magnetic material increases the amount of nonmagnetic substance which reduces the overall magnetization of the material.

The cytotoxicity study of both, bare and coated nanoparticles, was done on L929 cell line with different concentrations of nanoparticles. The L929 cell line was incubated with nanoparticles for 24 h with the concentrations of 0.1, 0.5, 1.0, 1.5, and 2.0 mg mL^{-1} at 37°C in a 5% CO atmosphere. The relative cell viability (%) compared with the control well containing cells without nanoparticles is calculated by the equation:

$$[A]_{tested}/[A]_{control}*100.$$

Over 93% and 94% cell viability was still obtained after 24-h incubation with 2 mg mL^{-1} concentration of bare Fe_3O_4 and coated nanoparticles, respectively. This shows that the coating of nanoparticles did not affect cytotoxicity much even up to 2 mg mL^{-1} concentration after 24 h. In this regards, this much concentration for both cytotoxicity and heating induction ability is studied in order to use it for hyperthermia therapy application. They have mentioned possible mechanisms of interactions of nanoparticles with the cells in our earlier report.

Finally, the goal focus of that work was developing a simple technique for the preparation of Fe_3O_4 nanoparticles and demonstrating their applicability in the biomedical field. From the present study, it is concluded that pure and stable phase of Fe_3O_4 nanoparticles can be obtained using $FeCl_2$ as the sole source and without using any other oxidant. Hence, the conventional method is more simplified and made cost-effective in the present work. The obtained particles are of polycrystalline nature. TEM images showed that the particles are round-shaped having a size 20.573.8 nm and 22.273.4 nm in cases of bare and coated particles, respectively. The resulting Fe_3O_4 nanoparticles were super paramagnetic at room temperature with a high saturation magnetization value. CH can be efficiently coated on the surface of Fe_3O_4 nanoparticles using a simple technique of ultrasonication. Cross-linking of amino groups in CH using GLD increases the stability of coating. Potential and DLS measurements of both the particles prove to have higher colloidal stability for the coated nanoparticles which is important in order to use them for *in vivo* applications. The preliminary results obtained from magnetic, hyperthermia, and cytotoxicity studies are highly encouraging due to monodispersivity, high saturation magnetization values, high SAR values, and low cytotoxicity.

6.3.2 Cobalt-Based Core/Shell Nanoparticles (CBCSN)

6.3.2.1 FeCo@C core/shell nanoparticles

Core–shell nanoparticles have been the subject of intensive research due to their potential applications from electromagnetic wave absorption to biomedicine [2]. Due to rapid development of wireless technology, it is necessary to develop advanced electromagnetic wave absorbers with light weight, strong absorption, and wide range of working frequency. Soft ferromagnetic nanoparticles (e.g., FeCo) are good candidates for this purpose because of their high saturation magnetization, higher snoek's limit, and high permeability at frequencies in the gigahertz range. In this research, they have prepared FeCo@C core–shell nanoparticles (FeCo nanoencapsulates) as a high dielectric–high magnetic material and investigated the effect of the carbon shell on an Electromagnetic properties of FeCo nanoparticles including their microwave absorption properties. FeCo nanoencapsulates prepared using a two-stage process: (1) synthesis of FeCo nanoparticles using water in oil (reverse micelle) microemulsion and (2) carbon coating of FeCo nanoparticles using the ACCVD method. The microstructural results showed that, that synthesized nanoparticles are spherical in shape with a narrow size distribution showing diffraction from (110), (200), and (211) lattice type planes. In addition, it indicated the formation of single crystal α-bcc structured FeCo alloy nanoparticles. The mean size of FeCo nanoparticles was determined about 9 nm by inspecting about 50 nm from TEM micrographs. The spherical shape of nanoparticles is due to the nature of the surfactant (CTAB) and co-surfactant (1-butanol) used. In this quaternary system, the polar co-surfactant (1-butanol) makes ion–dipole interactions with the surfactant and forms spherical aggregates in which the polar (ionic) ends of the surfactant molecules orient toward the center. In FeCo nanoencapsulates, interfacial polarization of space charges at core/shell interface produces a nanoscale capacitor leading to an enhancement in the real part of permittivity. Also for FeCo nanoencapsulates, the value of "ε" is higher than that of FeCo nanoparticles. So, there are two parts of semicircles which suggest a dual Debye relaxation mechanism while each arc is related to a Debye relaxation mechanism. In fact, during the alternation of the applied field, the charges redistribute alternatively between FeCo cores and graphite shells. Therefore, in addition to the dielectric relaxation of carbon shells, an interfacial relaxation between FeCo cores and graphite shells is also produced because of the presence of core/shell interfaces. But FeCo nanoparticles show a clear single semicircle but with a very smaller radius compared with FeCo nanoencapsulates which could be due to the weak interfacial relaxation between FeCo core and its oxide shell and there is no redistribution of charges due to the very high resistance of the oxide shell. The magnetic specifications showed the hysteresis curves of FeCo nanoparticles/nanoencapsulates at 298 K, respectively. The effect of graphite shell on the electromagnetic wave absorption properties is clearly seen from the figure. The RL for FeCo nanoencapsulates increases toward higher frequencies with increasing the absorber thickness. The maximum RL of about 40 dB is obtained for 2.5 mm thickness. There are two distinct absorption peaks in each thickness which indicates both dielectric and magnetic loss mechanisms in nanoencapsulates that is consistent with the results from dielectric and magnetic loss tangents. FeCo@C core–shell nanoparticles were synthesized using a two-stage process. First, FeCo nanoparticles were synthesized using the microemulsion technique, and then, FeCo nanoencapsulates were prepared using alcohol catalytic chemical vapor deposition. The effect of carbon coating on the magnetic properties of FeCo nanoparticles was investigated in detail. VSM results of FeCo nanoencapsulates showed lower saturation magnetization (40 emu/g) and higher coercivity (240 Oe) compared with that of FeCo nanoparticles (Ms = 60 emu/g and H = 100 Oe). This is due to the nonmagnetic graphite coating of FeCo nanoencapsulates which decreases the saturation magnetization and larger size of FeCo nanoencapsulates compared to FeCo nanoparticles which results in higher coercivity. Incorporation of the graphite shell into the microstructure of FeCo nanoencapsulates enhances the complex permittivity and permeability of FeCo nanoparticles, respectively, due to the presence of highly dielectric carbon shell and protection of magnetic properties of FeCo cores at high frequencies. Also, the graphite shell dramatically improves the microwave absorption properties of FeCo nanoparticles due to activation of several dielectric–magnetic loss mechanisms including dual dielectric relaxation and magnetic ferromagnetic resonance. The maximum RL of –40 dB was obtained for 40 wt% FeCo nanoencapsulates/paraffin nanocomposite with an effective absorption band (RLo –20 dB) of 2.8 GHz.

6.3.2.2 FeCo@PEG core/shell nanoparticles

Magnetic nanoparticles with high saturation magnetization are considered to be very important for magnetic energy, data storage, magnetic separation, drug delivery, and imaging applications [115]. The powders of FeCo nanoparticles with different chemical compositions such as $Fe_{20}Co_{80}$, $Fe_{50}Co_{50}$, $Fe_{60}Co_{40}$, and $Fe_{70}Co_{30}$ were prepared by modified polyol method. At first, mixtures of $(FeCl_2.4H_2O)$ and $(Co(Ac)_2.4H_2O)$ in desired proportions with suitable amounts of PEG were prepared in round bottom flasks. For all the compositions, the pH of the solutions was adjusted in between 10 and 11 by adding NaOH prior to the reduction process. Then, the PEG-metal salts solution was gradually heated up to 300°C while stirring continuously using a magnetic stirrer, and refluxed at this temperature for 2 h. During this process, it was observed that the solution turned black in all the cases within a few minutes after reaching the refluxing temperature. The solutions were then cooled down to room temperature, ultrasonicated for about 20 min, washed several times using ethanol and water, and finally collected by using a magnet. The synthesized powders were subsequently annealed for 2 h at 600 1C in the presence of hydrogen before characterizations. The X-ray diffraction patterns of the synthesized FeCo nanoparticles are shown three characteristic peaks values of 44.82, 65.2, and 82.661 corresponding to the crystal planes of

(110), (200), and (211) were observed for all FeCo of different compositions. Typical EDX analysis of the annealed FeCo samples reveal that the materials are mainly composed of Fe and Co metals only while displaying a small percentage (2–3%) of oxygen for all the samples with different Fe/Co ratios. Typical TEM images of the as-prepared FeCo nanoparticles reveal that the as-synthesized nanoparticles are monodisperse and spherical in shape with average size of about 10 nm. The smaller size of the FeCo nanoparticles, in the present study, has been successfully obtained mainly by controlling the reaction conditions such as reaction time, PEG, Fe/Co ratio, optimization of the concentration of hydroxyl ions, and the ultrasonication procedure. The TEM images determined that, the annealed sample where the particles were still spherical in shape but with increased sizes to about 50–90 nm. The magnetization of FeCo alloy depends on the ratio of Fe and Co atoms. The magnetization curves of FeCo nanoparticles with different Fe/Co ratios measured at room temperature and 5 K. As expected, the samples measured at lower temperatures resulted in higher values of saturation magnetization. In the inset in each figure, the magnetization trend as a function of Fe/Co ratio has been shown. Synthesize high magnetization FeCo nanoparticles using facile modified polyol method without surfactant has been succeeded. The parameters of the reaction conditions, time, hydroxyl ion concentration, and PEG were used while varying the quantities of Fe and Co precursors to effectively control the size and the magnetization values of FeCo nanoparticles. Transmission electron microscopy (TEM) shows that the synthesized nanoparticles are spherical in shape and monodisperse with average particle size of 10 nm before annealing. After the samples were annealed at 600°C in the presence of hydrogen, the particle sizes were increased to about 50–90 nm. These synthesized nanospheres are expected to be suitable as labels for targeted drug delivery and biosensors due to their high saturation magnetization. Also this method is considered promising for easy synthesis of different kinds of nanoparticles.

6.4 Biomedical Application of Magnetic Nanoparticles (MNPs)

Nanotechnology is an enabling technology that deals with nanometer-sized objects. It is expected that nanotechnology could be developed for several applications: materials, devices, and systems. At present, the nanomaterial application is the most advanced one, both in scientific knowledge and in commercial applications. A decade ago, nanoparticles were studied because of their size-dependent physical and chemical properties. Now, they have entered a commercial exploration period. Magnetic nanoparticles offer some attractive possibilities in medicine. Living organisms are built of cells that are typically 10 μm in diameter. However, the cell parts are much smaller and in the submicron size domain [35]. First advantage in medicine is that nanoparticles have controllable sizes ranging from a few nanometers up to tens of nanometers, which places them at dimensions that are smaller than those of a cell (10–100 μm), or comparable to size of a virus (20–450 nm), a protein (5–50 nm), or a gene (2 nm wide and 10–100 nm length). This means that they can "get close" to a biological entity of interest. This simple size comparison gives an idea of using nanoparticles as very small probes that would allow us to spy at the cellular machinery without introducing too much interference. Indeed, they can be coated with biological molecules to make them interact with or bind to a biological entity, thereby providing a controllable means of "tagging" or addressing it. Second one, if nanoparticles are magnetic, they can be manipulated by an external magnetic field gradient. This "action at a distance" combined with the intrinsic penetrability of magnetic fields into human tissue opens up many applications involving the transport and immobilization of magnetic nanoparticles, or of magnetically tagged biological entities. In this way, they can be made to deliver a package, such as an anticancer drug, to a targeted region of the body, such as a tumor. Third one, the magnetic nanoparticles can be made to resonantly respond to a time-varying magnetic field, with advantageous results related to the transfer of energy from the exciting field to the nanoparticle. For example, the particle can be made to heat up, which leads to their use as hyperthermia agents, delivering toxic amounts of thermal energy to targeted bodies such as tumors; or as chemotherapy and radiotherapy enhancement agents, where a moderate degree of tissue warming results in more effective malignant cell destruction. These, and many other potential applications, are made available in biomedicine as a result of the special physical properties of magnetic nanoparticles. Understanding of biological processes on the nanoscale level is a strong driving force behind development of nanotechnology [35, 36]. Magnetic nanoparticles can be a promising tool for several applications *in vitro* and *in vivo*. In medicine, many applications were investigated for diagnostics and therapy and some practical approaches were chosen. Magnetic immunobeads, magnetic streptavidin, DNA isolation, cell immunomagnetic separation (IMS), magnetic resonance imaging (MRI), magnetic targeted delivery of therapeutics, or magnetically induced hyperthermia are approaches of particular clinical relevance. Investigations on applicable particles induced a variability of micro and nanostructures with different materials, sizes, and specific surface chemistry. The nanoparticles for medicine are useful for therapy, imaging, and diagnostics of cancer and other diseases leading an entrapped or bound therapeutic or diagnostic target material to the area of interest, e.g., a tumor. The destination—targeted delivery—may be found by physical forces (magnetic) or with surface-bound antibodies (cell/tissue-specific) [37]. Some present applications of nanomaterial in biology and medicine are fluorescent biological labels [38–40], drug and gene delivery [41, 42], biodetection of pathogens [43], detection of proteins [44], probing of DNA structure [45], tissue engineering [46], tumor destruction via heating (hyperthermia) [47], separation and purification of biological molecules and cells [48], MRI contrast enhancement [49], and phagokinetic studies [50]. As mentioned above, nanomaterials are suitable for biotagging or labeling because they are as the same size as proteins. The other sufficient feature to use nanoparticles as biological tags is their biosusceptibility. In order to interact with a biological target, a molecular linker should be attached to the nanoparticle, acting as a bioinorganic interface. Examples of biological coatings may include antibodies, biopolymers such as collagen [20], or molecule mono layers (amino acids, sugars) that make the nanoparticles biocompatible [15, 22–23]. In addition, as optical detection techniques are wide spread in biological research, it is better if nanoparticles show fluorescence or have other optical features. Nanoparticles usually form the core of nano biomaterials. It can be used as a convenient surface for molecular assembly and may be composed of inorganic or polymer materials. It can also be in the form of nanovesicle surrounded by a membrane or a layer. The shape is not automatically spherical but sometimes cylindrical or plate like. Even more complicated shapes are possible. The size and size distribution might be important in some cases, for example, if penetration through a pore structure of a cellular membrane is required. The size and size distribution are becoming extremely critical when quantum-sized effects are used to control material properties. A tight control of the average particle size and a narrow distribution of sizes allow creating very efficient florescent probes that emit narrow light in a very wide range of wavelengths. This helps creating biomarkers with many well-distinguished colors. The core itself might have several layers and be multifunctional. For example, by combining magnetic and luminescent layers, one can both detect and manipulate the particles [51]. Fundamental understanding of chemical, electrical, optical, and magnetic properties of nanomaterial (materials with dimensions less than 100 nm) has been the significant importance over the past two decades [52, 53]. Magnetic nanoparticles are a specific class of nanomaterial, composed of at least one magnetic element. These materials could be used in a variety of forms including the following: in solution as ferrofluids for audio speakers [53], as particle arrays in magnetic storage media [54–59], as surface functionalized particles for biosensing applications [60–68], as powder compacts for power generation, conditioning, and conversion, in medical applications including magnetic targeted drug delivery [69–74], and contrasting agents in magnetic resonance imaging [75–79].

Magnetic nanoparticles have been proposed for biomedical applications for several years [116–118]. In recent years, nanotechnology has developed to a stage that makes it possible to produce, characterize, and specifically tailor the functional properties of nanoparticles to applications. This shows considerable promise for applications in biomedical and diagnostic fields, such as targeted drug delivery, hyperthermic treatment for malignant cells, and magnetic resonance imaging (MRI) [117–120]. There are three reasons why magnetic nanoparticles are useful in biomedical applications. First, the size of magnetic nanoparticles can be controlled, ranging from a few nanometers up to tens of nanometers and are thus smaller in size than a cell (10–100 μm), a virus (20–450 nm), a protein (5–50 nm), or a gene (2 nm wide and 10–100 nm long). Magnetic nanoparticles can get close to cells and genes, and they can be coated with biomolecules to make them interact or bind with biological entities. Second, magnetic nanoparticles can be manipulated by an external magnetic field gradient. Magnetic nanoparticles can be used to deliver a package, such as an anticancer drug, to a targeted region of the body such as a tumor. Third, magnetic nanoparticles can also be made to respond resonantly to a time-varying magnetic field, with an associated transfer of energy from the field to the nanoparticles. Magnetic nanoparticles can be made to heat up, which leads to their use as hyperthermia agents, delivering toxic amounts of thermal energy to targeted bodies such as tumors or as chemotherapy [117, 118, 120]. For biomedical applications, magnetic nanoparticles must (1) have a good thermal stability; (2) have a larger magnetic moment; (3) be biocompatible; (4) be able to form stable dispersion so the particles could be transported in living system; and (5) response well to AC magnetic fields. Furthermore, better control of particle size and properties will be necessary to use these particles in biomedical applications, in which uniformity of the properties will ensure accurate doses and delivery [118]. In order to interact with biological target, a biological or molecular coating layer acting as a bioinorganic interface should be attached to the nanoparticle. Examples of biological coatings may include antibodies, biopolymers such as collagen

[121], or monolayers of small molecules that make the nanoparticles biocompatible [122]. Magnetic particles as carriers for therapeutic agents have been used in experimental animals and clinical applications in humans. Mostly, they have used in combination with either diagnostic imaging procedures, and/or oncological therapeutic regimes. The aim of most of the research is to investigate the possibility of these magnetic particles to be used in clinical applications of musculoskeletal disorders as well (cartilage, joint capsules, bone, tendons, and ligaments). Biocompatible magnetic nanoparticles *in vitro* experiments insignificantly influence the cell's survive. Biocompatibility is made possible through chemical modification of the surface of the magnetic nanoparticles, usually by coating with biocompatible molecules such as dextran, polyvinyl alcohol (PVA), and phospholipids—all of which have been used on iron oxide nanoparticles [123].

6.4.1 Bioimaging Application of MNPs

MR imaging, one of the most powerful noninvasive imaging methods utilized in clinical medicine, is based on the relaxation of protons in tissues [124]. Upon accumulation in tissues, SIONPs enhance proton relaxation of specific tissues compared with that in surrounding tissues, serving as an MR contrast agent [125]. For *in vivo* MR imaging applications, SIONPs should have long half-life time in blood circulation for the improved efficiency of detection, diagnosis, and therapeutic management of solid tumors. Because opsonin plasma proteins are capable of interacting with plasma cell receptors on monocytes and macrophages, opsonin absorbed SIONPs will be quickly cleaned by circulating monocytes or fixed macrophages through phagocytosis, leading to elimination of SIONPs from blood circulation. The smaller the particle and the more neutral and hydrophilic its surface, the longer its plasma half-life [126]. Therefore, the surface of SIONPs has been modified with hydrophilic polymers to prevent absorption of the circulating plasma proteins. The use of contrast agents and tracers in medical imaging has a long history [127–129]. Known collectively as imaging contrast agents, these molecules possess physical characteristics that increase the strength of the signal coming out of the body. MRI contrast agents containing the element gadolinium (or iron), for example, do so by altering the magnetic field in the body, which boosts the strength (or reduce) of the MRI signal. They provide important information for diagnosis and therapy, but for some desired applications, a higher resolution is required than can be obtained using the currently available medical imaging techniques. In addition, Computed tomography

(CT) provides a good imaging modality for studying anatomical details as against positron emission tomography (PET) and another modalities which focus on metabolic pathways using X-ray absorption spectra as a detection signal. Gamma rays used in PET analysis have lower energy levels and lower penetration power than X-ray used in CT. Thus, high-resolution CT is most constructive in tracking the site and loci of some metabolic event or analyzing its histological impact as X-ray emissions are differentially absorbed by tissues according to their X-ray attenuation coefficient, which gives a visual spectrum for image reconstruction [11]. To achieve high resolution, several non-NP-based contrasting agents based on iodine, barium, barium sulfate, etc., are in use, which selectively highlight the tissue of interest during CT analysis. Conventional CT contrast agents have generally high renal toxicity and suffer from low imaging time, because of rapid renal clearance. Low molecular weight nanoparticle systems comprising of gold/iron oxide core/shell nanoparticles have shown great potential to be used as CT contrasting agents, for their stability and optimum residence time in the tissues and versatility in multifunctional imaging mode. Recent developments are going on using multimodal nanoparticles contrasting agents especially using electron dense elements such as iodine [130–134], or bismuth which form well-defined dispersion spectra when impinged by electromagnetic waves. Biocompatible, dual mode contrasting core/shell nanoparticle based agents have been reported, recently. Molecules are the most common radiotracers used for dual imaging purposes in PET/CT [135]. Nanoparticles having a super paramagnetic iron core cross-linked with dextrin forming the corona bonded molecules empower them to work for PET, CT, and MRI.

6.4.2 Controlled Drug Delivery (TDD) Applications of MNPs

Drug delivery remains a challenge in the management of cancer and another illness. The focus is on targeted cancer therapy. The newer approaches to cancer treatment not only supplement the conventional chemotherapy and radiotherapy but also prevent damage to normal tissues and prevent drug resistance. Innovative cancer therapies are based on current concepts of molecular biology of cancer. These include antiangiogenic agents, immunotherapy, bacterial agents, viral oncolysis, targeting of cyclic-dependent kinases and tyrosine kinase receptors, antisense approaches, gene therapy, and combination of various methods. Important methods of immunotherapy in cancer involve use of cytokines, monoclonal antibodies, cancer vaccines, and immunogene therapy [135]. The innovative pharmaceutical treatments obviously require novel modern methods of administration. The possibility of using ferrofluids for drug localization in blood vessels and in hollow organs is a contemporary task for drug development [136]. Pure magnetic particles are not stable in water-based solutions and suspensions; therefore, they cannot be used for medical application without biocompatible coating. The choice of polymers for magnetic nanoparticles coating to prevent them from adhering toxicity occurred to be not an easy one-step task, because these compositions should satisfy both the requirements of biocompatibility and biodegradability. During the past two decades, research into hydrogel delivery systems has focused primarily on systems containing polyacrylic acid (PAA) backbones. PAA hydrogels are known for their super-absorbancy and ability to form extended polymer networks through hydrogen bonding. In addition, they are excellent bioadhesives, which mean that they can adhere to mucosal linings within the gastrointestinal tract for extended periods, releasing their encapsulated medications slowly over time.

6.4.3 Cancer Diagnosis and Treatment via Hyperthermia Method (CDT) Using MNPs

In recent years, nanotechnology has achieved a stage that makes it possible to produce, characterize, and specifically tailor the functional properties of nanoparticles for clinical applications. This has led to various opportunities such as improving the quality of MRI, hyperthermic treatment for malignant cells, site-specific drug delivery, and the manipulation of cell membranes. To this end, a variety of iron oxide particles have been synthesized. A common failure in targeted systems is due to the opsonization of the particles on entry into the bloodstream, rendering the particles recognizable by the body's major defense system. In some cases, the combination of gene therapy with effects of hyperthermia may be possible. Heat-induced therapeutic gene expression is highly desired for gene therapy to minimize side effects. Furthermore, if the gene expression is triggered by heat stress, combined therapeutic effects of hyperthermia and gene therapy may be possible [35]. Hyperthermia therapy is a type of cancer treatment in which the body tissue is exposed to high temperature of 42°C or higher, which is found to be more harmful to cancer cells than to normal healthy cells. Mild hyperthermia is performed at 41–46°C to simulate the immune response for nonspecific immunotherapy of cancers, while thermo ablation is performed at 46–56°C to kill cancer cells by direct cell necrosis, coagulation, or carbonization [148–149].

The challenge of this cancer therapy lies in controlling the heating effect to only the local tumor site so as to not harm the nearby healthy cells. To this end, magnetic hyperthermia has emerged as one of the most promising approaches for heat localization. Magnetic hyperthermia treatment is based upon the idea that magnetic nanoparticles heat up under oscillating magnetic field. For this therapy, biocompatible magnetic nanoparticles are introduced into the tumor site either by direct injection or by targeted delivery. Oscillating magnetic field, or magnetic field generated by sending alternating current through a coil, is applied and magnetic nanoparticles interact with this field to generate heat through various mechanisms. The concept of magnetic materials in hyperthermia was first proven in 1957 when Gilchrist and coworkers heated various tissue samples with 20–100 nm size particles of γ-Fe_2O_3 exposed to a 1.2 MHz magnetic field [150]. Much progress in the field of magnetic nanoparticles using various types and sizes of magnetic materials, magnetic field strengths and frequencies, methods of preparation, coatings, and nanoparticle delivery have been made [151–152]. In 2007, Jordan and coworkers have reported the first clinical study of magnetic hyperthermia, showing that aminosilane-coated super paramagnetic iron oxide nanoparticles could be safely applied for the treatment of brain tumors, achieving hyperthermic temperatures while being well tolerated by patients [153].

6.5 Conclusion

Diagnosis of diseases is very important in health care, which in turn not only enhance the effectiveness of medical treatment but also save human life where early diagnosis is crucial. However, in many cases, early diagnosis needs sophisticated biomedical instruments or improved techniques. The application of nanoparticles in the realm of biomedical engineering has ushered in a new era for the development for novel contrast agent and drug delivery vehicle, which has the potential to revolutionize in the area of health care [11]. Biomedical application of different nanoparticles showed in Table 6.3. The idealistic concept of a single platform for drug delivery to its monitoring of drug release seems to be feasible in the near future, because of the recent advances in the application of novel nanomaterials in this field. The versatility of nanoparticles has been applied in various studies related to disease diagnostics, early detection studies, and better contrast agents for improved imaging techniques. The development of new drug delivery vehicles has not

Table 6.3   Biomedical application of different nanoparticles

Core/Shell Nanoparticles	Surface Modification	Application
Fe_3O_4/SiO_2	Fluorescein isothiocyanate dye, chelated	MRI, amperometric sensor
$Fe_3O_4/PAH/Au$		
$Fe_3O_4/chitosan$ or oleic acid	Hemoglobin for H_2O_2 detection/enzyme, nucleotides	MRI, optical imaging and drug delivery
$Fe_3O_4/silica/Au$		
Fe_2O_3/PEG or PEI		MRI
$Fe_2O_3/2methacryloyloxyethyl$		X-ray
$Fe_2O_3/SiO_2/Au$	PEG, amino acid, FTIC, antibody conjugation	MRI, biolabeling optical imaging, drug delivery
$Fe_3O_4/CaCo_3/PMMA/MnO$	PEG, glucuronic acid	MRI, cell labeling
Fe oxide or Fe_3O_4/Au	DNA ligase enzyme	Piezometric and optic sensor
Fe_3O_4 embedded in Poly (D-lactic)/PLA/PVP		MRI, ultrasound
Fe/CNP	Poly acrylic acid) (PAA), polyvinyl pyrrolidone (PVP), poly (2-acetoxyethyl methacrylate) (PAEMA)	MRI
FeCo/Au	PNA oligomers	MRI, optical sensor
FeCo/C	PAA	MRI, hyperthermia cancer treatment
FeCo	Hydroxy apatite	MRI, hyperthermia cancer treatment

only reduced the payload of the drugs but has also improved the efficacy of the drug in the system because of improved bio- and cytocompatibility along with increased circulation time. Thus, the advent of nanoparticles has influenced all the spheres pertaining to medical biotechnology and biomedical engineering, improving and enhancing the already existing techniques along with the experimentation of new and advanced techniques for drug delivery and its monitoring. In this article, two general fields of applications namely diagnosis (analytical—biosensor/nucleotide interactions or visual—bioimaging) and transportation (drug delivery and gene transfection) are discussed. The review clearly shows, except bioimaging, that most other areas of bio-application are concentrated on nanoparticles of noble metals or magnetic materials.

Acknowledgement

This work was supported financially by Materials and Biomaterials Research Center (MBMRC). The authors are thankful to Associate Prof. Mehdi Hadavi, Director of MBMRC, and Assistant Prof. Salman Afghahi, Director of Advanced Materials Institute.

References

[1] Amirsadegh Rezazadeh Nochehdehi, Sandri, M., and Moham- madzadeh, A. (2015). "Bio magnetic Nano Particles (BMNPs) used for cancer treatment via Hyperthermia method", *World Congress on Medical Physics and Biomedical Engineering*, Volume 51 of the series IFMBE Proceedings pp. 827–827, Toronto, Canada, June 7–12.

[2] Afghahi, S. S. S., and Shokuhfar, A. (2014). "Two step synthe- sis, electromagnetic and microwave absorbing properties of FeCo@C core–shell nanostructure", *J. Magn. Magn. Mater.* 370, 37–44.

[3] WHO Report (2004). "Priority Medicines for Europe and the World", commissioned by the Government of the Netherlands, 18 November 2004.

[4] Bertil, A. (2005). "Forward Look report on Nanomedicine", ESF & EMRC.

[5] Kyler, J. C. (2010). "Core-shell Nanoparticles: Synthesis, Design, and Characterization", July.

[6] Bai, X., Son, S. J., Zhang, S. X., Liu, W., Jordan, E. K. Frank, J. A., Venkatesan, T., and Lee, S. B. (2008). *Nanomedicine*.

[7] Gupta, A. K., and Gupta, M. (2005). *Biomaterials*. 26, 3995.

[8] Maurer, T., Ott, F., Chaboussant, G., Soumare, Y., Piquemal, J. Y., and Viau, G. (2007). *Physics App. Biomater. Physics Letter.*

[9] Yan, J.-M., Zhang, X.-B., Akita, T., Haruta, M., Xu, Q. J. (2010). *Am. Chem. Soc.*

[10] Srivastava, C., Nikles, D. E., and Thompson, G. B. (2008). *J. Appl. Phys.* 104, 104314.

[11] Chatterjee, K., Sreerupa Sarkar, K. Jagajjanani Rao, and Santanu Paria (2014). "Core/shell nanoparticles in biomedical applications", *Adv. Colloid Interface Sci.* 209, 8–39.

[12] Liu G, Swierczewska M, Lee S, and Chen X. (2010). "Functional nanoparticles for molecular imaging guided gene delivery", *Nano Today* 5, 524–539.

[13] de la Isla, A., Brostow, W., Bujard, B., Estevez, M., Rodriguez, J. R., Vargas, S., and Castano, V. M. (2003). *Materials Res. Innovation,* 7–110.

[14] Yoshida, J., and Kobayashi, T. (1999). *J. Magn. Magn. Mater.* 194, 176.

[15] Molday, R. S., and MacKenzie, D. (1982). *J. Immunol. Methods* 52, 353.

[16] Weissleder, R., Elizondo, G., Wittenburg, J., Rabito, C. A., Bengele, H. H., and Josephson, L. (1990). *Radiology,* 175, 489.

[17] Parak, W. J., Boudreau, R., Gros, M. L., Gerion, D., Zanchet, D., Micheel, C. M., Williams, S. C., Alivisatos, A. P., and Larabell, C. A. (2002). *Adv. Mater.* 14, 882.

[18] Sinani, V. A., Koktysh, D. S., Yun, B. G., Matts, R. L., Pappas, T. C., Motamedi, M., Thomas, S. N., and Kotov, N. A. (2003). *Nano Lett.* 3, 1177.

[19] Zhang, Y., Kohler, N., and Zhang, M. (2002). *Biomaterials* 23, 1553.

[20] Gutwein, L. G., and Webster, T. J. (2002). *J. Nanoparticle. Res.* 4, 231.

[21] Gutwein, L. G., and Webster, T. J. (2004). *Biomaterials* 25, 4175.

[22] Sangregorio, C., Wieman, J. K., Connor, J. O., and Rosenzweig, Z. (1999). *J. Appl. Phys.* 85, 5699.

[23] Pardoe, H., Chua-anusorn, W., St Pierre, T. G., and Dobson, J. (2001). *J. Magn. Mag. Mater.* 225, 41.

[24] Burns, M. A., and Graves, D. J. (1985). *Biotechnol. Progress,* 1, 95.

[25] Mehla, R. V., Upadhyay, R. V., Charles, S. W., and Ramchand, N. (1997). *Biotechnol Teclm,* 11, 493.

[26] Koneracka, M., Kopcansky, P., Antalk, M., Thabo, M., Ramchand, N., Lobo, D., Mehta, R. V., and Upadhyay, R. V. (1999). *J. Magn. Magn. Mater.* 201, 427.

[27] Koneracka, M., Kopcansky, P., Timko, M., Ramchand, C. N., de Sequeira, A., and Trevan, M. (2002). *J. Mol. Catalysis A Enzymatic,* 18, 13.

[28] Babincova, M., Leszczynska, D., Sourivong, P., and Babinec, P. (2000). *Med. Hypoth.* 54, 177.

[29] Davis, S. S. (1997). *Trend Biotechnol.* 15, 217.

[30] Brightman, M. W. (1965). *Am. J. Anat.* 117, 193.

[31] Gaur, U., Sahoo, S. K., De, T. K., Ghosh, P. C., Maitra, A., and Ghosh, P. E. (2000). *Int J Pharm,* 202, 1.

[32] Liberti, P. A., Rao, G., and Terstappen, L. (2001). *J. Magn. Mater.* 225, 301.

[33] Paul, F., Melville, D., Roaih, S., and Warhurst, D. (1981). *IEEE Trans. Magn.* 17, 2822.

[34] Seesod, N., Nopparat, P., Hedrum, A., Holder, A., Thaithong, S., and Uhlen, M. (1997). *J. Lundeberg, Am. J. Tropical Med. Hygiene* 56, 322.

[35] Gubin, S. P. (2009). *Magnetic Nanoparticles*. Weinheim: Wiley-VCH Verlag GmbH & Co., ISBN: 978-3-527-40790-3.

[36] Nikiforov, V. N., and Filinova, E. Yu. (2009). *Biomedical Applications of Magnetic Nanoparticles*. Weinheim: Wiley-VCH Verlag GmbH & Co., ISBN: 978-3-527-40790-3.

[37] Alexiou, C., Jurgons, R., Seliger, C., and Iro, H. (2006). *J. Nanosci. Nanotechnol.* 6, 2762.

[38] Bruchez, M., Moronne, M., Gin, P., Weiss, S., and Alivisatos, A. P. (2013). *Science* 281.

[39] Chan, W. C. W., and Nie, S. M. (2016). *Science* 281.

[40] Wang, S., Mamedova, N., Kotov, N. A., and Chen, W. (2002). *J. Stud. Nano Letters,* 2, 817.

[41] Mah, C., Zolotukhin, I., Fraites, T. J., Dobson, J., Batich, C., Byrne, B. J. (2000). *Mol. Therapy*, 1, 239.

[42] Panatarotto, D., Prtidos, C. D., Hoebeke, J., Brown, F., Kramer, E., Briand, J. P., Muller, S., Prato, M., and Bianco, A. (2003). *Chem. Biol.* 10, 961.

[43] Edelstein, R. L., Tamanaha, C. R., Sheehan, P. E., Miller, M. M., Baselt, D. R., Whitman, L. J., and Colton, R. J. (2000). *Biosens Bioelectron*, 14, 805.

[44] Nam, J. M., Thaxton, C. C., and Mirkin. C. A. (2003). *Science*, 301, 1884.

[45] Mahtab, R., Rogers, J. P., and Murphy, C. J. (1995). *J. Am. Chem. Soc.* 117, 9099.

[46] Ma, J., Wong, H., Kong, L. B., and Peng, K. W. (2003). *Nanotechnology* 14, 619.

[47] Weissleder, R., Elizondo, G., Wittenburg, J., Rabito, C. A., Bengele, H. H., and Josephson, L. (1990). *Radiology*, 175, 489.

[48] Parak, W. J., Boudreau, R., Gros, M. L., Gerion, D., Zanchet, D., Micheel, C. M., Williams, S. C., Alivisatos, A. P., and Larabell, C. A. (2002). *Adv. Mater.* 14, 882.

[49] Sinani, V. A., Koktysh, D. S., Yun, B. G., Matts, R. L., Pappas, T. C., Motamedi, M., Thomas, S. N., and Kotov, N. A. (2003). *Nano Lett.* 3, 1177.

[50] Zhang, Y., Kohler, N., and Zhang, M. (2002). *Biomaterials* 23, 1553.

[51] Gref, R., Minamitake, Y., Peracchia, M. T., Trubetskoy, V., Torchilin, V., and Langer, R. (1994). *Science* 263, 1600.

[52] Hadjipanayis, G. C. (1999). Nanophase hard magnets. *Magn. Magn. Mater.* 200, 373–391.

[53] Willard M. A., Kurihara, L. K., Carpenter, E. E., Calvin, S., and Harris, V. G. (2004). "Chemically prepared magnetic nanoparticles", *Int. Mater. Rev.* 49, 125–170.

[54] Sun S., Murray, C. B., Weller, D., Folks, L., and Moser, A. (2000). "Monodisperse FePt Nanoparticles and Ferromagnetic FePt Nanocrystal Superlattices", *Science* 287, 1989–1992.

[55] Misra, R. D. K., Ha, T., Kadman, Y., Powell, C. J., Stiles, M. D., and McMichael, R. D. (1995). "STM studies of GMR spin valves", *MRS-Proceeding*, 384, 373–383.

[56] Meny, C., Panissod, P., Humbert, P., Nozienes, J. P., Speriosu, V. S., Gurney, B. A., and Zehringer, R. (1993). "Structural study of Cu/Co/Cu/NiFe/FeMn spin valves by nuclear magnetic resonance", *J. Magn. Magn. Mater.* 121, 406–408.

[57] Huang, T. C., Nozieres, J. P., Speriosu, V. S., Gurney, B. S., and Lefakis, H. (1992). "Effect of annealing on the interfaces of giant-magnetoresistance spin-valve structures", *Appl. Phys. Lett.*, 62, 1478–1480.

[58] Soeya, S., Tadokoro, S., Imagawa, T., Fuyama, M., and Narishige, S. (1993). "Magnetic exchange coupling for bilayered Ni81fe19/NiO and trilayered Ni81Fe19/NiFeNb/NiO films", *J. Appl. Phys.* 74, 6297–6301.

[59] Egelhoff Jr., W. F., Chen, P. J., Powell, C. J., Stiles, M. D., McMichael, R. D., Lin, C.-L., Sivertsen, J. M., Judy, J. H., Takano, K., and Berkowitz, A. E. (1996). "The tradeoff between large GMR and small GMR and small coercivity in symmetric spin-valves", *J. Appl. Phys.* 79, 2491.

[60] Miller, M. M., Prinz, G. A., Cheng, S. F., and Bounnak, S. (2002). "Detection of a micron-sized magnetic sphere using a ring-shaped anisotropic magnetoresistance-based sensor: A model for a magnetoresistance-based", *Appl. Phys. Lett.* 81, 2211–2214.

[61] Blanc-Beguin, F., Nabily, S., Gieraltowski, J., Turzo, A., Querellou, S., Salaun, P. Y. (2009). "Cytotoxity and GMI bio-sensor detection of maghemite nanoparticles internalized into cells", *J. Magn. Magn. Mater.* 321, 192–197.

[62] Cavalli, G., Banu, S., Ranasinghe, T., Broder, G. R., Martins, H. F. P., Neylon, C., Morgan, H., Bradley, M., and Roach, P. L. (2007). "Multistep synthesis on SU-8: combining microfabrication and solid-phase chemistry on a single material", *J. Combined Chem.* 9, 462–472.

[63] Choi, J. W., Oh, K. W., Thomas, J. H., Heineman, W. R., Halsall, H. B., Nevin, J. H., Helmicki, A. J., Henderson, H. T., and Ahn, C. H. (2002). "An integrated microfluidic biochemical detection system for protein analysis with magnetic bead-based sampling capabilities Lab Chip", 2, 27–30.

[64] Golub, T. R., Slonim, D. K., Tamayo, P., Huard, C., Gaasenbeek, M., Mesirov, J. P., Coller, H., Loh, M. L., Downing, J. R., Caligiuri, M. A., Bloomfield, C. D., and Lander, E. S. (1999). "Molecular classification of cancer: class discovery and class prediction by gene expression monitoring", *Science* 286, 531–537.

[65] Graham, D. L., Ferreira, H. A., Bernardo, J., Freitas, P. P., and Cabral, J. M. S. (2003). "Single magnetic microsphere placement and detection on-chip using current line designs with integrated spin valve sensors: biotechnological applications", *J. Appl. Phys.* 91, 7786–7788.

[66] Togawa, K., Sanbonsugi, H., Sandhu, A., Abe, M., Narimatsu, H., Nishio, K., Handa, H. (2005). "High sensitivity InSb Hall effect biosensor platform for DNA detection and biomolecular recognition using functionalized magnetic nanobeads", *J. J. Appl. Phys.* 44, 1494–1497.

[67] Mihajlović, G., Aledealat, K., Xiong, P., Moler, S., Field, M., and Sullivan, G. J. (2007). "Magnetic characterization of a single superparamagnetic bead by phase-sensitive micro-Hall magnetometry", *Appl. Phys. Lett.* 91, 172518–172521.

[68] Shen, W., Mathison, L. C., Petrenko, V. A., and Chin, B. A. (2010). "A pulse system for spectrum analysis of magnetoelastic biosensors", *Appl. Phys. Lett.* 96, 163502–163505.

[69] Gertz, F., Azimov, R., and Khitun, A. (2012). "Biological cell positioning and spatially selective destruction via magnetic nanoparticles", *Appl. Phys. Lett.* 101, 013701–013704.

[70] Lubbe, A. S., Bergemann, C., Riess, H., Schriever, F., Reichardt, P., Possinger, K., Matthias, M., Dorken, B., Herrmann, F., Gurtler, R., Hohenberger, P., Haas, N., Sohr, R., Sander, B., Lemke, A. J., Ohlendorf, D., Huhnt, W., and Huhn, D. (1996). "Clinical experiences with magnetic drug targeting: a phase I study with 4'-epidoxorubicin in 14 patients with advanced solid tumors", *Cancer Res.* 56, 4686–4693.

[71] Zhang, C., Wangler, B., Morgenstern, B., Zentgraf, H., Eisenhut, M., Untenecker, H., Kruger, R., Huss, R., Seliger, C., Semmler, W., and Kiessling, F. (2002). "Silica- and alkoxysilane coated ultrasmall superparamagnetic iron oxide particles: a promising tool to label cells for magnetic resonance imaging", *Langmuir* 23, 1427–1434.

[72] Gref, R., Luck, M., Quellec, P., Marchand, M., Dellacherie, E., Harnisch, S., Blunk, T., and Muller, R. H. (2000). "'Stealth' corona-core nanoparticles surface modified by polyethylene glycol (PEG): influences of the corona (PEG chain length and surface density) and of the core composition on phagocytic uptake and plasma protein adsorption", *Colloids Surf. B Biointerfaces*, 18, 301–313.

[73] Mikhaylova, M., Jo, Y. S., Kim, D. K., Bobrysheva, N., Andersson, Y., Eriksson, T., Osmolowsky, M., Semenov, V., and Muhammed, M. (2004). "The effect of biocompatible coating layers on magnetic properties of superparamagnetic iron oxide nanoparticles", *Hyperfine Interact.* 156, 257–263.

[74] Veiseh, M., Gabikian, P., Bahrami, S. B., Veiseh, O., Zhang, M., Hackman, R. C., Ravanpay, A. C., Stroud, M. R., Kusuma, Y., Hansen, S. J., Kwok, D., Munoz, N. M., Sze, R. W., Grady, W. M., Greenberg, N. M., Ellenbogen, R. G., and Olson, J. M. (2007). "Tumor paint: a chlorotoxin: Cy5.5 bioconjugate for intraoperative visualization of cancer foci", *Cancer Res.* 67, 6882–6888.

[75] Glover, P., and Mansfield, P. (2001). "Limits to magnetic resonance microscopy", *Rep. Prog. Phys.* 65, 1489–1511.

[76] Kooi, M. E., Cappendijk, V. C., Cleutjens, K., Kessels, A. G. H., Kitslaar, P., Borgers, M., Frederik, P. M., Daemen, M., and van Engelshoven, J. M. A. (2003). "Accumulation of ultrasmall superparamagnetic particles of iron oxide in human atherosclerotic plaques can be detected by in vivo magnetic resonance imaging", *Circulation* 107, 2453–2458.

[77] Trivedi, R. A., U-King-Im, J. M., Graves, M. J., Cross, J. J., Horsley, J., Goddard, M. J., Skepper, J. N., Quartey, G., Warburton, E., Joubert, I., Wang, L. Q., Kirkpatrick, P. J., Brown, J., and Gillard, J. H. (2004). "In vivo detection of macrophages in human carotid atheroma—temporal dependence of ultrasmall superparamagnetic particles of iron oxide enhanced MRI", *Stroke* 35, 1631–1635.

[78] Schulze, K., Koch, A., Schopf, B., Petri, A., Steitz, B., Chastellain, M., Hofmann, M., Hofmann, H., and von Rechenberg, B. (2004). "Intraarticular application of superparamagnetic nanoparticles and their uptake

by synovial membrane—an experimental study in sheep", *J. Magn. Magn. Mater.* 293, 419–432.

[79] Varallyay, P., Nesbit, G., Muldoon, L. L., Nixon, R. R., Delashaw, J., Cohen, J. I., Petrillo, A., Rink, D., and Neuwelt, E. A. (2002). "Comparison of two super paramagnetic viral-sized iron oxide particles ferumoxides and ferumoxtran-10 with a gadolinium chelate in imaging intracranial tumors", *Am. J. Neuroradiol.* 23, 510–519.

[80] Hong, R., Li, J., Wang, J., and Li, H. (2007). "Comparison of schemes for preparing magnetic Fe_3O_4 nanoparticles", Chinese Society of Particuology and Institute of Process Engineering, Chinese Academy of Sciences. Elsevier B.V.

[81] Bonnemann, H., Brand, R. A., Brijoux, W., Hofstadt, H.-W., Frerichs, M., Kempter, V., Maus-Freidrichs, W., Matoussevitch, N., Nagabhushana, K. S., Voights, V., and Caps, V. (2005). "Air stable Fe and Fe Co magnetic fluids—synthesis and characterization", *Appl. Organomet. Chem.* 19, 790–796.

[82] Chaubey, G. S., Barcena, C., Paudyal, N., Rong, C., Gao, J., Sun, S., and Liu, J. P. (2007). "Synthesis and stabilization of FeCo nanoparticles", *Am. Chem. Soc.* 129, 7214–7215.

[83] Fievet, F., Lagire, J. P., Blin, B., Beaudoin, B., and Figlarz, M. (1989). "Nucleation and growth of bimetallic CoNi and FeNi monodisperse particles prepared in polyols", *Solid State Ionics* 32/33, 198–205.

[84] Joseyphus, R. J., Mastumoto, T., Takahashi, H., Kodama, D., Tohij, K., and Joseyphus, B. (2007). "Designed synthesis of cobalt and its alloys by polyol process", *Solid State Chem.* 180, 3008–3018.

[85] Repetsky, S. P., Melnyk, I. M., Tatarenko, V. A., Len, E. G., and Vyshivanaya, I. G. (2009). *J. Alloys Compd.* 480, 13.

[86] Moulas, G., Lehnert, A., Rusponi, S., Zabloudil, J., Etz, C., Ouazi, S., Etzkorn, M., Bencok, P., Gambardella, P., Weinberger, P., and Brune, H. (2008). *Phys. Rev. B, PRB* 78, 214424.

[87] Elmen, G. W. (1929). "Magnetic material and appliance", US patent No. 1739752.

[88] Desvaux C., Amiens, C., Fejes, P., Renaud, P., Respaud, M., Lecante, P., Snoeck, E., and Chaudret, B. (2005). "Multimillimetre-large superlattices of air-stable iron–cobalt nanoparticles", *Nature Mater.* 4, 750–753.

[89] Jones, N. J., McNerny, K. L., Wise, A. T., Sorescu, M., McHenry, M. E., and Laughlin, D. E. (2010). "Observation of oxidation mechanism

and kinetics in faceted FeCo magnetic nanoparticles", *Appl. Phys.* 107, 09A3041–09A3043.

[90] Thirumal, E., Prabhu, D., Chattopadhyay, K., and Ravichandran, V. (2010). "Magnetic, electric, and dielectric properties of FeCo alloy nanoparticles dispersed in amorphous matrix", *Physica Status Solidi A.* 207, 2505–2510.

[91] Seo, W. S., Lee, J. H., Sun, X.; Suzuki, Y., Mann, D., Liu, Z., Terashima, M., Yang, P. C., Mcconnell, M. V., Nishimura, D. G., and Dai, H. (2006). "FeCo/graphitic-shell nanocrystals as advanced magnetic-resonance-imaging and near-infrared agents", *Nature Mater.* 5, 971–976.

[92] Zhang, L., and Manthiram, A. (1996). *J. Appl. Phys.* 80, 4534.

[93] Shin, S. J., Kim, Y. H., Kim, C. W., Cha, H. G., Kim, Y. J., and Kang, Y. S. (2007). *Curr. Appl. Phys.* 7, 404.

[94] Cho, U., Lee, Y., Kumar, S., Lee, C., and Koo, B. (2009). *Sci. China Series E: Technol. Sci.* 52, 19.

[95] Glavee, G. N., Klabunde, K. J., Sorensen, C. M., and Hadjapanayis, G. C. (1992). *Langmuir*, 8, 771.

[96] Glavee, G. N., Klabunde, K. J., Sorensen, C. M., and Hadjipanayis, G. C. (1993). *Inorg. Chem.* 32, 474.

[97] Glavee, G. N., Klabunde, K. J., Sorensen, C. M., and Hadjipanayis, G. C. (1995). *Inorg. Chem.* 34, 28.

[98] Glavee, G. N., Klabunde, K. J., Sorensen, C. M., Hadjipanayis, G. C., Tang, Z. X., and Yiping, L. (1993). *Nanostruct. Mater.* 3, 391.

[99] Klabunde, K. J., Zhang, D., Glavee, G. N., Sorensen, C. M., and Hadjipanayis, G. C. (1994). *Chem. Mater.* 6, 784.

[100] Glavee, G. N., Klabunde, K. J., Sorensen, C. M., and Hadjipanayis, G. C. (1993). *Langmuir* 9, 162.

[101] Ghazanfari, M., Johar, F., and Yazdani, A. (2014). "Synthesis and characterization of Fe_3O_4@Ag core-shell: structural, morphological, and magnetic properties", *J. Ultrafine Grained Nanostruct. Mater.* 47(2), 97–103.

[102] Mikhaylova, M., et al. (2004). "Superparamagnetism of magnetite nanoparticles: dependence on surface modification", *Langmuir* 20(6), 2472–2477.

[103] Suto, M., et al. (2009). "Heat dissipation mechanism of magnetite nanoparticles in magnetic fluid hyperthermia", *J. Magn. Magn. Mater.* 321(10), 14931496.

[104] Wan, J., et al. (2007). "Monodisperse water-soluble magnetite nanoparticles prepared by polyol process for high-performance magnetic resonance imaging", *Chem. Commun.* 47, 5004–5006.

[105] Jain, P. K., et al. (2009). "Surface plasmon resonance enhanced magneto-optics (SuPREMO): Faraday rotation enhancement in gold-coated iron oxide nanocrystals", *Nano Lett.* 9(4), 1644–1650.

[106] Patil, R. M., Shete, P. B., Thorat, N. D., Otari, S. V., Barick, K. C., Prasad, A., Ningthoujam, R. S., Tiwale, B. M., and Pawar, S. H. (2013). "Superparamagnetic iron oxide/chitosan core/shells for hyperthermia application: Improved colloidal stability and biocompatibility", *J. Magn. Magn. Mater.* Dec.

[107] Wang, C. Y., Zhou, Y., Mo, X., Jiang, W. Q., Chen, B., and Chen, Z. Y. (2000). *Mater. Res. Bull.* 35, 755–759.

[108] Todaka, Y., Nakamura, M., Hattori, S., Tsuchiya, K., and Umemoto, M. (2003). *Mater. Trans.* 44, 277–284.

[109] Sasaki, T. Terauchi Koshizaki, S. N., and Umehara, H. (1998). *Appl. Surf. Sci.* 127–129, 398–402.

[110] Makovec, D., and Košak, A. J. (2005). *J. Magn. Magn. Mater.* 289, 32–35.

[111] Chin, S. F., Pang, S. C., and Tan, C. H. (2011). *J. Mater. Environ. Sci.* 2, 299–302.

[112] Sairam, M., Naidu, B. V. K., Nataraj, S. K., Sreedhar, B., Aminabhavi, T. M. (2006). *J. Membr. Sci.* 283, 65–73.

[113] Cornell, R. M., and Weinheim, U. S. (1991). *Iron Oxides in the Laboratory: Preparation and Characterization.* Germany: VCH Publishers.

[114] Chen, S., Feng, J., Guo, X., Hong, J., and Ding, W. (2005). *Mater. Lett.* 59, 985–988.

[115] Mohamed Abbas, Md. Nazrul Islam, B. Parvatheeswara Rao, Tomoyuki gawa, Migaku Takahashi, and CheolGi Kim (2013). "One-pot synthesis of high magnetization air-stable FeCo nanoparticles by modified polyol method", *Mater. Lett.* 91, 326–329.

[116] Poudyal N. (2008). "Fabrication of Superparamagnetic and Ferromagnetic Nanoparticles", The University of Texas at Arlington, Dec.

[117] Berry, C. C., and Curtis, A. S. G. (2003). "Fictionalization of magnetic nanoparticles for application in biomedicine", *J. Physics. D: Appl. Phys.* 36, R198.

[118] Pankhurst, Q. A., Connolly, J., Jones, J. S. K., and Dobson J. (2003). "Applications of magnetic nanoparticles in biomedicine". *J. Phys. D: Appl. Phys.* 36, R167.

[119] Willard, M. A., Kulrihara, L. K., Carpenter, E. E., Calvin, S., and Harris, V. G. (2004). "Chemically prepared magnetic nanoparticles", *Int. Mater. Rev.* 49, 3.

[120] Chastellain, M., Petri, A., Hofmann, M., and Hofmann, H. (2002). "Synthesis and patterning of magnetic nanostructures", *Eur. Cells Mater.* 3, 1, 11.

[121] Van Tomme, S. R., Storm, G., Hennink, W. E. (2008). "In situ gelling hydrogels for pharmaceutical and biomedical applications", *Int. J. Pharm.*, 355, 1–18.

[122] Chen, S., Wang, L., Duce, S. L., Brown, S., Lee, S., Melzer, A., et al. (2010). "Engineered biocompatible nanoparticles for in vivo imaging applications", *J. Am. Chem. Soc.* 132, 15022–15029.

[123] Ghosh Chaudhuri, R., and Paria, S. (2011). Core/shell nanoparticles: classes, properties, synthesis mechanisms, characterization, and applications. *Chem. Rev.* 112, 2373–433.

[124] Oh, J. K., and Park, J. M. (2011). "Iron oxide-based super paramagnetic polymeric nanomaterials: Design, preparation, and biomedical application", *Prog. Polymer Sci.* 36, 168–189.

[125] Sun, C., Lee, J. S. H., and Zhang, M. (2008). "Magnetic nanoparticles in MR imaging and drug delivery", *Adv. Drug Delivery Rev.* 60, 1252–1265.

[126] Mornet, S., Vasseur, S., Grasset, F., Duguet, E. (2004). "Magnetic nanoparticle design for medical diagnosis and therapy", *J. Mater. Chem.* 14, 2161–2175.

[127] Gleich, B., and Weizenecker, J. (2005). *Nature*, 435, 1214.

[128] Saini, S., Stark, D. D., Hahn, P. F., Wittenberg, J., Brady, T. J., and Ferrucci, J. T. (1987). *Radiology*, 162, 211.

[129] Hengerer, A., and Grimm, J. (2006). *Biomed. Imaging Int. J.* 2, 238.

[130] Ho Kong, W., Jae Lee, W., Yun Cui, Z., Hyun Bae, K., Gwan Park, T., and Hoon Kim, J. et al. (2007). "Nanoparticulate carrier containing water-insoluble iodinated oil as a multifunctional contrast agent for computed tomography imaging", *Biomaterials* 28, 5555–5561.

[131] Goh, V., Bartram, C., and Halligan S. (2009). "Effect of intravenous contrast agent volume on colorectal cancer vascular parameters as measured by perfusion computed tomography", *Clin. Radiol.* 64, 368–372.

[132] Xu, C., Tung, G. A., Sun, S. (2008). "Size and concentration effect of gold nanoparticles on X-ray attenuation as measured on computed tomography", *Chem. Mater.* 20, 4167–4169.

[133] Aydogan, B, Li, J., Rajh, T., Chaudhary, A., Chmura, S. J., Pelizzari, C., et al. (2010). "AuNP-DG: deoxyglucose-labeled gold nanoparticles as X-ray computed tomography contrast agents for cancer imaging", *Mol. Imaging. Biol.* 12, 463–467.

[134] Kojima, C., Umeda, Y., Ogawa, M., Harada, A., Magata, Y., and Kono, K. (2010). "X-ray computed tomography contrast agents prepared by seeded growth of gold nanoparticles in PEGylated dendrimer", *Nanotechnology* 21, 245104.

[135] McCall, K. C., Barbee, D. L., Kissick, M. W., and Jeraj, R. (2010). "PET imaging for the quantification of biologically heterogeneous tumours: measuring the effect of relative position on image-based quantification of dose-painting targets", *Phys. Med. Biol.* 55, 2789.

[136] Ruuge, E. K., and Rusetski, A. N. (1993). *J. Magn. Magn. Mater.* 122, 335.

[137] Martina, M.-S., Nicolas, V., Wilhelm, C., Menager, C., Barratt, G., and Lesieur, S. (2007). *Biomaterials* 28, 4143.

[138] Allen, C., Dos Santos, N., Gallagher, R., Chiu, G. N. C., Shu, Y., Li, W. M., Johnstone, S. A., Janoff, A. S., Mayer, L. D., Webb, M. S., and Bally, M. B. (2002). *Biosci. Rep.* 22, 225.

[139] Hattrup, C. L., Gendler, S. J. (2006). *Breast Cancer Res.* 8, R37.

[140] Moase, E. H., Qi, W., Ishida, T., Gabos, Z., Longenecker, B. M., Zimmermann, G. L., Ding, L., Krantz, M., and Allen, T. M. (2001). *Biochim. Biophys. Acta.* 1510, 43.

[141] Park, J. W. (2002). *Breast Cancer Res.* 4, 95.

[142] Mailander, V., and Landfester, K. (2009). "Interaction of nanoparticles with cells", *Biomacromolecules* 10, 2379–2400.

[143] Kasthuri, J., Poornima, S., Dinseh, J. P. (2011). "Evaluation of the interaction of gold nanoparticles with CT- DNA using spectroscopic and electrophoretic techniques", *J. Biosci. Res.* 2, 1–4.

[144] Sun, L., Zhang, Z., Wang, S., Zhang, J., Li, H., Ren, L., et al. (2009). "Effect of pH on the interaction of gold nanoparticles with DNA and application in the detection of human p53 gene mutation", *Nanoscale Res. Lett.* 4, 216–220.

[145] Yang, J., Lee, J. Y., and Too, H.-P. (2006). "Chow G-M. Inhibition of DNA hybridization by small metal nanoparticles", *Biophys. Chem.* 120, 87–95.

[146] Robinson, I., Tung, L. D., Maenosono, S., Wlti, C., and Thanh, N. T. (2010). "Synthesis of core–shell gold coated magnetic nanoparticles and their interaction with thiolated DNA", *Nanoscale* 2, 2624–2630.

[147] Li, X., Zhang, J., and Gu, H. (2011). "Adsorption and desorption behaviors of DNA with magnetic mesoporous silica nanoparticles", *Langmuir* 27, 6099–6106.

[148] Peeraphatdit, C. (2010). *Magnetic Nanoparticles for Applications in Oscillating Magnetic Field.* Iowa State University.

[149] Hergt, R., Dutz, S., Müller, R., and Zeisberger, M. (2006). *J. Phys. Condens. Mat.* 18, S2919.

[150] Pankhurst, Q. A. (2003). *J. Phys. D Appl. Phys.* 36, 167.

[151] Higai, K., Masuda, D., Matsuzawa, Y., Satoh, T., and Matsumoto, K. (1999). *Biol. Pharmaceut. Bullet.* 22, 333–338.

[152] Perry, J. D., James, A. L., Morris, K. A., Oliver, M., Chilvers, K. F., Reed, R. H., and Gould, F. K. (2006). *J. Appl. Microbiol.* 101, 977–985.

[153] Horst, M. A. v. d., Stalcup, T. P., Kaledhonkar, S., Kumauchi, M., Hara, M., Xie, A., Hellingwerf, K. J., and Hoff, W. D. (2009). *J. Am. Chem. Soc.* 131, 17443–17451.

[154] Kumar, C. S. S. R., and Mohammad, F. (2011). "Magnetic nanomaterials for hyperthermia-based therapy and controlled drug delivery", *Adv. Drug Delivery Rev.* 63, 789–808.

7

Comparative Study on Cytotoxic and Bactericidal Effect of Nanoscale Zero Valent Iron Synthesized through Chemical and Biological Methods

Sharath R.[1], Harish B. G.[2], Chandraprabha M. N.[1], Samrat K.[1], Nagaraju Kottam[3], Hari Krishna R.[3], Rakesh G. Kashyap[4] and Muktha H[1]

[1]Department of Biotechnology, M. S. Ramaiah Institute of Technology, Bangalore, 560054, India
[2]Department of MCA, Visvesvaraya Institute of Advanced Technology, Muddenahalli, Chikkaballapur, 562101, India
[3]Department of Chemistry M. S. Ramaiah Institute of Technology, Bangalore, 560054, India
[4]Department of Nanotechnology, Visvesvaraya Institute of Advanced Technology, Muddenahalli, Chikkaballapur, 562101, India

Abstract

The improper application of conventional antimicrobials has made them unavailing which have led to the antibiotic resistance crisis over the world. Nanotechnology is a key to address this pressing problem as it proffers improvements and upswing in therapeutics. Especially, the nanoparticles with novel properties have a wide range of potential applications in biomedical fields, in textile fabrics, and in water treatment as disinfectants or antibiofilm agents and have great potential for early detection, accurate diagnosis, and tailored treatment of cancer. Nanoscale zero valent iron (nZVI), a promising material which plays a very important role in environmental remediation of heavy metals such as Pb, As, Cr, and Cd can be removed easily from water bodies using magnetic property nZVI. nZVI also acts as very good bactericidal agent. In the present study, synthesis of nZVI by chemical and biological method using homeopathic extract of Phytolacca Berry (*Phytolacca Decandra*) was carried out to compare its effect on microbes and cancer cell lines. The morphology of the nanoparticles was determined through scanning electron microscopy (SEM) images and the crystalline structure was identified using X-ray diffraction (XRD) studies. The biologically synthesized nZVI particles were found to be more stable. The screening of antibacterial activity of synthesized nZVI particles against Gram-negative, Gram-positive bacteria was performed. Both chemically and biologically synthesized nZVI showed significant activity compare to that of standard drug ciprofloxacin. The nZVI synthesized by biological method was found to be more effective in inhibiting Gram-negative and Gram-positive bacteria. Cytotoxic activity of nZVI was screened against MCF-7 (human breast adenocarcinoma) cell lines, and it was found that biologically synthesized nanoparticles had more cytotoxic effect on MCF-7 cell lines.

7.1 Introduction

The most common causes of human mortality are infectious diseases and internal malfunctions of the body. Infectious diseases kill more people worldwide than any other single cause. Infectious diseases such as malaria, tuberculosis, and AIDS are caused by germs which include bacteria, viruses, fungi, and protozoa. These microbes pose a serious threat to human health and biodiversity. The improper use of antimicrobial agents has led to the development of new resistance mechanisms followed by the global spread of resistant organisms, which threatens the effective treatment of common infectious diseases. It is important to develop new antimicrobials from various sources as alternative agents.

Internal malfunctions of the body include cancer, failure of the vital organs, and cardiovascular diseases. Cancer is observed as the most dangerous class of disease categorized by uncontrolled cell growth [1, 2]. There is a marginal increase in cancer cases in the last few years, and most of the time, it ends up with taking life [3–5]. In many types of cancer, we are yet to find a satisfactory medicine or carrier of medicine as in case of drug delivery to be used as a satisfactory chemotherapeutic agent.

The mode of action of antimicrobial agents includes inhibition of cell wall synthesis, protein synthesis, DNA replication, and alteration of intermediary metabolism. If the target for antimicrobial action is located within the cell wall of bacteria, then antimicrobials agents must be able to penetrate to the site of location. But penetration through the cell wall is usually achieved by passive or active transport mechanism. However, microbes have developed antimicrobial resistance due to improper usage and remarkable capability to survive, adopt, and evolve. Because of which effective treatment for infectious diseases has become difficult.

The antimicrobial resistance may be caused due to inactivation of the drug, reduced binding capacity, antimicrobial effect of the drug, and decreased permeability due to modification of the binding sites, metabolic pathways, and reduced intracellular accumulation of antimicrobial agents, respectively [6, 7].

Advances in nanotechnology especially synthesis of metal and metal–oxide nanoparticles with potential applications in industry, food packaging, textiles, medicine, and therapeutics have raised exciting opportunities for specific drug delivery for cancer and have novel antimicrobial agents/nanocomposites to overcome antimicrobial resistance [8].

Nanoparticles, typically 0.2–100 nm in size, having a high surface-to-volume ratio are emerging as a class of therapeutics for cancer treatment. Nanoparticles can be composed of several functional molecules simultaneously, such as small molecule drugs, peptides, proteins, and nucleic acids. By using both passive and active targeting strategies, nanoparticles can increase the intracellular concentration of drugs in cancer cells while minimizing toxicity in normal cells, thereby enhancing anticancer effects and reducing systemic toxicity simultaneously, when compared with the therapeutic entities they contain. Furthermore, nanoparticles offer the potential to overcome drug resistance, since nanoparticles can bypass the P-glycoprotein efflux pump, one of the main drug resistance mechanisms, leading to greater intracellular accumulation [9, 10].

Cancer has become one of the leading causes of human mortality in recent years. Rapid and unbridled cell division leading to tumour growth (abnormal growth of mass of cells) is a symptomatic feature of cancer. Currently, chemotherapy, radiation therapy, and surgery are widely accepted to be effective cancer treatment methods but these methods have not been a downright solution for its cure. The reason being that is these treatment methods greatly damage the normal cells making patients weak and exhausted; over to this, they present a fair risk of metastasis or cancer reoccurrence. However, they do much damage than any good.

Despite advancements in anticancer drugs (paclitaxel and cisplatin), specificity in effectively targeting cancer cells has been barely achieved.

That is, anticancer drugs may have been advanced in tumour regression or blocking the cell proliferation, but non-specificity of these drugs will still pose a deleterious effect on normal cells. In addition to this, sometimes, chemotherapeutic drugs become useless because of cancerous cells becoming less sensitive or exhibit resistance overtime in a condition what is called multidrug resistance (MDR); thus, this further adds complications in treatment procedures.

In pursuit of achieving specificity and minimizing the side effects of anticancer drugs, a lot of research are being carried out. At present, nanotechnology is being looked upon as a potential solution in treatment and diagnosis of cancer. The nanoscale dimension enables to strategically design and develop the drug molecules suitably to target the diseased site with specificity. Nanotechnology particularly nanoparticles cannot only be useful in the treatment of cancer but they can also be designed to be way more sensitive to detect cancer cells even at small percentages, thus enabling early detection of cancer.

As nanoparticles offer various advantages compared to the bulk materials, many different particles of metallic or non-metallic and polymeric origins in nanoscale dimension range are being immensely explored for their potential use in cancer treatment. Drug permeability and retention are important factors for effective treatment in cancer. However, with particles in the nanoscale range, enhanced permeability and retention effect could be achieved while passively targeting nanoparticles via leaky neovasculatures and absence of lymphatic drainage in tumour will lead to accumulation and retention of nanoparticles.

Tumour-specific targeting could be achieved by suitably functionalizing the surface of the nanoparticles to have ligand molecules that bind specifically to tumour biomarkers. Also, coating the surface of the nanoparticles with polyethylene glycol (PEG) could enhance retention time or make nanoparticles circulate for longer time. Drug loading capacity and drug release time can be invariably controlled using nanoparticles. Preferably, nanoparticles can be made non-toxic or biocompatible through surface modifications. Thus, nanoparticles offer wide range of flexibility to tailor its properties to get the desired effect.

Among the various metal-based nanoparticles, iron oxide nanoparticles are being investigated extensively for their super magnetic properties which can be utilized to destroy tumour cells. The magnetic property of iron oxide nanoparticles is used in 'hyperthermia' treatment of cancerous tumour.

Generally, hyperthermia treatment involves application of slightly high temperature across an exposed tumour region which is enough to kill or destroy cancerous tumour cells. There are various approaches followed in hyperthermia treatment while making use of microwaves or ionizing radiation or laser as an external energy source to generate heat. These techniques may fairly do the work, but the radiation or laser intensity used may also damage the neighbouring cells.

Iron oxide nanoparticle with super magnetic property allows to generate in situ heat using external magnetic field of non-hazardous frequency under an oscillating magnetic field. That is, on application of alternating magnetic field, iron oxide nanoparticles localized in tumour region absorb magnetic energy and start emanating heat which allows increase in temperature in the region. Consequently, the iron oxide nanoparticles can be well controlled by applying or removing external magnetic field. The magnetic field frequency used in hyperthermia falls in the range of kHz to 1 MHz.

This range of magnetic field frequency is safe and is quiet enough to achieve penetration depths viable to reach tissues or organs in the human body. As cancerous cells are highly sensitive to change in temperature, any increase in temperature above 42°C would naturally kill tumour cells, thus enabling selective obliteration of diseased cells only [11]. Nevertheless, there are few important factors to be taken into consideration for this technique to produce desired results. Nanoparticles should have the following properties for their use in hyperthermia treatment.

- Zero remanent magnetization.
- High saturation magnetization.

It is necessary for particles to have zero residual magnetization, else particles agglomerate under residual magnetism or remanent magnetization leading to embolus formation which will occlude blood circulation, thereby creating serious health implications [12].

Higher saturation magnetization is required to produce more heat without the loss of magnetic property of the material. As size increases, saturation magnetization increases, which works efficiently for hyperthermia treatments, but undesirable for biomedical application because as particle size increases, cell permeation decreases. With less nanoparticle penetrating the tumour site, on application of external magnetic field, only the nanoparticles located exterior to the tumour will get activated; this will have less effect on the tumour [13]. However, it is important to have smaller sized particles for them to be able to cross the membrane barriers and get distributed uniformly inside the tumour to induce desired heating effect and kill tumour cells.

Magnetic properties of a particle are highly size-dependent, so it is important to ascertain both size distribution and corresponding change in magnetic property of nanoparticles for them to be applicable in biomedical field. Biocompatibility of nanoparticle is another such prerequisite for their use in hyperthermia treatment. Since nanoparticles are foreign substances, they naturally trigger immune responses and get subjected for opsonization. Also, iron oxide nanoparticles may react with hydrogen peroxide produced in the body and lead to fenton reaction, resulting in free radical production which is toxic to cells.

Therefore, the iron oxide nanoparticles need to be biocompatible for them to be any applicable in biomedical use. Nanoparticles can be made to be biocompatible by modifying or functionalizing their surface by using biomolecules such as antibodies, protein molecules, and viruses or coating their surface with polymeric coating agents. Surface coating the nanoparticles will make them less toxic and ensure longer retention time. Functionalizing the surface with target-specific ligand molecule will enhance specificity and damage to the neighbouring normal cells can be greatly reduced.

Iron oxide nanoparticles can also be used as anticancer drug carriers. Anticancer drugs such as doxorubicin (DOX) and paclitaxel rapamycin can be loaded into encapsulation material coated on to the nanoparticle surface. Preferably, drug-loaded magnetic nanoparticles are administered through intravenous (IV) route. Once the magnetic nanoparticles are introduced into the circulatory system, external magnetic field is used to guide these nanoparticles to the target site. Another strategy of delivering chemotherapeutic agents is integration of magnetic nanoparticles into thermoresponsive polymer jackets.

Magnetic nanoparticles within the thermoresponsive polymers can be loaded with anticancer therapeutics and guided to the target site through the influence of external magnetic field. Once these magnetic–polymer nanoparticles reach the target site, alternating magnetic field is applied, and magnetic nanoparticles start to emanate heat which actuates thermoresponsive polymers to expand and release anticancer drugs into the target site. Hence, using nanoparticles, there can be various ways in designing smart drug delivery systems which can efficiently and more effectively deliver drugs to the targeted sites [14].

Apart from advanced treatment strategies, cancer diagnosis is also of prime importance which gives a window for early therapeutic interventions which can considerably avoid advancements in tumour growth. Iron oxide nanoparticles are studied extensively for their use as contrast agent for MRI [15]. Iron oxide nanoparticles with functionalized surface have been formulated to enhance contrast in imaging. Functionalized nanoparticles with contrast enhancing ability can selectively aggregate at tumour site and help in assessing extent of tumour growth.

Passive targeting and active targeting are different approaches in delivering the contrast enhancing nanoparticles at the tumour site. The leaky vasculature and absence of lymphatic drainage of the tumour enables permeation and retention of nanoparticles at the site of tumour which is called passive targeting. But passive targeting is only effective when there are leaky vasculatures which are not always found to be distributed uniformly across the tumour site. However, with active targeting, contrast nanoparticles can be delivered to the specific tumour site. Attachment of target-specific ligand molecules on to the nanoparticles will enable nanoparticles to aggregate in the tumour site.

Apart from passive and active targeting approaches, magnetically guided targeting is being accepted to be simple and reliable. Unlike passive and active targeting, magnetically guided targeting requires external magnetic field to guide the contrasting enchasing iron oxide nanoparticles to the targeted site [16]. At present, all three approaches are being used in formulating iron oxide nanoparticles as contrasting agents. Currently, the use of ferumoxytol iron oxide nanoparticles as contrast agent is under clinical trials [17].

Numerous methods have been used to synthesize nanoparticles such as physical, chemical, and biological. But physical and chemical methods use high radiation and concentrated reducing and stabilizing agents that are harmful to human health and environment. Therefore, as biological synthesis is a single-step bioreduction method and uses less energy, it is used to synthesize eco-friendly nanoparticles [18].

Different parts of plants such as leaves, roots, flowers, fruits, stem, and peel have been used for synthesis of metallic nanoparticles in various size and shapes [19, 20]. Hence, in the present study, fruit extract of Phytolacca Berry (*Phytolacca Decandra*) was used to synthesize nanoscale zero valent iron (nZVI) particles. These biologically synthesized metallic nanoparticles have potential applications in the field of treatment, development surgical nanodevices, and diagnosis. So, eco-friendly synthesis would be considered as building blocks for future generations to control chronic diseases [21, 22].

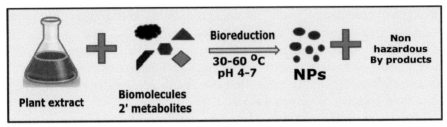

Figure 7.1   Mechanism of biosynthesis of nanoparticles.

Nanoscale zero valent iron (nZVI), a promising material, plays a very important role in environmental remediation. Heavy metals such as Pb, As, Cr, and Cd can be removed easily from water bodies using nZVI. Recent studies have shown that it can also act as effective bactericidal agent.

The bioreduction mechanism of ferrous sulphate ($FeSO_4 \cdot 7H_2O$) with plant extract leads to the formation of nZVI by following reaction [23].

$$FeSO_4 \cdot 7H_2O + Plantmetabolites \longrightarrow Fe^0 \text{ NPs} + by\text{-}products$$

The proposed mechanism of biosynthesis of nanoparticles is described as follows (Figure 7.1).

Several secondary metabolites promote the formation of metallic nanoparticles form the ionic precursors. The bioreduction reaction involves biomolecules such as sugars, proteins, organic compounds, pigments, and resins. Chemical compounds such as saponins, terpenoids, polyphenols, and alkaloids are produced by plants participating in defence mechanisms against controlling various acute diseases. These secondary metabolites play important role in biological synthesis of metallic nanoparticles [20].

As a whole, synthesis of metallic nanoparticles using plant extract involves three main steps: 1) activation phase is the first step in which reduction of metal ions and nucleation of the reduced metal atoms occurs; 2) in the second step (i.e. growth phase), heterogeneous nucleation and growth of adjacent nanoparticles into larger size particles take place which further promote metal ion reduction accompanied by increase in the thermodynamic stability of nanoparticles; and 3) the last step is the termination phase which defines the nanoparticles shape [24]. The schematic representation of formation of metallic nanoparticles using plant extract is shown in Figure 7.2. Aggregation nanoparticles occur if the extent of growth phase increases, leading to form various irregular shapes such as nanotubes and nanoprisms. Termination step is strongly influenced by plant extract ability to stabilize metal nanoparticles to acquire energetically stable conformation [25, 26].

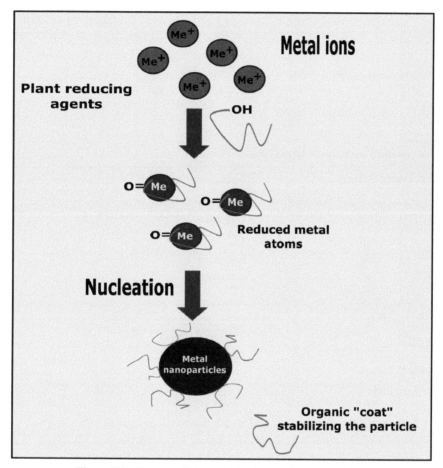

Figure 7.2   Process of biosynthesis of metallic nanoparticles.

Phytolacca Berry (*Phytolacca Decandra*) is commonly called as poke-weed, poke—root is an herbaceous plant belongs to Phytolaccaceae family and is widely used as homeopathic medicine to treat several diseases. The berries contain phytolaccic acid, which is gummy, non-deliquescent, soluble in water and alcohol with acid reaction, and hardly soluble in ether. The berries are commonly used as a remedy for cancer and rheumatism. It is found throughout North America, the Azores, North Africa, and China.

In India, it is extensively used for treating diarrhoea and chronic dysentery nausea and for muscular weakness, double vision, weight loss, etc. The presence of terpenoids and saponins such as phytalaccosides A, B, C, and

D shows effect on chronic rheumatism, obesity and granular conjunctivitis, and skin diseases, and it cures leucorrhoea, anti-inflammatory activity, and also antiviral effect. Phytolacca is also used to treat the people suffering from thyroid disorder and metabolic problems [27].

This study reveals the cytotoxic and bactericidal effect of nanoscale zero valent iron (nZVI) synthesized through chemical and biological methods.

7.2 Materials and Methods

7.2.1 Materials

Ferrous sulphate heptahydrate [$FeSO_4 \cdot 7H_2O$], ethylene diamine tetra acetic acid disodium salt [$C_{10}H_{16}N_2O_8$] (EDTA), sodium borohydride [$NaBH_4$], and ethanol were obtained from S D Fine Chemical Limited. In addition to above chemicals, ethanolic extract of Phytolacca Berry was procured from SBL PVT. LTD.

7.2.2 Methods

7.2.2.1 Synthesis of nanoscale Zero Valent Iron (nZVI) by chemical and biological methods

Synthesis by Chemical Method

In this method, the particles were synthesized in aqueous phase by chemical reduction of ferrous sulphate solution (0.1 M) using sodium borohydride (0.75 M) in the presence of EDTA as a stabilizing agent. In a typical set, 0.1 M of $FeSO_4$ was dissolved in 75 ml of nitrogen purged Millipore water and the solution was heated up to 60°C with constant stirring. To the above solution, 50 ml of 0.1 M EDTA solution was added and allowed to stir for 5 min. The aqueous solution of $NaBH_4$ (0.75 M) was then added drop wise to it under constant stirring until colour changes to dark black, indicating the formation of nanoscale zero valent iron (nZVI) [28]. The solution containing the nanoparticles was transferred to centrifuge tubes immediately and was centrifuged at 10,000 rpm for 10 min. The pellet containing the nanoparticles was ethanol washed for two times. Then, the particles were dried in hot air oven at 85°C for 10 min.

Synthesis by Biological Method

A total of 50 ml of 0.1 M of $FeSO_4$ solution was heated up to 60°C with constant stirring. To this solution, 50 ml of Phytolacca Berry extract was then added under constant stirring. The resulting solution was stirred constantly for 1 h until colour changes to dark black, indicating the formation of zero valent iron nanoparticles. The solution containing nanoparticles was immediately centrifuged and washed with ethanol. The particles were then dried in hot air oven at 85°C for 10 min [29, 30].

7.2.3 Characterization Studies

Characterization studies such as X-ray powder diffractometer (PXRD, Rigaku Ultima IV, Japan), Fourier transform infrared spectrometer (FT-IR, Bruker-Alpha FTIR, Germany), and field emission scanning electron microscopy (FESEM, Carl Zeiss-ULTRA 55, Germany) were used to analyse the structure and the composition of functional groups present in the sample and surface morphology, respectively, of synthesized nZVI using chemical and biological methods.

7.2.4 Screening of Bactericidal and Cytotoxic Activity

Bactericidal Activity

Well-diffusion assay was carried out to establish the bactericidal activity of nZVI synthesized by chemical and biological method against six clinical strains [Gram-negative strains – *Escherichia coli, Kelbsiella pneumonia, Pseudomonas aeruginosa, and Salmonella paratyphi;* and Gram-positive strains – *Bacillus subtilis and Staphylococcus aureus*]. These strains were maintained on agar slants at 4°C and subcultured before use. Then, the strains were swabbed on the agar plates and four wells were punched. One milligram of nanoparticles were added to 1 ml of DMSO and sonicated for 10 mins so that the nanoparticles get dispersed in DMSO. The dispersed solutions were loaded into the wells. The plates were incubated at 37°C for 24 h. Ciprofloxacin (50 µg) and DMSO were used as standard and control, respectively. The zone of inhibition around each well after the incubation period was recorded [31].

Cytotoxic Activity

The cytotoxic activity of both chemically and biologically synthesized nZVI was screened on MCF-7 (human breast adenocarcinoma) cell lines by MTT assay. MTT assay is a colorimetric assay used for the determination of cell proliferation and cytotoxicity, based on reduction of the yellow-coloured water-soluble tetrazolium dye MTT to formazan crystals. Mitochondrial lactate dehydrogenase produced by live cells reduces MTT to insoluble for-mazan crystals, which upon dissolution into an appropriate solvent exhibits purple colour, the intensity of which is proportional to the number of viable cells and can be measured spectrophotometrically at 570 nm.

7.3 Results and Discussion

7.3.1 Synthesis of Nanoscale Zero Valent Iron (nZVI) by Chemical and Biological Methods

Nanoscale zero valent iron (nZVI) synthesized by chemical reduction and biological methods was primarily confirmed by colour change of the reaction mixture from pale yellow to black and light green to black clearly indicating the formation of nZVI, respectively [29, 32].

7.3.2 Characterization of nZVI Particles

The crystalline nature of nZVI synthesized through biological and chemical routes was studied by X-ray diffraction (XRD) analysis. The XRD pattern (Figure 7.3) of nZVI synthesized through biological and chemical routes showed diffraction peak at 2 Θ value 44.25° and 45.62°, respectively, cor-respond to the (111) phase of body-centred cubic structure (BCC) of iron (JCPDS card file no. 00–006–0696) [28, 29, 33].

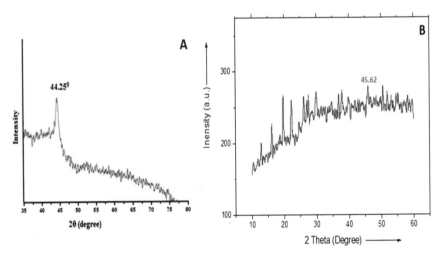

Figure 7.3   XRD pattern of nZVI synthesized by (A) chemical and (B) biological methods.

FTIR analysis provides information about vibrational state of absorbed molecule and hence the nature of surface complexes. The FTIR spectra of nZVI particles synthesized form chemical route indicate peaks at 3420.35, 2924.04, 1626.21, 1384.30, 1190.23, 1110.84, and 619.87 cm^{-1}. The absorption band at 3420.35 and 2924.04 cm^{-1} was due to weak stretching vibration of surface hydroxyl group and bulk hydroxyl stretch. The peaks at 1626.21, 1384.30, 1190.23, 1110.84, and 619.87 cm^{-1} correspond to absorption of water, asymmetric stretching vibrations of –COO– functional group, and metal nanoparticles with oxygen (i.e. nZVI), respectively (Figure 7.4A).

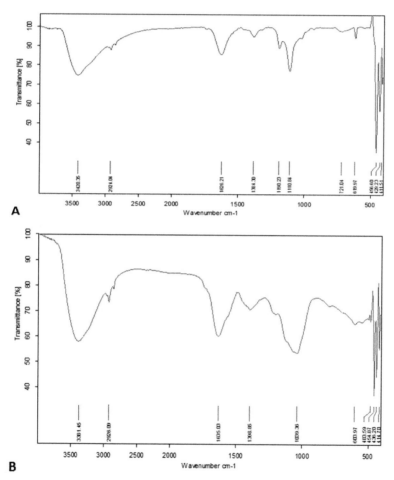

Figure 7.4 FT-IR Spectrum of nZVI synthesized by (A) chemical and (B) biological methods.

Similarly, the characteristic absorption bands at 3381.45, 2926.91, 1635.03, 1398.85, 1039.36, and 603.97 cm^{-1} of nZVI particles synthesized form biological route indicate stretching vibrations for O-H, C=C aromatic ring stretching vibration, C-N stretching vibration for aromatic amines, C–N stretching vibration of aliphatic amines, and metal nanoparticles with oxygen stretches, respectively. The above bands are due to oxidized polyphenols, amines, carboxyl, and carbonyl groups which help increase the stability of nZVI and protect from oxidizing (Figure 7.4B) [33–36].

The surface morphology of synthesized nZVI by chemical and biological routes was studied by SEM. The scanning electron micrograph of nZVI revealed spherical morphology with particle size in the range of 80–90 nm and 60–80 nm, respectively (Figure 7.5A and B).

Figure 7.5   Scanning electron micrograph (A) magnification of 50.00 K X of nZVI particles synthesized by chemical method and (B) magnification of 50.00 K X of nZVI synthesized by biological method.

7.3.3 Screening of Bactericidal and Cytotoxic Activity

Bactericidal Activity

Bactericidal potential of nZVI synthesized by chemical and biological routes was assessed in terms of zone of inhibition of bacterial growth against six clinical isolates [Gram-negative strains – *Escherichia coli, Kelbsiella pneumonia, Pseudomonas aeruginosa, and Salmonella paratyphi;* and Gram-positive strains – *Bacillus subtilis and Staphylococcus aureus*]. The results of the antibacterial activity are shown in Table 7.1 and Figure 7.6. NZVI synthesized by biological route showed significant bactericidal activity compared chemical route against Gram-negative strains. Furthermore, nZVI synthesized through both methods showed moderately significant bactericidal activity against all bacterial strains as compared to the standard drug. Thus, bactericidal effect of nZVI was due to oxidative stress generated by reactive oxygen species (ROS) including superoxide radicals (O^{2-}), hydroxyl radicals (–OH), hydrogen peroxide (H_2O_2), and singlet oxygen ($^1O_2$) can cause damage to proteins and DNA in bacteria or damage cell membranes, leading to cellular injury and death [33, 37, 38].

Cytotoxic Activity

The cytotoxic activity of both chemically and biologically synthesized nZVI was screened on MCF-7 (human breast adenocarcinoma) cell lines by MTT assay. The percentage viability graph of cancer cells at different concentrations of biologically and chemically synthesized nZVI is shown in Figure 7.7A and 7.7B. The initial concentration of 100 ug/ml of the chemically synthesized nanoparticles showed 40% viability and % viability decreased with increase in the concentration of the nanoparticles. On the other hand, biologically synthesized nanoparticles at the concentration 100 ug/ml showed 33% viability. This indicates that the biologically

Table 7.1 Zone of inhibition (mm) of nZVI synthesized by chemical and biological methods

Clinical Isolates	Drug (mm)	nZVI (Chemical Method) (mm)	nZVI (Biological Method) (mm)
Escherichia coli	40.00 ± 1.39	33.00 ± 0.77	38.00 ± 1.02
Bacillus subtilis	48.00 ± 0.77	40.00 ± 0.38	43.00 ± 0.38
Salmonella paratyphi	43.33 ± 0.80	37.67 ± 2.56	39.33 ± 1.82
Klebsiella pneumonia	38.33 ± 0.62	28.33 ± 0.91	32.67 ± 0.29
Staphylococcus aureus	43.33 ± 1.53	31.00 ± 0.67	33.67 ± 0.88
Pseudomonas aeruginosa	39.00 ± 0.22	34.67 ± 0.44	36.00 ± 0.22

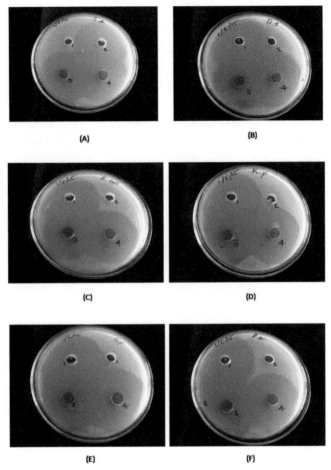

Figure 7.6 Screening of antibacterial activity of synthesized nZVI against (A) Staphylococcus *aureus,* (B) *Bacillus subtilis,* (C) *Escherichia coli,* (D) *Klebsiella pneumonia,* (E) *Salmonella paratyphi, and* (F) Pseudomonas aeruginosa – (1) ciprofloxacin, (2) DMSO, (3) biological method, and (4) chemical method.

synthesized nanoparticles were effective at minimum concentration compared to the chemically synthesized nanoparticles. The increased cytotoxicity of the biologically synthesized nanoparticles may be due to the presence of phytoconstants such as phytolaccic acid, punicotannic acid, gallic acid, and alkaloids such as pelletierine, methyl-pelletierine, and pseudo-pelletierine and iso-pelletierine in the phytolacca berry extract [39].

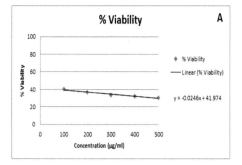 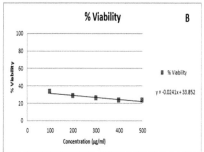

Figure 7.7 Graph of % viability of cancer cells treated with (A) chemically synthesized nZVI and (B) biologically synthesized nZVI.

7.4 Conclusion

The stability of nanoscale zero valent iron particles was found to be increased by several folds in biological synthesis as compared to the chemical synthesis and was successfully synthesized. Improved antibacterial activity was seen for nZVI synthesized using biological method against Gramnegative (*Escherichia coli, Kelbsiella pneumonia, Pseudomonas aeruginosa, Salmonella paratyphi*), Gram-positive (*Bacillus subtilis, Staphylococcus aureus*) bacterial strains. The MTT assay on MCF-7 (human breast adenocarcinoma) cell lines treated with the nanoparticles showed effective anticancer activity. The biologically synthesized nanoparticles showed better cytotoxicity towards MCF-7 cell lines due to the presence of additional photocomposes of Phytolacca Berry extract which suggests the advantages of biologically synthesized nanoparticles over chemical synthesis. Above its beneficial effects, biological synthesis method is cost-effective, environmental friendly, and free of hazardous chemicals.

References

[1] Chow, A. Y. (2010). Cell cycle control by oncogenes and tumor suppressors: driving the transformation of normal cells into cancerous cells. *Nat. Educ.* 3, 7.

[2] Suriamoorthy, P., Zhang, X., Hao, G., Joly, A. G., Singh, S., Hossu, M., et al. (2010). Folic acid-CdTe quantum dot conjugates and their applications for cancer cell targeting. *Cancer Nanotechnol.* 10, 65.

[3] Dite, G. S., Whittemore, A. S., Knight, J. A., John, E. M., Milne, R. L., AndrulisI, L., et al. (2010). Increased cancer risks for relatives of very

early-onset breast cancer cases with and without BRCA1 and BRCA2 mutations. *Br. J. Canc.* 103, 1103–1108.

[4] Parveen, S., and Sahoo, S. K. (2010). Evaluation of cytotoxicity and mechanism of apoptosis of doxorubicin using folate-decorated chitosan nanoparticles for to retinoblastoma. *Cancer Nanotechnol.* 1, 47–62.

[5] Smith, R. A., Cokkinides, V., Brooks, D., Saslow, D., and Brawley, O. W. (2011). A review of current American Cancer Society guidelines and issues in cancer screening. *CA Cancer J. Clin.* 60, 99–119.

[6] Awad, H. M., Kamal, Y. E. S., Aziz, R., Sarmidi, M. R., and El-Enshasy, H. A. (2012). Antibiotics as microbial secondary metabolites: Production and application. *J. Teknol.* 59, 101–111.

[7] Schmieder, R., and Edwards, R. (2012). Insights into antibiotic resistance through metagenomic approaches. *Future Microbiol.* 7, 73–89.

[8] Akhtar, M., Swamy, M. K., Umar, A., Sahli, A., and Abdullah, A. (2015). Biosynthesis and characterization of silver nanoparticles from methanol leaf extract of *Cassia didymobotyra* and assessment of their antioxidant and antibacterial activities. *J. Nanosci. Nanotechnol.* 15, 9818–9823.

[9] Ravishankar Rai, V., Jamuna Bai, A. (2011). "Nanoparticles and their potential application as antimicrobials," in *Science against Microbial Pathogen: Communicating Current Research and Technological Advances*, ed. A. Méndez-Vilas, 2, 197–209.

[10] Wang, X., Wang, Y., Chen, Z. G., and Shin, D. M. (2009). Advances of cancer therapy by nanotechnology. *Cancer Res. Treat.* 41, 1–11.

[11] Andra W., and Nowak, H. (1998). *Magnetism in Medicine: A Handbook.* First Edn. Germany: Wiley-VCH.

[12] Hofmann-Amtenbrink, M., von Rechenberg, B., and Hofmann H. (2009). "Superparamagnetic nanoparticles for biomedical applications," In, editor. *Nanostructured Materials for Biomedical Applications*, ed. M. C. Tan, 119.

[13] Rivas, J., Bañobre-López, M., Piñeiro-Redondo, Y., Rivas, B., and López-Quintela, M. A. (2012). *J. Magn. Magn. Mater.* 324:3499.

[14] Dionigi, C., Piñeiro, Y., Riminucci, A., Bañobre, M., Rivas, J., and Dediu V. (2013). Regulating the thermal response of PNIPAM hydrogels by controlling the adsorption of magnetite nanoparticles. *Appl. Phys. A.*

[15] Hao, R., Xing, R., Xu, Z., Hou, Y., Gao, S., and Sun, S. (2010). Synthesis, functionalization, and biomedical applications of multifunctional magnetic nanoparticles. *Adv. Mater.* 22, 2729–2742. doi: 10.1002/adma.201000260

[16] Subbiahdossa, G., Sharifia, S., Grijpmaa, D. W., Laurentc, S., van der Meia, H. C., Mahmoudid, M., et al. (2012). Magnetic targeting of surface-modified superparamagnetic iron oxide nanoparticles yields antibacterial efficacy against biofilms of gentamicin-resistant staphylococci. *Acta Biomater.* 8, 2047–2055.

[17] M. D. Anderson Cancer Center. (2000). *Ferumoxytol – Iron Oxide Nanoparticle Magnetic Resonance Dynamic Contrast Enhanced MRI, National Library of Medicine, USA, Bethesda, MD, 2015 ClinicalTrials.gov [Internet].* Available at: http://clinicaltrials.gov/show/NCT0189 5829 NLM Identifier: accessed on 26.10.16.

[18] Sathishkumar, M., Sneha, K., Won, S. W., Cho, C. W., Kim, S., and Yun, Y. S. (2009). *Cinnamon zeylanicum* bark extract and powder mediated green synthesis of nano-crystalline silver particles and its bactericidal activity. *Colloid Surf. B* 73, 332–338.

[19] Chandran, S. P., Chaudhary, M., Pasricha, R., Ahmad, A., and Sastry, M. (2006). Synthesis of gold nanotriangles and silver nanoparticles using Aloe vera plant extract. *Biotechnol. Prog.* 22, 577–583.

[20] Dubey, S. P., Lahtinen, M., Sarkka, H., and Sillanpaa, M. (2010). Bioprospective of *Sorbus aucuparia* leaf extract in development of silver and gold nanocolloids. *Colloid Surf. B* 80, 26–33.

[21] Bar, H., Bhui, D. K., Sahoo, G. P., Sarkar, P., De, S. P., and Misra, A. (2009). Green synthesis of silver nanoparticles using latex of *Jatropha curcas. Colloids Surf. A*, 339, 134–139.

[22] Cruz, D., Fale., P. L., Mourato, A., Vaz, P. D., Serralheiro, M. L., Lino, and A. R. L. (2010). Preparation and physicochemical characterization of Ag nanoparticles biosynthesized by *Lippia citriodora* (Lemon Verbena). *Colloid Surf. B*, 81, 67–73.

[23] Kuppusamy, P., Yusoff., M. M., Maniam., G. P., and Govindan., N. (2014). Biosynthesis of metallic nanoparticles using plant derivatives and their new avenues in pharmacological applications – An updated report. *Saudi Pharma. J.* 24, 473–484.

[24] Si, S., and Mandal, T. K. (2007). Tryptophan-based peptides to synthesize gold and silver nanoparticles: a mechanistic and kinetic study. *Chemistry* 13, 3160–3168.

[25] Kim, J., Rheem, Y., Yoo, B., Chong, Y., Bozhilov, K. N., Kim, D., et al. (2010). Peptide-mediated shape- and size-tunable synthesis of gold nanostructures. *Acta Biomater.* 6, 2681–2689.

[26] Makarov, V. V., Love, A. J., Sinitsyna, O. V., Makarova, S. S., Yaminsky, I. V., Taliansky, M. E., et al. (2014). Green nanotechnologies: Synthesis of metal nanoparticles using plants. *Acta Naturae* 6, 35–44.

[27] Monkiedjea, A., Englandeb, A. J., James H. Wallc., and Anderson, A. C. (2008). A new method for determining concentrations of Endod-S (*Phytolacca dodecandra*) in water during mollusciciding. *J. Environ. Sci. Health* 25, 777–786.

[28] Allabaksh, M. B., Mandal, B. K., Kesarla, M. K., Kumar, K. S., and Reddy, P. S. (2010). Preparation of stable zero valent iron nanoparticles using different chelating agents. *J. Chem. Pharm. Res.* 2, 67–74.

[29] Pattanayak, M., and Nayak, P. L. (2013). Green synthesis and characterization of zero valent iron nanoparticles from the leaf extract of *Azadirachta indica* (Neem). *World J. Nanosci. Technol.* 2, 6–9.

[30] Bhattacharya, S. S., Das, J., Das, S., Samadder, A., Das, D., De, A., et al. (2012). Rapid green synthesis of silver nanoparticles from silver nitrate by a homeopathic mother tincture *Phytolacca Decandra*. *J. Chin. Med.* 10, 546–554.

[31] Behera, S. S., Patra, J. K., Pramanik, K., Panda, N., and Thatoi, H. (2012). Characterization and evaluation of antibacterial activities of chemically synthesized iron oxide nanoparticles. *World J. Nano Sci. Eng.* 2, 196–200.

[32] Shah, S., Chandraprabha, M. N., and Samrat, K. (2014). Synthesis and characterization of zero valent iron nanoparticles and assessment of its antibacterial activity. *Int. Rev. Appl. Biotechnol. Biochem.* 2, 145–151.

[33] Ravikumar, K. V. G., Kumar, D., Rajeshwari, A., Madhu, G. M., Mrudula, P., Chandrasekaran, N., et al. (2016). A comparative study with biologically and chemically synthesized nZVI: applications in Cr (VI) removal and ecotoxicity assessment using indigenous microorganisms from chromium-contaminated site. *Environ. Sci. Pollut. Res.* 23, 2613–2627.

[34] Jing-Feng, G., Hong-Yu. L., Kai-Ling, P., and Chun-Ying, S. (2016). Green synthesis of nanoscale zero-valent iron using a grape seed extract as a stabilizing agent and the application for quick decolorization of azo and anthraquinone dyes. *RSC Adv.* 6, 22526–22537.

[35] Das, R. K., Borthakur, B. B., and Bora, U. (2010). Green synthesis of gold nanoparticles using ethanolic leaf extract of Centella Asiatic. *Mater. Lett.* 64, 1445–1447.

[36] Kumar, K. M., Mandal, B. K., Kumar, K. S., Reddy, P. S., and Sreedhar, B. (2013). Biobased green method to synthesise palladium and iron

nanoparticles using *Terminalia chebula* aqueous extract. Spectrochim. *Acta* 102, 128–133.

[37] Chaudière, J., and Ferrari-Iliou R. (1999). Intracellular antioxidants: from chemical to biochemical mechanisms. *Food Chem. Toxicol.* 37, 949–962.

[38] Espitia, P. J. P., Soares, N. F. F., Coimbra, J. S. R., de Andrade, N. J., Cruz, R. S., and Medeiros, E. A. A. (2012). Zinc oxide nanoparticles: synthesis, antimicrobial activity and food packaging applications. *Food Bioprocess Technol.* 5, 1447–1464.

[39] Clarke., J. H. (2005). *A Dictionary of Practical Materia Medica.* Uttar Pradesh: B Jain Publishers Pvt. Ltd., 2, 802–812.

8

Simulation Studies of Nanomotors Based on Carbon Nanotubes for Nanodelivery Systems

Sunita Negi

Amity University, Manesar, Haryana, India

Abstract

The application of nanotechnology in the biological field is of great importance because of the ample application in the treatment of diseases and application of drugs. The nanomotors, for example, could be attached to a biological entity for such applications. In this chapter, we discuss such applications and the techniques which could be used in this direction. The compatibility of a carbon nanotube (CNT) in particular with the calmodulin (CaM) protein would be discussed.

As a first step, we simulated a nanomotor based on a double-walled carbon nanotube (DWNT) and studied its response using classical molecular dynamics (MD) simulations [1, 2]. Pure motor like behavior is observed for a particular range of amplitude and frequency of the applied electric field. In our works, we proceed with the simulation of a CNT with the calmodulin protein acting as a first step toward biological nanomotors. The behavior of the biological nanomotor in different environmental conditions would be studied later. Our previous work in the field of protein simulations would be beneficial in this direction [3–6].

8.1 Introduction

Carbon nanotubes (CNT) have attracted a lot of interest in the past few decades from the point of view of application of them in many research areas such as medical, electronics, and nanotechnology. Carbon nanotubes are simple tube-like structures which are formed from a single sheet of graphite. These tubes are found in arm-chair, zig–zag, and chiral forms and can be single-walled nanotubes (SWNT), double-walled nanotubes (DWNT), or multi-walled nanotubes (MWNT) depending on the number of layers of carbon atoms.

Biological motors are one of the most remarkable products of evolution as they are capable of performing biological tasks with much higher efficiency. Many researchers are working in the field of biological motors as these could be a boon to the medical science for the applications such as drug delivery. These motors coupled to the other nanocomponents such as carbon nanotubes could work as a very efficient machine/device which could be used to perform complex biological tasks in a very efficient way.

Micro- and nanomotors are propelled by the application of fuel such as hydrogen peroxide and hydrazine as some kind of propulsion is required for the operation of these motors. A laser excitation is used in [1, 2] to give a start to the nanomotor. But the biological motors, on the other hand, are not compatible with these kinds of fuels because of the compatibility issues. Also the excitation given in terms of laser would be critical to handle as it might have some adverse effect on the biological counterpart of the motor.

Simulations of carbon nanotubes are important from the point of view of application of them in nanodevices such as nanosensors or nanomotors [1, 2]. R. E. Tuzan [1] and S. Negi [2] reported the application of these single- and double-walled carbon nanotubes as nanomotors. These nanomotors are important from the point of view of application of these for the targeted drug delivery inside the human body. Unidirectional motor-like behavior is observed on applying an electric field of suitable amplitude and frequency as reported in [2].

8.2 Nanomotor

A nanomotor is a nanoscale device made from various entities such as carbon nanotube. A molecular dynamics (MD) study is performed on simulating these nanomotors by Tuzan et al. and Negi et al. [1, 2]. The simulations are performed for 100 ps by placing a shaft (inner CNT) inside a sleeve (outer CNT) as seen in Figure 8.1.

Tuzan et al. used the empirical Lenard Jones potential, whereas Negi et al. used the Brenner potential to simulate the interactions of carbon nanotube. An extra Nordlund term [3] is used to take into account the carbon nanotube

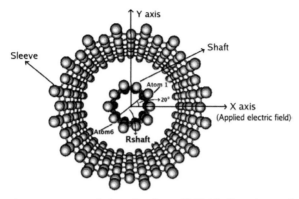

Figure 8.1 A nanomotor consisting of an inner CNT (shaft) and outer CNT (sleeve).

interaction with the graphite surface. Tuzan reported oscillatory pendulum-like behavior of the shaft nanotube inside the sleeve nanotube, whereas Negi et al. could work out a parameter space in the amplitude and frequency region of the applied field where full motor-like behavior is achieved. This parameter space would depend upon the configuration of the two nanotubes comprising the nanosystem and the spacing between the two. For a shaft and sleeve of configuration as shown in Figure 8.1, a useful parameter space could be obtained.

8.3 Protein Structure

Proteins are biological entities which are found inside the cells. There are various types of protein which differ on the basis of their function inside our body. The calmodulin protein is one of the proteins which is found in the eukaryotic cells. This protein helps in calcium signaling inside the human body. The basic structure of calmodulin protein is as shown in Figure 8.2. It consists of N- and C-lobes and the central linker shown in green, cyan, and purple color, respectively. The flexibility of the central linker part is mainly responsible for the overall conformation change and dynamics of the protein as discussed in [4].

We have earlier studied the conformation change in this protein at different pH values [4]. At lower pH values, this protein is observed to show a much compact form as compared to its structure at a physiological value of pH, i.e. 7.0. Figure 8.3 shows the root mean square deviation (RMSD) for the same. The protein is observed to show a higher value of RMSD at lower pH

Figure 8.2   Fully loaded calmodulin (CaM) protein with four calcium ions; N-lobe shown in green, the linker in purple, and C-lobe in cyan.

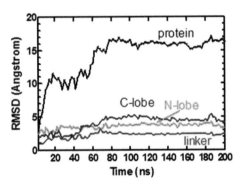

Figure 8.3   Root mean square deviation (RMSD) of the N-lope, C-lobe, linker, and protein at a lower pH of 5.0.

value, indicating more dynamics of the protein in this case. The main reason for this compactness is observed due to the salt bridge and hydrophobic patch formation. The lower pH value helps in triggering the ionic interaction between the residues of the different parts of the protein.

We have also studied the behavior of this protein at different environmental conditions such as calcium ion concentration, temperature, and ionic strength as in [5]. Figure 8.4 shows the behavior of CaM at different calcium

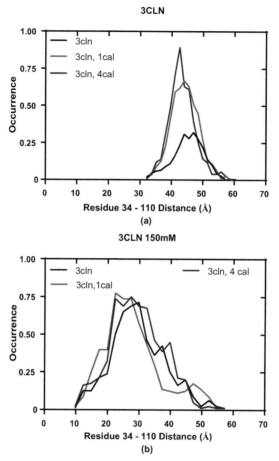

Figure 8.4 Normalized distance distribution (in Å) measured between residues 34 and 110, observed at the end of 100 ns in (a) ionic strength corresponding to a physiological value (b) ionic strength equal to 150 mM.

ion concentrations. The respective normalized distance distributions are as seen in Figure 8.4 with different stages of the calcium ion removal from the protein shown in different colors. The overall distance distributions do not show a significant shift on the removal of calcium ions in each case. Only the distribution peak is observed to acquire a higher value in the case of physiological ionic strength on the removal of calcium ions from the extended form as seen in Figure 8.4 (a). This implies that in this case, the calcium ion removal just acts as an attribute to add flexibility to the overall

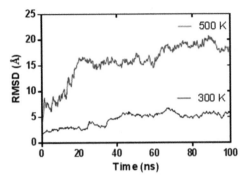

Figure 8.5   RMSD of the protein as a function of time at different temperatures. A significant deviation can be observed at 500 K as compared to 300 K implying a much greater flexibility at 500 K.

protein structure rather than giving any overall compactness. In the case of lower ionic strength of 150 mM, the normalized distance distribution is not observed to show any significant change. This is in agreement with our earlier result that the calcium ion removal does not make any significant changes in the conformation/flexibility of the protein at a lower ionic strength [5].

The behavior of the CaM at different temperature values is as seen in Figure 8.5. A higher RMSD of value nearly 15–20 Angstrom is observed at 500 K. The average value of the RMSD increases from 5 Å to 18 Å on increasing the temperature from 300 K to 500 K. Thus, a significant shift in the average value of the RMSD is observed with an increase in the temperature. This increase in the RMSD value would be the result of kinetic energy imparted to the protein residues at a higher temperature. The kinetic energy imparted to the protein would affect the stability of the protein to a much greater extent and confirms the previous observation of the change in the conformation on changing the temperature [5].

8.4 Simulation Method

In our previous works, the CaM protein is first solvated in a water box of dimension 10 Angstrom. This solvation is done using VMD plugin "solvate" [6]. Package "namd" [7] is then used to perform the simulation of CaM protein dissolved in a water box. The protein is observed to undergo a compact conformation on performing a 100-ns simulation. Then, various environmental conditions are varied one by one keeping the others constant and the effect of temperature is observed to be the most effective on changing the overall conformation of the protein.

In the present work, we simulate the calmodulin protein in combination with a single-walled carbon nanotube. The structure is simulated using NAMD package keeping a minimum distance of 6 Angstrom between the two. Both the structures are equilibrated together under atmospheric temperature and pressure conditions. Usual relaxation of the protein along with the CNT is observed in this case. This work is important from the point of view of application of these types of structures in the biological nanomotors.

8.5 Summary and Future Scope

Biological motors are one of the most remarkable products of evolution as they are capable of performing biological tasks with much higher efficiency. Many researchers are working in the field of biological motors as these could be a boon to the medical science for the applications such as drug delivery. These motors coupled to the other nanocomponents such as carbon nanotubes could work as a very efficient machine/device which could be used to perform complex biological tasks in a very efficient way.

Micro and nanomotors are propelled by the application of fuel such as hydrogen peroxide and hydrazine as some kind of propulsion is required for the operation of these motors. A laser excitation could be used to give a start to these nanomotors. But the biological motors, on the other hand, are not compatible with these kinds of fuels because of the compatibility issues.

In this work, we have performed the simulations of a single-walled carbon nanotube in combination with a calmodulin protein for the future application of these as a biological nanomotor. The simulation is performed at room temperature 300 K and atmospheric pressure. The SWNT and the calmodulin protein are observed to be showing a usual relaxation dynamics at these environmental conditions without showing any absurd behavior confirming the possibility of such simulations. The further simulations of such kind at different temperature conditions are in progress.

References

[1] R. E. Tuzan., D. W. Noid, and B. G. Sumpter. *Nanotechnology* **6**, 52–63 (1995).
[2] S. Negi, M. Warrier, K. Nordlund and S. Chaturvedi. *Comput. Mat. Sci.* **44**, 979–987 (2009).

[3] K. Nordlund, J. Keinonen and T. Mattila. *Phys. Rev. Lett.* **77**, 699 (1996).

[4] S. Negi, A. Ozlem Aykut, A. Rana Atilgan and C. Atilgan. *J. Phys. Chem. B.* **116**, 7145–7153 (2012).

[5] S. Negi. *Biophys. J.* **2014**, 329703 (2014).

[6] J. C. Gordon, J. B. Myers, T. Folta, V. Shoja, L. S. Heath and A. Onufriev. *Nucleic Acids Res.* **33**, W368–W371 (2005).

[7] C. M. Shepherd and H. J. Vogel. *Biophys. J.* **87**, 780–791 (2004).

9

Synthesis and Characterization of Lipid-Conjugated Carbon Nanotubes for Targeted Drug Delivery to Human Breast Cancer Cells

**Jawahar Natarajan, Surendra Ekkuluri and
Veera Venkata Satyanarayana Reddy Karri**

Department of Pharmaceutics, JSS College of Pharmacy, Ootacamund,
Jagadguru Sri Shivarathreeswara University, Mysuru, India

Abstract

This study deals with the investigation carried out on the formulation and evaluation of lipid-conjugated carbon nanotubes containing raloxifene hydrochloride (RH), to improve its targeting efficiency to the human breast cancer cells and its oral bioavailability. Preformulation studies like saturation solubility studies and compatibility of the selected excipients and RH were carried by FT-IR peak matching method. Carbon nanotubes were prepared by sonication method, and the effect of certain process and formulation variables such as stirring speed and stirring time on particle size was studied. Crystallinity studies were carried out for pure drug and optimized formulations by DSC. Lipid-conjugated carbon nanotubes containing RH were evaluated for zeta potential, drug loading efficiency, SEM studies, polydispersity, and *in vitro* release studies. Cytotoxicity studies were done by using MCF-7 breast cancer cells, to know the cytotoxic effect of formulation on the human breast cancer cells. In solubility studies of drug in different solvents, the solubility was found to be highest in pH 4.0 phosphate buffer. Selected lipid and other excipients were found to be compatible with RH based on FT-IR peak matching method and DSC studies. Better adsorption of drug on to the carbon nanotubes was observed in sonication method. The study on various process variables revealed that all the variables are important in preparation of carbon nanotubes. Batch prepared with 5.0 g graphite powder, 3 h stirring time, 2500 rpm stirring speed and 25 g potassium chlorate, 25 ml nitric acid, and 50 ml sulfuric acid showed minimum particle size and was identified as an ideal batch. Batches prepared using phosphatidyl choline and folic acid gave ideal zeta potential value. The entrapment efficiency and drug loading were found to be higher in formulation F3 and F4 compared to other formulations. The formulated lipid-conjugated RH-loaded carbon nanotubes showed a significant increase in cytotoxicity on breast cancer cells compared to pure drug suspension.

9.1 Introduction

Breast cancer becomes the most common causes of deaths worldwide. Thirty-three percentage of cancer cases diagnosed in women are due to breast cancer [1]. Chemotherapy in addition to the primary surgical removal of tumors is a necessary treatment for breast cancer. Widespread use of adjuvant chemotherapy in breast cancer has led to dramatic improvements in survival. RH, a highly effective drug for the treatment of invasive breast cancer and osteoporosis in post-menopausal women, shows poor oral bioavailability of 2%. Further, its non-selectivity in tumor cells causes various adverse effects. Carbon nanotube (CNT), discovered by Iijima in 1991, is a novel-type synthetic nanomaterials with distinct hollow, cylindrical structure. CNT can be viewed as rolled from layers of graphene sheets. These CNTs posses specific properties such as drug carrier, *in vivo* detection, and imagining [2]. Further, biodistribution studies of CNTs have shown high tumor accumulation of functionalized SWNT [3]. Due to these interesting properties, CNT has been widely investigated for delivery of antitumor agents, including DNA [4], siRNA [5], peptides [6], and drugs [7]. Hence, in this research, RH has been loaded into CNTs to target tumor cells in breast cancer.

9.2 Experimental Part

9.2.1 Preformulation Studies

Preformulation may be described as a phase of the research and development process where the formulation scientist characterizes the physical, chemical, and mechanical properties of a new drug substance, in order to develop stable, safe, and effective dosage forms. Ideally, the preformulation phase begins early in the discovery process such that appropriate physical, chemical data are available to aid in the selection of new chemical entities that enter the development process. During this evaluation, possible interaction with various inert ingredients intended for use in final dosage form is also considered.

9.2.1.1 Solubility studies

The solubility of the RH was determined in various solvents by adding an excess amount of drug to 10 ml of solvents in conical flasks. The flasks were kept at $25 \pm 0.5°C$ in isothermal shaker for 72 h to reach equilibrium. The equilibrated samples were removed from the shaker and centrifuged at 4000 rpm for 15 min. The supernatant was taken and filtered through 0.45-μm membrane filter. The concentration of RH was determined after suitable dilution by using UV–visible spectrophotometry at 286 nm.

9.2.1.2 Standard calibration curve

A stock solution of RH was prepared by dissolving 10 mg of drug in 10 ml methanol and with different buffers viz. pH 4.0 and pH 7.4 to give a stock solution of concentration 1 mg/ml, filter through 0.45-μm PVDF membrane filter. From the stock solution, different aliquots were taken in series of 10 ml volumetric flasks and volume made up with methanol to get a series of working standard solutions of concentrations, 2, 4, 6, 8, and 10 μg/ml. The absorbance of samples was obtained spectrophotometrically against the reagent blank at 286 nm. The calibration curves were constructed by plotting drug concentration versus the absorbance value at 286 nm, and the regression equation was computed.

9.2.1.3 Compatibility study

Compatibility of drug and excipients was studied using Fourier transform infrared spectroscopy (FT-IR) and differential scanning calorimetry (DSC).

9.2.2 Preparation of Carbon Nanotubes

A total of 5.0 g of graphite powder (99.99% purity, Aldrich) was slowly added to a mixture of fuming nitric acid (25 ml) and sulfuric acid (50 ml) for 30 minute. After cooling the mixture down to 5°C in an ice bath, 25.0 g of potassium chlorate was slowly added to the solution while stirring for 30 min. Since a lot of heat produced while adding potassium chlorate into the mixture, we have taken special care during this step. The solution is heated up to 70°C for 24 h and placed in the air for 3 days. Most graphite precipitates on the bottom but some reacted carbons will float. The floating carbon materials are transferred into DI (deionized) water (1 l). After stirring it for 1 hour, the solution is immediately filtered and the sample is vacuum dried [8].

9.2.2.1 Study on the effect of formulation process variables

The effect of formulation on process variables such as stirring time and stirring speed and on the particle size was studied. To investigate the effect of formulation on process variables, each time, one parameter was varied, keeping the others constant. From the results obtained, optimum level of those variables was selected and kept constant in the subsequent evaluations.

(A) Effect of Stirring Time

Four different batches of carbon nanotubes were prepared corresponding to 1, 2, 3, and 4 h stirring time keeping the following parameters constant.

Graphite powder	: 5.0 g
Sulfuric acid	: 50 ml
Stirring time	: 1, 2, 3, and 4 h
Potassium chlorate	: 25.0 g
Nitric acid	: 25 ml

(B) Effect of Stirring Speed

Four different batches of carbon nanotubes were prepared corresponding to 1000, 1500, 2000, and 2500 rpm stirring speed keeping the following parameters constant.

Graphite powder	: 5.0 g
Sulfuric acid	: 50 ml
Stirring speed (rpm)	: 1000, 1500, 2000, and 2500
Potassium chlorate	: 25.0 g
Nitric acid	: 25 ml

9.2.3 Purification, Cutting, and Oxidation of CNTs

CNTs were processed overnight with 1 M HNO3 at room temperature. They were then filtered through a membrane with a pore size of 200 nm, washed with purified water until a pH of 7 was reached, and vacuum dried. Subsequently, the following procedures were performed to obtain the purification, cutting, and oxidation of the CNTs: The CNTs were allowed to react in an ultrasonic bath at 45°C in a strong acid solution (3:1 mixture of 18.4M H_2SO_4

and 4M HNO$_3$) for 4 h, filtered and washed with purified water until a pH of 7 was reached, and finally vacuum dried [9].

9.2.4 Particle Size, Zeta Potential, and Polydispersity Index (PDI)

Particle size and zeta potential of the carbon nanotubes were measured by photon correlation spectroscopy using a Malvern Zetasizer Nano ZS90 (Malvern Instruments, Worcestershire, UK), which works on the Mie theory. All size and zeta potential measurements were carried out at 25°C using disposable polystyrene cells and disposable plain folded capillary zeta cells, respectively, after appropriate dilution with original dispersion preparation medium. In order to investigate the effect of stirring speed and stirring time on particle size.

9.2.5 Surface Morphology by Scanning Electron Microscopy (SEM)

The morphology of carbon nanotubes was determined by scanning electron microscopy (SEM) (JB.Kayes, Pharmaceutics: The Science of Dosage Form Design, 1st edn, 1999). The samples were examined at suitable accelerating voltage, usually 20 kV, at different magnifications. Image analysis software (Amrita University, Cochin) was employed to obtain an automatic analysis result of the shape and surface morphology.

9.2.6 Preparation of CNTs-RH-Folic Acid (CNTs-RH-FA)

RH (20 mg) was mixed with CNTs (10 mg), PVP K30 (110 mg), and folic acid (10 mg) in methanol (5 ml) using an ultrasonic bath for 10 min and then rotary vacuum evaporated to dryness to obtain CNTs complexed with RH. The complex was suspended with 10 ml of an aqueous solution of PC (150 mg) and PVP K30 (10 mg) by probe sonication for 30 min. The resulting suspension was centrifuged at 5000 rpm for 10 min to remove the excess RH and insoluble CNTs, and then, the supernatants were lyophilized to produce the final formulation of CNT-RH-FA (Table 9.1) [10].

9.2.7 Characterization of CNT-RH-FA

9.2.7.1 Size distribution and zeta potential

A 100 μl aliquot of CNTs-RH-FA was diluted in 1.9 ml ultrapure H$_2$O. The size and zeta potential of the CNT-RH-FA were determined. The instrument

Table 9.1   Formulation of RH-CNTs-FA

Formulations	RH (mg)	CNTs (mg)	PVP K30 (mg)	PEG 4000 (mg)	Phosphatidyl Choline (mg)	Folic Acid (mg)
Fl	20	10	120	–	150	–
F2	20	10		120	150	–
F3	20	10	120	–	150	10
F4	20	10	–	120	150	10
F5	20	10	120	–	–	10
F6	20	10	–	120		10

used is the Malvern Zetasizer Nano, Series ZEN1002 (Malvern) in cuvette DTS0012 with a 532-nm green laser and a scattering angle of 173°C.

9.2.8 Scanning Electron Microscopy (SEM)

External surface morphology of lyophilized drug-loaded carbon nanotubes was recorded using SEM (FEI QUANTA 200 SEM/EDAX, UK) at 20 kV as an accelerating voltage. Weighed amount of samples (5–7 mg) were mounted on an aluminum stub with double-sided adhesive tape. The tape was firmly attached to the stub, and lyophilized sample was scattered carefully over its surface. The stub with the sample was then sputter-coated with a thin layer of gold to make the sample conductive. Processed sample was subjected to SEM analysis. The images were captured and recorded.

9.2.9 Determination of Loading Efficiency

A 0.5 ml volume of CNTs-RH-FA was diluted with 4.5 ml of methanol, sonicated for 1 h, and then centrifuged at 10,000 rpm for 10 min to remove the excess CNTs in the formulation. The supernatant solution was used to determine the RH concentration using an UV–visible spectrophotometry (Shimadzu) with a wavelength at 286 nm.

9.2.10 *In Vitro* Drug-Release Studies

The pattern of drug release from targeted carrier was studied at the physiological temperature of 37°C and pH of 7.4, and 4.0 in PBS. The proper amount of loaded carriers containing 10 mg of RH was placed in dialysis tubing (12 kDa) and immersed in 900 ml of the release media and then placed in an incubator shaker set at 100 rpm for 3 days. At predetermined time intervals, 5 ml of the release medium was removed and replaced with fresh medium and the 5 ml of the collected sample was centrifuged, and the concentration of released RH in the supernatant was estimated by UV–visible spectrophotometry at 286 nm. The concentration of drug released at a given time was calculated using a standard curve for RH [11].

9.2.11 Determination of Mitochondrial Synthesis by MTT Assay

The monolayer cell culture was trypsinized, and the cell count was adjusted to 1.0×105 cells/ml using MEM/DMEM medium containing 10% FBS/NBCS. To each well of a 96-well microtitre plate, 100 μl of the diluted cell suspension (approximately 10,000 cells/well) was added. After 24 h, when a partial monolayer was formed, the supernatant was flicked off, the monolayer was washed once with medium, and 100 μl of different sample concentrations prepared in maintenance media were added per well to the partial monolayer in microtitre plates. The plates were then incubated at 37°C for 3 days in 5% CO_2 atmosphere, and microscopic examination was carried out and observations were recorded every 24 h. After 72 h, the sample solutions in the wells were discarded and 50 μl of MTT (2 mg/ml) in MEM-PR (MEM without phenol red) was added to each well. The plates were gently shaken and incubated for 3 h at 37°C in 5% CO_2 atmosphere. The supernatant was removed, 50 μl of propanol was added, and the plates were gently shaken to solubilize the formed formazan. The absorbance was measured using a microplate reader at a wavelength of 540 nm. The percentage growth inhibition was calculated using the following formula and concentration of drug or test samples needed to inhibit cell growth by 50% values were generated from the dose–response curves for each cell line [12].

$$\% \text{ Growth Inhibition} = 100 - \frac{\text{Mean OD of individual test group}}{\text{Mean OD of Control group}}$$

9.3 Results and Discussion

9.3.1 Preformulation Studies

9.3.1.1 Solubility studies

The saturation solubility of RH was determined in different solvents viz. 0.1 N HCl, phosphate buffer pH 4.0, pH 6.8, and pH 7.4. RH was found to be highly soluble in pH 4.0. The solubility in 0.1 N HCl, phosphate buffer pH 6.8, and pH 7.4 was found to be 130.3 ± 6.1 μg /ml, 243.4 ± 4.1 μg/ml, and 275.46 ± 0.92 μg/ml, respectively.

9.3.1.2 Development of calibration curve

Calibration curve of the drug was developed to found out the linearity between concentration of drug in the solution and its optical density. It was concluded that the perfect linearity between the concentration and absorbance was observed when the concentration range was from 2 µg/mL to 10 µg/mL. Figures 9.1 and 9.2 show the calibration curve of RH using phosphate buffer pH 4.0 and pH 7.4. The "Slope (K)" and "Intercept (β)" value was found to be 0.020 and 0.0013, and 0.012 and 0.0023, respectively.

9.3.1.3 Crystallinity study by using DSC

The results of DSC are shown in Figures 9.3 and 9.4. These studies revealed that there was obscene of the drug peak in the formulations. From the results, it was concluded that the drug was changed its nature from crystalline to polymorphic form.

9.3.1.4 Compatibility studies using FT-IR

A physical mixture of drug and excipients was prepared and mixed with suitable quantity of potassium bromide. This mixture was compressed to form a transparent pellet using a hydraulic press at 15 tons pressure. It was scanned from 4000 to 400 cm^{-1} in a FTIR spectrophotometer (FTIR 8400 S, Shimadzu). The IR spectrum of the physical mixture was compared with

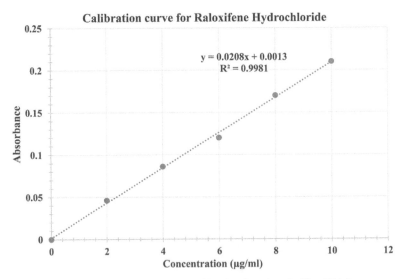

Figure 9.1   Calibration curve of RH in phosphate buffer pH 4.0.

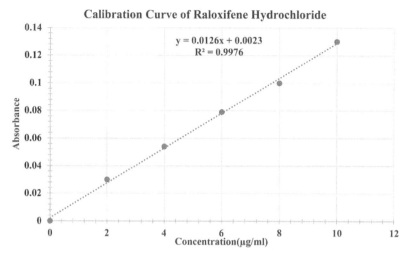

Figure 9.2 Calibration curve of RH in phosphate buffer pH 7.4.

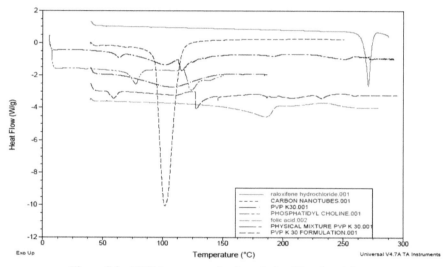

Figure 9.3 DSC thermogram for formulation F3 and excipients.

those of pure drug and excipients, and peak matching was done to detect any appearance or disappearance of peaks.

The results for FTIR studies are shown in Figures 9.5–9.8. These results of FTIR indicate that there was no physical and chemical interaction in

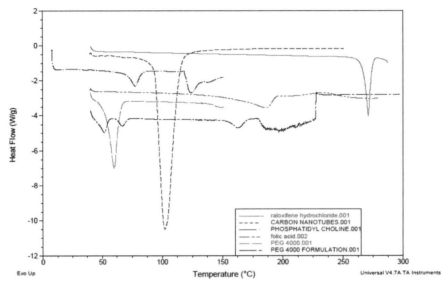

Figure 9.4   DSC thermogram for formulation F4 and excipients.

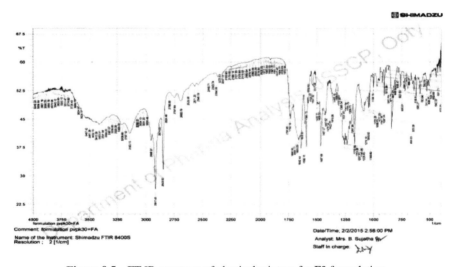

Figure 9.5   FT-IR spectrum of physical mixture for F3 formulation.

between drug and studied excipients. The frequencies of functional groups of drug remained intact in physical mixture containing different excipients. So it was concluded that there was no major interaction occurred.

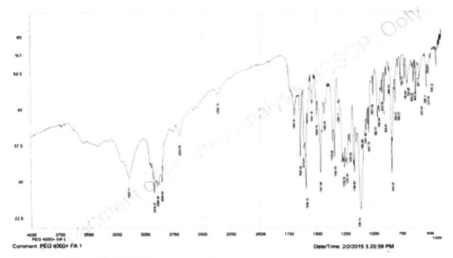

Figure 9.6 FT-IR spectrum of physical mixture for F4 formulation.

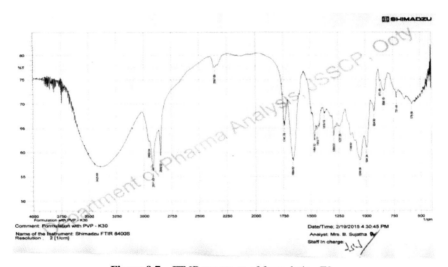

Figure 9.7 FT-IR spectrum of formulation F3.

9.3.2 Preparation of Carbon Nanotubes

Effect of Process Variables on Formulation: Various batches of carbon nanotubes were prepared. The influence of formulation/process variables viz. stirring time and stirring speed on particle size was studied.

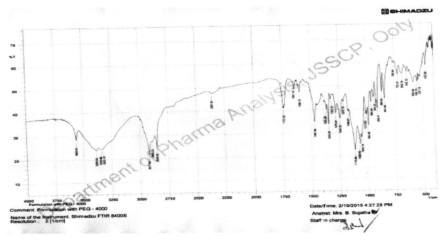

Figure 9.8   FT-IR spectrum of formulation F4.

Effect of stirring time: The effect of stirring time on particle size is given in table. It was observed that particle size was in the range of 1.9–3.2 μm. It is obvious that with increase in stirring time, there was decrease in particle size up to 4 h. When the stirring was continued up to 5 h, there was no significant decrease in particle size. Thus, 3 h time was taken as optimum time for the preparation of further batches.

Effect of stirring speed: The effect of stirring speed on particle size is given in Table 9.2. It was observed that particle size was in the range of 1.9–3.2 μm. It is obvious that with increase in stirring speed, there was decrease in particle size up to 2500 rpm. When the stirring was done at 3000 rpm, there was no significant decrease in particle size. Thus, 2000 rpm stirring speed was taken as optimum time for the preparation of further batches.

Table 9.2   Zeta potential, particle size, and PDI

Formulations	Zeta Potential (Mv)	Particle Size (nm)	Polydispersity Index
F1	26.1	291.9	0.411
F2	33.0	217.8	0.526
F3	24.7	234.2	0.237
F4	20.1	197.7	0.342
F5	15.4	352.4	0.517
F6	15.2	320.6	0.470

9.3.3 Particle Size Distribution and Zeta Potential

Zeta potential is the potential difference between the dispersion medium and the stationary layer of fluid attached to the dispersed particle. The significance of zeta potential is that its value can be related to the stability of colloidal dispersions. The zeta potential indicates the degree of repulsion between adjacent, similarly charged particles in dispersion. For molecules and particles that are small enough, a high zeta potential will confer stability, i.e., the solution or dispersion will resist aggregation. When the potential is low, attraction exceeds repulsion and the dispersion will break and flocculate.

Thus, two batches were prepared without phosphatidyl choline keeping other formulation variables as constant. It was found that batch prepared without phosphatidyl choline had zeta potential of 15.2 Mv and 15.4 Mv, which might be due to negative charge distributed at surface of CNTs, while batch prepared with phosphatidyl choline had zeta potential of 24.7 Mv. Table 9.2 shows the zeta potential of all the formulations. Figures 9.9–9.12 show the particle size distribution and zeta potential for optimized formulations.

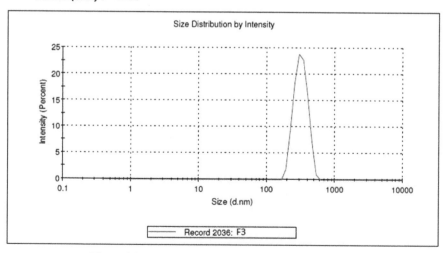

		Size (d.nm):	% Intensity:	St Dev (d.n...
Z-Average (d.nm): 234.2	Peak 1:	319.0	100.0	70.74
Pdl: 0.237	Peak 2:	0.000	0.0	0.000
Intercept: 0.910	Peak 3:	0.000	0.0	0.000
Result quality : Good				

Figure 9.9 Particle size distribution of formulation F3.

Results

		Mean (mV)	Area (%)	St Dev (mV)
Zeta Potential (mV): 24.7	Peak 1:	24.7	100.0	5.86
Zeta Deviation (mV): 5.86	Peak 2:	0.00	0.0	0.00
Conductivity (mS/cm): 0.226	Peak 3:	0.00	0.0	0.00
Result quality : Good				

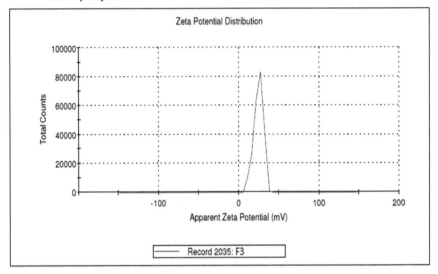

Figure 9.10   Zeta potential of formulation F3.

9.3.4 Scanning Electron Microscopy (SEM)

The external morphological studies (SEM) revealed that the outer diameter of the CNTs increased in size and the surfaces of the CNTs became non-uniform after the CNTs were non-covalently functionalized with PC, PVP K30, and FA (Figure 9.13), confirming the non-covalent functionalization of the CNTs. These results suggest that the increase in the dispersibility of the CNTs might be partially due to the PC and PVP K30 surrounding the CNTs (Figure 9.14), which help prevent aggregation. And Figures 9.13 and 9.14 shows the morphology of purified CNTs.

9.3.5 Drug Loading Efficiency

The loading of RH onto the CNTs was determined by the analysis of the supernatant for free drug using a UV–vis spectrophotometer after centrifugation of the RH-loaded CNTs. Table 9.3 shows the drug loading efficiency

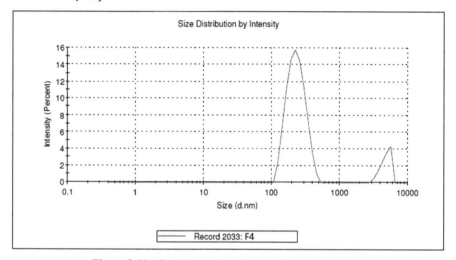

		Size (d.nm):	% Intensity:	St Dev (d.n...
Z-Average (d.nm): 197.7	**Peak 1:**	204.3	88.0	71.24
PdI: 0.342	**Peak 2:**	4748	12.0	750.1
Intercept: 0.928	**Peak 3:**	0.000	0.0	0.000
Result quality : Good				

Figure 9.11 Particle size distribution of formulation F4.

for all formulated batches. From the results, we observed that the lipid core was found to affect the extent of drug loading efficiency. As observed, the formulations (F3 and F4) formulated with phosphatidyl choline show better drug loading efficiency compared to formulations (F5 and F6) and the maximum entrapment efficiency was 74.2% (F3) and 71.2% (F4).

9.3.6 *In Vitro* Drug-Release Studies

The drug-release profile of RH from the RH-loaded CNTs of both optimized formulations (F3 and F4) was studied at 37°C in PBS at two different pH conditions 7.4 and 4.0 with continuous shaking at 100 rpm for 72 h. The temperature of 37°C was selected for drug-release response because it is close to the physiological temperature. The pH of 7.4 corresponds to physiological pH, and pH of 4.0 corresponds to lysosomal pH of cancer cells. Tables 9.1 and 9.2 show the drug-release profile of optimized formulations, and the drug-release curves (Figure 9.15) indicate that the release of RH from the CNTs is pH-triggered, and the drug-release studies were carried out till it

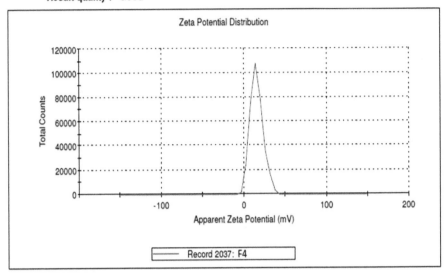

	Mean (mV)	Area (%)	St Dev (mV)
Zeta Potential (mV): 20.1	Peak 1: 20.1	100.0	7.72
Zeta Deviation (mV): 7.72	Peak 2: 0.00	0.0	0.00
Conductivity (mS/cm): 0.201	Peak 3: 0.00	0.0	0.00
Result quality : Good			

Figure 9.12   Zeta potential of formulation F4.

reached the stationary phase. At pH 7.4, the drug-release curve shows that RH loaded on CNTs is released at a very low and slow rate, with very minimum drug release. However, at pH 4.0, the RH-release rate was significantly enhanced during the initial 6 h. We observed an initial faster drug release up to 24 h, followed by a sustained-release pattern till 72 h. These results can be ascribed to the hydrogen-bonding interaction between RH and CNTs, which is stronger in neutral conditions, resulting in a controlled release. However, the drug-release pattern under acidic media indicates a higher amount of RH release than at neutral conditions. This efficient loading and release of RH indicate strong $\pi-\pi$ stacking interaction between CNTs and RH. Around 92% of the drug was released from F3 formulation within 72 h in pH 4.0 buffer, whereas 87% of the drug was released from formulation F4 in pH 4.0 buffer. And 40% of the drug was released from F3 formulation within 72 h in pH 7.4 buffer, whereas 37% of the drug was released from formulation F4 in pH 7.4 buffer, indicating a higher percentage of release under acidic conditions.

Figure 9.13　Scanning electron micrograph of purified CNTs.

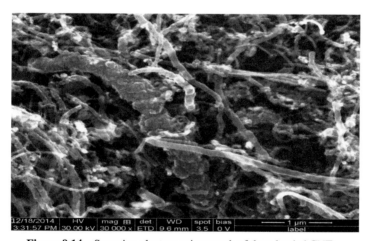

Figure 9.14　Scanning electron micrograph of drug-loaded CNTs.

Table 9.3　Drug loading efficiency

Formulations	Drug Loading Efficiency (%)
F1	69.2
F2	68.6
F3	74.2
F4	71.2
F5	54.8
F6	54.2

9.3.7 *In Vitro* Cytotoxicity Studies

The cytotoxicity studies were carried out by MTT assay for optimized formulations (F3&F4). The *in vitro* cytotoxicity profile of the RH-CNTs-FA in comparison with free RH was studied using MTT assay. MCF-7 breast cancer cells were used for the cytotoxicity analysis. Figures 9.16, 9.17, and 9.18 show the *in vitro* cytotoxicity studies for the optimized formulations and pure drug. Five different concentrations each of the RH-CNTs-FA and RH as test sample were used. The assays were carried out for 72 h, and the absorbance readings were taken for analysis. The 50% cytotoxic concentration (CC_{50}) was calculated. The results showed that formulation F3 (CC_{50} = 43.57305 µg/ml) is more cytotoxic than the formulation F4 (CC_{50} = 57.5487 µg/ml). The results were compared with the standard drug RH HCL (CC_{50} = 41.4825 µg/ml) (Figures 9.16, 9.17, 9.18, 9.19, 9.20, and 9.21).

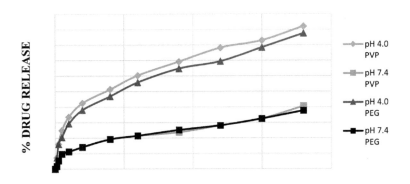

Figure 9.15   *In vitro* drug-release profile.

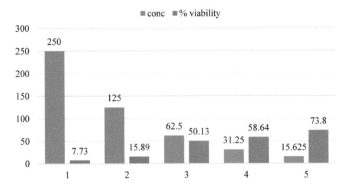

Figure 9.16   Cytotoxicity studies of formulation F3.

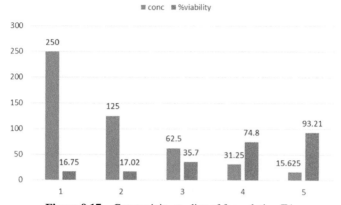

Figure 9.17 Cytotoxicity studies of formulation F4.

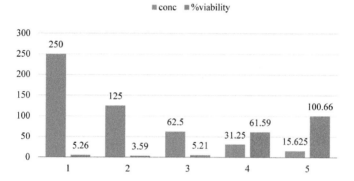

Figure 9.18 Cytotoxicity studies of pure drug.

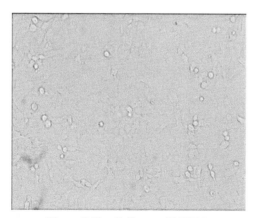

Figure 9.19 Cell control MCF-7.

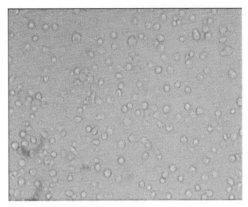

Figure 9.20   100% cytotoxic culture at highest concentration 250 µg/ml.

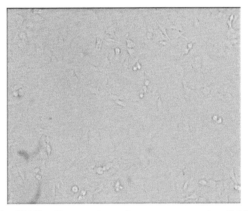

Figure 9.21   100% cell protection at lowest concentration (15.625 µg/ml).

9.4 Conclusion

In conclusion, sonication method was suitable for drug adsorption on to the carbon nanotubes. Lipophilic drugs such as RH can be successfully loaded into the lipids. The formulated lipid-conjugated RH-loaded carbon nanotubes showed a significant increase in cytotoxicity on breast cancer cells compared to pure drug suspension. The formulated carbon nanotubes may be preferred as drug carriers for anticancer drugs to overcome the drug-related adverse effects and to minimize the toxic effects of the drug on the normal cells and targeted to cancer cells.

References

[1] Lester, J. (2007). Breast cancer in 2007: Incidence, risk assessment, and risk reduction strategies. *Clin. J. Oncol. Nurs.* 11, 619–622.

[2] Kostarelos, K., et al. (2007). Cellular uptake of functionalized carbon nanotubes is independent of functional group and cell type. *Nat. Nanotech.* 2, 108–113.

[3] Liu, Z., et al. (2007). In vivo biodistribution and highly efficient tumour targeting of carbon nanotubes in mice. *Nat. Nanotech* 2, 47–52.

[4] Liu, Y., et al. (2005). Polyethylenimine-grafted multiwalled carbon nanotubes for secure noncovalent immobilization and efficient delivery of DNA. *Angewandte Chemie* (International ed. in English), 44, 4782–4785.

[5] Cheung, W., et al. (2010). DNA and carbon nanotubes as medicine. *Adv. Drug Del. Rev.* 62, 633–649.

[6] Zhang, Z., et al. (2006). Delivery of telomerase reverse transcriptase small interfering RNA in complex with positively charged single-walled carbon nanotubes suppresses tumor growth. *Clin. Cancer Res.* 12, 4933–4939.

[7] Varkouhi, A. K., et al. (2011). SiRNA delivery with functionalized carbon nanotubes. *Int. J. Pharm.* 416, 419–425.

[8] Zhao, D., et al. (2011). Carbon nanotubes enhance CpG uptake and potentiate antiglioma immunity. *Clin. Cancer Res.* 17, 771–782.

[9] Prakash, S., et al. (2011). Polymeric nanohybrids and functionalized carbon nanotubes as drug delivery carriers for cancer therapy. *Adv. Drug Del. Rev.* 63, 1340–1351.

[10] Bethune, D. S., Kiang, C. H., De-Vries, M., Gorman, G., and Savoy, R. (1993). Cobalt-catalyzed growth of carbon nanotubes with single-atomic-layer walls. *Nature* 363, 605–607.

[11] Heister, E., Neves, V., Lamprecht, C., Silva, S. R. P., and Coley, H. M. (2012). Drug loading, dispersion stability, and therapeutic efficacy in targeted drug delivery with carbon nanotubes. *Carbon* 50, 622–632.

[12] Li, H., Zhang, N., Hao, Y., Wang, Y., Jia, S., Zhang, H., et al. (2013). Formulation of curcumin delivery with functionalized single-walled carbon nanotubes: characteristics and anticancer effects in vitro. *Drug Deliv. Dove Press* 9, 234–256.

10

Phytosynthesis of Silver Nanoparticles and Its Potent Antimicrobial Efficacy

Soumya Soman and Joseph George Ray

Laboratory of Ecology and Eco-technology, School of Biosciences, Mahatma Gandhi University, Kottayam, Kerala, India

Abstract

Green synthesis of metallic nanoparticles has become a promising field of research in recent years. Owing to its nature of importance, biosynthetic approach using plants has become an alternative for physical and chemical methods. In phytosynthesis, plant biomolecules mediate reduction of metals to metal nanoparticles quite rapidly and it is an eco-friendly synthetic process. Due to the large number of bioconstituents involved in the plants, the exact mechanism of reduction is not well known. However, certain interplay of phytochemicals in the reduction process of metals to metal nanoparticles is already understood. In phytosynthesis, the various functional groups associated with the phytochemicals act both as reducing and capping agent for the nanoparticles formed. Phytosynthesis of metal nanoparticles is now considered as a green chemical method of high application in biomedical field. Among diverse metal nanoparticles, phytosynthesized silver nanoparticles are used as novel antibacterial agents that have various biomedical applications. In the last few decades, research out put on various aspects of phytosynthesis of nanoparticles, which explain the methodology of synthesis, mechanism of formation of nanoparticles, and the diverse biomedical applications of the same have appeared in large numbers. Reviews on all these aspects have already published. However, reviews with specific focus on antibacterial applications of phytosynthesized nanoparticles are quite rare, especially on its recent advancements. In the present review, emphasis is given on the antibacterial nature of the phytosynthesized nanoparticles, especially silver nanoparticles and its bacterial toxicity. An overview of recent findings on the mechanism of bacterial toxicity, which is an important aspect in light of evolution of silver as an antimicrobial agent, is also examined. Additionally, publications on the mode of interaction of bacteria with nanoparticles using the TEM studies are also critically examined. Finally, the application of phytosynthesized nanoparticles, its limitations, and possible future studies are also explained.

10.1 Introduction

Nanobiotechnolgy, is a multidisciplinary field that involves the technological developments in the field of biotechnology, nanotechnology, physics, chemistry, and material science [1]. Nanoparticles (NPs) are getting high scientific interest in modern times because they bridge the gap between the bulk materials and its atomic/molecular counter parts. They are of great interest due to its extremely small size and large surface to volume ratio, which bring about change both in their physical and chemical properties such as mechanical properties, catalytic activity, sterical and biological properties, optical absorption and melting point compared to their bulk counter parts [2–4]. NPs exhibit size- and shape-dependent characteristics due to which they are being used in the applications ranging from biosensing and catalysts to optics, chemical sensors, electrometers, and computer transistors and electronic, logic, and memory schemes. They have many different applications in diverse fields such as medical imaging, nanocomposites, filters, drug delivery, hyperthermia of tumors [5–8], and novel antimicrobial agents.

Nanoparticles are commonly synthesized by physical, chemical, and biological methods. In the physicochemical methods, the use of toxic or potentially hazardous solvents, stabilizers, and precursors are inevitable, whereas biological methods do not involve the use of such toxic materials. Biogenic reactants such as plant extracts, microorganisms, fungi, algae, and actinomycetes are used in the biological methods for shaping or stabilizing the NPs with advanced antimicrobial properties. Most of these are not only cost effective, water soluble, hazard free, and biocompatible reductants and stabilizers, but also have additional advantages such as enhancing the properties of the NPs. Therefore, the green synthesis of antimicrobial NPs has become highly significant. Figure 10.1 explains the overview of different applications of NPs formed through the biological way.

In recent years, efficient green chemistry approaches for the fabrication of commercially viable noble metallic NPs have become a major focus of researchers. The three main criteria of green chemical preparation of NPs

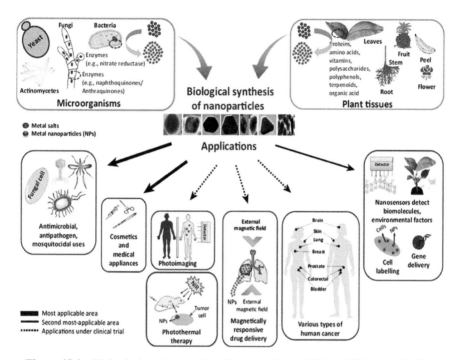

Figure 10.1 Biological synthesis and applications of metal NPs in different fields [9].

include choice of the solvent medium, environmentally benign reducing agent, and nontoxic material for the stabilization of the NPs [10]. Phytosynthesis of NPs using plant extracts satisfies the above-mentioned choices, because it uses water as solvent medium instead of the organic solvents. Moreover, phytochemicals in the plant extract act as both reducing and stabilizing agents for the rapid production of stable NPs, which have additional antimicrobial properties. Therefore, phytosynthesis of antimicrobial NPs using biological sources has been receiving tremendous research attention these days. The present review focuses on synthesis of NPs using plant extracts as plants itself act as a nanofactory.

10.1.1 Advantages of Phytosynthesis

The use of living plants or their extract for reducing metal salts to metal nanoparticle [11–14] is one of the most prominent advantages of phytosynthesis. Diverse kinds of plants of different medicinal properties are common in all environments, which are easily available and safe to handle as it does not involve any toxic solvents or other chemicals. Naturally, phytosynthesis becomes more eco-friendly and it offers production of NPs of diverse medicinal properties. Compared to physicochemical methods, it is quite a rapid process as well. It is more advantageous than the microbial synthesis, because it does not involve costly microbial cultures and its maintenance or the associated biological risks. Moreover, the processing of NPs after synthesis is quite cumbersome in the microbial synthetic process [15, 16]. In phytosynthesis using plant extracts, NPs are synthesized extracellularly and the conditions for synthesis can be easily controlled so as to obtain NPs of desired shape and size. All these advantages of phytosynthesis have led to the evolution of a new terminology for the synthesis of NPs using plant resources as phyto-nanotechnology, which mainly focuses on synthesis of NPs of variable shapes, sizes, and stability over time without the use of any capping and stabilizing agents [9].

Plant extracts may act as both reducing and capping agents during the process of synthesis. The phytochemicals in plants are known to have influence on NP characteristics [12] and its properties. Since the nature and number of chemicals vary with different plants, the process is complex in nature [17]. Moreover, the surface of NPs selectively absorbs biomolecules in the plant extracts forming a corona, which ultimately influence the properties of the synthesized particles, the reason for an additional efficacy for the biosynthesized NPs [18, 19]. In the physicochemical synthesis, various additional steps are required for the capping and stabilizing of the NPs, which can be avoided in phytosynthesis [20]. Phytosynthesis of NPs also helps in controlling the shape and size of NPs. The reaction parameters that control phytosynthesis such as the volume of plant extract used and the temperature during the reaction or pH of the medium, and concentration of the metal salts [21] are very critical to obtain the NPs of desired size [22] and shape. Therefore, phytosynthesis is mainly concerned with the potential use of NPs for human benefits. In addition to all these advantages, phytosynthesis also influences scalability, biocompatibility, and the medical applicability of synthesized NPs, in which water is the reducing medium [23].

10.1.2 Major Objective

It was in 2007, a comprehensive account on green chemical approach to nanosynthesis in the development of new materials and associated applications appeared as a review (Dahl et al. [24]). The major focus in the review was on the development of high precision, low waste management methods of nanomanufacturing, which was crucial for commercialization. All the methods, which involved wet chemical approaches to the production, functionlaization, purification, and assembly of NP building blocks, are explained in the review. The authors attempted to identify and illustrate the concepts and techniques of green nanosynthesis that help research to transform the nanoscience area from discovery phase to a production phase. Later, Lin et al. [25] briefly looked into the details of publications on various methods of preparation of AgNPs using polyoxometallates, polysaccharides, tollens, irradiation, and by plants and microorganisms, as well as few applications of AgNPs. Parikh et al. [26] provided an overview of biological methods (yeasts, plants, algae, fungi, bacteria, and viruses) of synthesis of metallic NPs by both intracellular and extracellular ways, where they emphasized the low-cost, energy-efficient, and nontoxic nature of the phytosynthetic process.

Narayanan and Shakthivel in 2011 [15] gave an in-depth account on synthesis of NPs using plants, algae, diatoms, and biocompatible agents. They had explained in detail about the fabrication of different NPs such as cobalt, copper, silver, gold, bimetallic alloys, silica, palladium, platinum, iridium, magnetite, and quantum dots from the above-mentioned sources. In the study entitled "greener synthesis of metal NPs using plants" [13], the author gave a detailed account of various plant biomass or living plants used in metal NPs. Some of the recent reviews give detailed account on different plants being used as an alternative source for NP synthesis; in addition, they also looked at the various factors, which influence the formation of NP during phytosynthesis. However, no attempts have so far been focused on the use of plant extracts as a source for AgNP synthesis and the possible mechanism behinds the phytosynthesis of such antibacterial. Therefore, the objective of the present review is to discuss specifically the role of phytosynthesized AgNPs as antibacterial agent and the mechanism of its bacterial toxicity. The antibacterial activity of the phytosynthesized AgNPs and its role as an alternative to be a nano-antimicrobial is analyzed in detail.

10.2 Phytosynthesis of AgNPs

Phytosynthesis of NPs includes three different categories: (1) intracellular synthesis of metals and conversion inside the plant cells into NPs, (2) extracellular biosynthesis of NPs by plant biomass, and (3) extracellular biosynthesis by plant extracts. In the present review, major emphasis is given on the recent advancement of the third category, which occurred in the last 13 years.

Shakthivel et al. [15] have observed the drawbacks of methods 1 and 2 and have reviewed the disadvantages in the intracellular synthesis of NPs as well as in extracellular biosynthesis using biomass. According to the authors, the difficulties include involvement of complex biochemical pathways in the intracellular synthesis of metal NPs so that localization of NPs is based on the presence of enzymes or proteins associated with the recovery of the NPs. Therefore, the biosynthesis of NPs using such methods is found tedious and expensive and needs enzymes to degrade the cellulosic materials. Therefore, the major focus on phytosynthesis of NPs is now in the extracellular methods. It is also observed that the extracellular synthesis has greater potential in commercial application [27].

10.2.1 Extracellular Synthesis of AgNPs Using Plant Extracts: A Few Case Studies

Shankar et al. in 2003 [28] successfully synthesized highly stable and crystalline AgNPs using geranium leaf extract and silver ions as substrates. Thus, phytosynthesis as an alternative method of more rapid and reproducible process was established. Later, many authors followed similar procedures using diverse plants and reiterated the utility of phytosynthesis. Shankar et al. [29] reported the synthesis of AgNPs using *Azadirachta indica* in which the reduction process for AgNP formation was complete within 4 h. Chandran et al. [30] and Li et al. [31] illustrated the synthesis of AgNPs using *Aloe vera* and *Capsicum annum*, respectively. They observed the formation of well-dispersed spherical AgNPs with size of 15 ± 4 nm. AgNPs of size 10–45 nm was synthesized using *Capsicum annum* leaf extract at room temperature where the authors recorded the XPS spectrum, and according to them, silver ions in the extract reduced into the metallic form and protein in the extract further capped the same [31].

Song et al. [27] made a comparison of extracellular synthesis of AgNPs in five different plant extracts. Among the five plants, magnolia leaf broth was found to be the plant that took the lowest window period of 11 min for formation NPs. According to their observation, the size of the particle can be controlled by changing the reaction temperature, leaf broth concentration, and $AgNO_3$ concentration. Huang et al. [32] in their synthesis of AgNPs using Cinnamomum leaf applied a continuous-flow tubular microreactor for biosynthesis of AgNPs to scale up its production; however, the method did not find acceptance as a superior one any more in the literature. Kasthuri et al. [33] reported the ability of apiin in the leaf extract of henna to aid formation of gold and silver NPs. Secondary hydroxyl and carbonyl groups of apiin are found responsible for the reduction. According to Tripathy et al. [34], different variables such as reductant concentration, reaction pH, mixing ratio of reactants, and interaction time have certain effect on the morphology and size of the AgNP particles while synthesis is carried out using aqueous leaf extract of *Azardirachta indica*. The researchers explained the tuning of various bioprocess parameters for NPs of desired shape and size.

Christensen et al. in 2011 [35] synthesized AgNPs (10–25 nm) using *Murraya koeniggi* leaf extract within 15 min of reaction. According to the authors, increase in concentration of the broth is directly proportional to rate of reduction, whereas the same is inversely proportional to the particle size and agglomeration of NPs. Similar observation was obtained by Annamalai et al. in phytosynthesis of Ag NPs using leaf extract of *Phyllanthus amarus* Schum and Thonn [36] and by Zayed et al. [37] for AgNPs fabricated using *Malva parviflora*. Further evidences to these facts are found in reports of Dubey et al. in synthesis of AgNPs using *Sorbus aucuparia* leaf extract, in *Murraya Koeniggi*, in *Ocimum sanctum,* and in *Mangifera indica* [38–40]. All these investigations not only reiterated the role of biomolecules present in the leaf extract as reducing and capping agent during the synthesis, but also confirmed that formation of the NPs is quite rapid and the particles show high stability in the phytosynthesis. One of the major focuses in phytosynthesis was faster synthesis of stable AgNPs. Prarthana et al. (2011) [41] achieved synthesis of AgNPs within 10 min using *Citrus limon* and inferred that citric acid synergistically with bio-organics in the lemon juice caused the reduction and capping process.

Various modern techniques are applied in the identification and characterization of the NPs or the environment conditions during the formulation of the same. Yilmaz et al. in their synthesis of AgNPs using *Stevia rebaubiana* [42] leaf extract, applied proton nuclear magnetic resonance spectrum to assess the extract after the accumulation of AgNPs, which revealed the existence of aliphatic CH_2, alcoholic CH_3 groups as well as aromatic compounds in the medium, but without any sign of aldehydes or carboxylic acids. In another study by Metz et al. [43], SERS analysis demonstrated that organic constituents of coffee were present as adlayers on metal nanoparticle surfaces, and in addition, TGA measurements stated the presence of organic residues bound to the carbon microsphere scaffolds. This study brings out the importance to validate methods to perform careful analysis nanomaterials synthesized using greener methods.

In the selection of plants for phytosynthesis, one of the major focuses is antioxidant richness in the plant. Nadagouda et al. [44] became successful in synthesis of AgNPs with antioxidant coatings by using extracts of blackberry, blueberry, pomegranate and turmeric, rich in antioxidants such as polyphenols, catechin, quercetin, anthocynanins, and curcumin. Thus, it became established that phytosynthesis using antioxidant-rich plants enables formation of AgNPs with antioxidant coatings for better delivery of the same in therapeutic applications.

It may be noted that phytosynthesis using plants rich in amino acids, citric acid, flavonoids, heterocyclic compounds, terpenoids, phenolic compounds, and alkaloids enables fast formation of stable particles because such biomolecules act as reducing agents in the process. In the conversion of metal ions to nanoparticles, reduction, nucleation, and subsequent growth are important steps. Moreover, the synthesized particles will get capped by the various functional groups such as alcohols, aldehydes, amines, ketones, or extracellular proteins, which make the nanoparticles stable. Plants vary in the amounts of phytochemicals, which is likely to have diverse influence on the reduction process and stabilization of the nanoparticles preventing aggregation. Therefore, trials using diverse plants and environment conditions are significant to unravel the mysteries regarding the formation of wonderful nanoparticles using phytosynthesis.

In the last decade, researchers have used extracts of different plant parts for phytosynthesis of the NPs. Plant parts utilized include whole plant [45], bark [46, 47], latex [48], seed [49], fruit peel [50], inflorescence [51], corn [52], tuber [53], fruit [41], flower [54], rhizome [55], stem and root [16], gum [56], and even bran [57]. Researchers have also tried animal and plant products such as gelatine and glucose [58] and starch [59]. In addition to all these, combinations of plant extracts such as D-Sorbitol along with leaf extract of *Polyathia longifolia* (Kaviya et al. [50]) is used to increase the stability of AgNPs, and the synthesized NPs were found highly toxic to the gram-negative bacteria.

In general, research output indicates phytosynthesis requires lesser time than that for physiochemical methods. For instance, some extracts produce silver nanoparticles within 2 min, 5 min, 45 min, 1 h, and 2 h [18, 60–62]. The rate of reduction is much faster than using microorganisms [13, 63, 64]. AgNPs formed by plant extracts remain stable for longer times which can be analyzed using the TEM and DLS studies. Polydispersity index (PDI < 0.5) obtained from DLS measurements is a good sign to analyze the stability of particles [65, 66] along with TEM images showing dispersed particles after a storage period of 6 months–1 year. In most of the studies, the involvement of functional groups of phytochemicals is revealed spectroscopically, i.e., by FT/IR studies. The shape and size of particles are analyzed mainly using TEM, SEM, AFM, XRD, and DLS. Figure 10.2 describes different parameters which are needed for homogenous production of NPs so as to produce them in large scale.

10.2.2 Effect of Environment Parameters Influencing Phytosynthesis

It may be noted that the efficiency of phytosynthesis is dependent, mainly on three parameters namely pH, temperature, and incubation time, which influence the morphology and size of nanoparticles. Therefore, the main

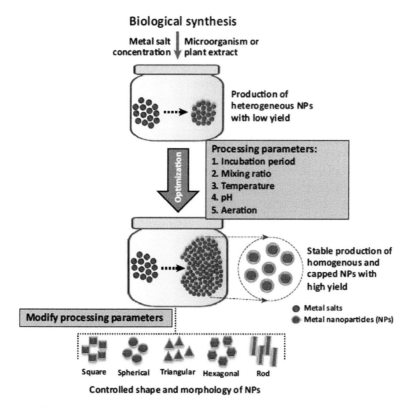

Figure 10.2   Different parameters which are involved in the production of homogenous NPs [9].

challenge associated with biosynthesis is the method to control the shape and size of particles so as to obtain monodisperse nanoparticles. Many researchers have focused on constraining the diverse physicochemical environment factors and the influence of the same on biomolecules in the extract acting as reducing or capping agents [60, 67].

The role of pH in phytosynthesis of NPs is crucial as observed by previous researchers. The pH changes can lead to changes in the charge of phytochemicals present in the plant, which further affects their binding ability and reduction of metal ions during NP synthesis. Many researchers have observed that the size or shape or both and yield of NPs vary with the change in pH [11, 68, 69]. Phytosynthesis using *Cinnamomum zeylanicum* has shown that acidic pH favours aggregation [47]. The latter author observed that at higher pH, availability of a large number of functional groups helps the binding of higher number of Ag^+ to bind and subsequently form large number of NPs. It was also observed that phytosynthesis leads to a drop in pH of the medium [70]. However, influence of alkaline pH in the formation of hexagonal and triangular gold nanoparticles in 2010 [71] and AgNPs in the extract of *Curcuma longa* [72] is also observed. According to the authors, more negatively charged functional groups capable of efficient binding and reducing of silver ions in the alkaline pH make possible synthesis of NPs.

Zeta potentials or the charge of the nanoparticles is also pH dependent. In 2010, Prabha et al. [73] could synthesize spherical AgNPs at room temperature within 10 min into the reaction. They were found to have a high negative zeta potential value and were stable under a wide pH range. Dubey et al. [74] reported that AgNPs show a negative zeta potential value at strong acidic pH compared to alkaline pH so that the NPs possesses high stability and small size at basic pH. The authors used the SPR spectra at various pHs to substantiate this fact. Contradictory to this finding, Andreescu et al. [75] observed rapid and complete reduction of silver at elevated alkaline pH with NPs having a negative zeta potential. They reported that an increase in pH of the medium is due to an increase in negative zeta potential that led to the formation of highly dispersed nanoparticles. In contrast, Dwivedi and Gopal [76] revealed that silver and gold particles are stable in a wider range of pH as they observed very slight variation in the zeta potential between pH 2–10 in their study using extracts of *Chenopodium album*.

Temperature is also positively correlated to the rate, shape, and size of NPs [77]. These authors observed increase in the rate of formation of gold nanoparticles of different morphology at high temperatures. They could differentiate the synthesis of nanorod and platelet-shaped particles at higher temperature, while spherical particles were seen at lower temperatures. Therefore, the above studies implied temperature as one crucial factor dominating the shape and size of NPs. Soman and Ray [66] also reported similar observation about formation of AgNPs at higher temperature. According to Sathishkumar et al. [72] an increase in surface plasmon resonance with an increase in temperature is responsible for the positive correlation between yield of particles and temperature. In general, the size of NPs decreases toward increase in temperature [78, 79]. In the study using tansy fruit extract, Prabha et al. [69] observed an increase in yield of AuNPs and AgNPs with an increase in temperature from 25 to 150°C. It may be noted that with an increase in temperature, the rate of reaction also increases, which enhances the synthesis of particles [76, 80]. The increase in sharpness of absorbance peak obtained in the UV spectrum reiterates the lowering in the size of synthesized nanoparticles along with increase in temperature. Quite contradictory to the general finding, Lengke et al. [81] observed larger particle formation at higher temperature due to the efficient crystal growth of (111) NPs.

Incubation time also affects the synthesis of nanoparticles. It is the time required to complete the steps of reaction. Shankar et al. [29] observed that within 4 h of reaction of leaf extract and silver nitrate, 90% reduction of silver salt to nanoparticles is completed. According to the authors, reduction of AgNPs takes more time than that of AuNPs; because the redox potential of silver ions to metallic silver is lower than gold. Li et al. [31] demonstrated synthesis of AgNPs using *Capsicum annuum* extract at room temperature; the authors suggested that the size of NPs increases with reaction time. Increase in incubation time causes phase change from polycrystalline to single crystalline nature of the NPs. Tripathy et al. [34] observed AgNP, which was synthesized using aqueous leaf extract of *Azardirachta indica*. They identified presence of organics in the neem leaf broth that has control on both nucleation and growth of AgNPs. The spectral studies revealed that, after a period of 24 h, the action of organics no more continues and, by that time, a desired dimension of NPs is reached. Prabha et al. [73] were able to synthesize spherical AgNPs (12 nm) at room temperature within 10 min into the reaction, and reaction was completed in a day.

Increase in sharpness of UV spectra reveals continuation of NP formation. Increase in incubation time has no influence on formation of NPs from 20 min to 2 h in AgNP synthesis using Chenopodium leaf extract, but slight variation in the rate of formation of NPs takes place after 2 h [76]. In the Tansy fruit-mediated synthesis of Ag and Au particles, Dubey et al. also observed a similar influence of incubation time on phytosynthesis. An incubation time of 60 min was observed as the optimum incubation time in the formation of AgNPs using Mangosteen leaf extract [70] and *Rosa damascene* [82]. All these studies point to the importance of different environment parameters essential to tailor successful synthesis of nanoparticles of desired size and shape in phytosynthesis.

10.3 Probable Mechanism for AgNP Formation

The three main components for the formation of metallic NPs are the reducing agent, stabilizing agent, and a solvent medium which solubilize the metal of interest [83]. As discussed in the introduction part, the synthesis of NPs using plant extracts is strictly a green nanotechnology method which follows the green chemistry principles to the core because here the extract itself acts as the reducing and stabilizing agent. Phytosynthesis abides to the principles of green chemistry by the use of aqueous medium instead of organic solvents which is apparently more eco-friendly and cost effective.

Tanhn et al. [84] reviewed the major process of nucleation and growth of silver nanoparticles. The authors have presented details of identification of the two phases involved in the formation of nanoparticles. The first stage is initiation of autocatalytic reduction nucleation (as proposed by Finke-Watzky [85]). In the second phase, increase in size of particles takes place by agglomeration of small particles so that total number of particles decreases in the medium of synthesis, which is known as Ostwald ripening (in agreement with LSW theory [86]). Shields et al. [87] also describe initial nucleation (follows burst nucleation, LaMer mechanism) followed by coalescence as the process of growth of particles.

Lukman et al. [88] have identified three distinct stages of biosynthetic process, i.e., a short induction period, a growth phase, and a termination period for the silver nanoparticles. Always particles rearrange themselves to achieve a morphology which is more stable. The growth rate of phytosynthesized NPs is slower than the chemical and physical methods. Gan and Li [89] state that when strong ligands are absent in the solution, metallic ions associate with extract through ionic binding with the reducing agents present in the plant extracts. They attribute the presence of pi electrons and carbonyl groups in their molecular structures for absorption of reducing agents to the surface of metallic NPs. Figure 10.3 illustrates the different steps in nanoparticle formation.

NPs of desired shape and size can be obtained by adjusting the reaction conditions such as pH, temperature, and metal concentration. Nanoparticles

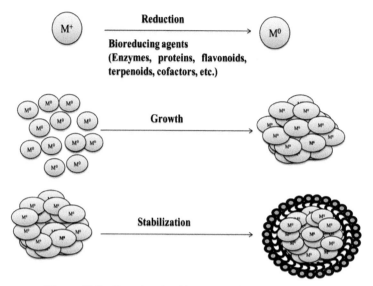

Figure 10.3   Steps involved in nanoparticle synthesis [90].

with different morphology are formed as a result of coalescence of small particles in the solution. For example, the shapes such as nanorods, nanowires, and nanoprisms are formed by aggregation. The stability of fabricated nanoparticles is determined by the different composition of biomolecules present in the plants and also the reaction condition. Different plants have different biomolecules; therefore, it is quite difficult to maintain and reproduce the shape and size of NPs in phytosynthesis [91].

10.4 Importance of Antibacterial Activity of Phytosynthesized AgNPs

Antibiotic resistance has become a growing global threat due to the resistance developed by microorganisms against antibiotics due to its overuse and misuse [92, 93]. The real meaning of a resistant bacteria is "they respond, but only at a higher concentration" [94, 95]. Bacteria resist antibiotics as a result of chromosomal mutation or inductive expression of a latent chromosomal gene or by exchange of genetic material through transformation [96], transduction (Bacteriophage) [97], conjugation by plasmids (extra chromosomal DNA) [98], and efflux [99]. *Pseudomonas aeruginosa*, methicillin resistance *Staphylococcus aureus* (MRSA), and Enterococcus are the most common drug-resistant bacteria which cause severe illness [100, 101].

Bacteria are often successful in developing resistant mechanism to new drugs or there analogs as accounted [102]. The action on bacteria is by the simultaneous processes which include electrostatic interaction with the cell membrane, production of reactive oxygen species, ion release, and internalization [103]. This makes the resistant mechanism expressed by antibiotics ineffective. The surface area and chemical reactivity of nanomaterials are higher or exceedingly large than its bulk counter parts. Therefore, at a low concentration, they provide high bioactivity [102] and can be used alone or in combination against bacteria and microbes.

During 1990s, medical fraternity was in need of an effective antimicrobial coating for the medical devices. The medical field uses appliances ranging from orthopedic pins, plates, implants, urinary catheters, and wound dressings, and an inexpensive antimicrobial coating was needed so as to improve the antibacterial efficacy of the above-mentioned devices. The antimicrobial efficacy of metallic ions such as Ag, Au, Pt, Ir, Zn, Cu, Sn, Sb, and Bi [104] were already established. Among them, silver was used to treat and prevent microbial infections due to its oligodynamic action. Silver salts (like silver nitrate) were used in ancient medical systems for the disinfection of wounds and surgical devices and also in first water filtering systems [105, 106] due to its antimicrobial action. The use of nanoparticulate silver in consumer products is the greatest application of the nanomaterials use till date, and the products vary from wound dressings, socks, other textiles, air filters, toothpaste, baby products, vacuum cleaners, and washing machines [107]. There were several failures for the available metallic silver coatings. The most probable alternative strategies to confront this failure was either to try activation processes along with coating or use of AgNPs [103].

The production of AgNPs by chemical reduction (using hydrazine hydrate, sodium borohydride, DMF, and ethylene glycol) may lead to the absorption of harsh toxic chemicals on to the surfaces of NPs raising the question about the toxicity of thus-synthesized particles. FDA has approved the use of AgNPs in making wound-healing material due to its antibacterial activity where either metallic AgNP or ionic silver can be used. Thus, AgNPs synthesized using plants are an alternative which can be exploited for its antimicrobial efficiency. Aqueous silver nitrate solution on reaction with plant leaf extract results in the formation of highly stable, crystalline NPs which assembled in the reaction medium into particles with different morphologies. The shape of particles depends on the reaction parameters optimized for different plants. The size control of NPs is shown to be a function of time. It has been widely demonstrated that the bioactive properties of AgNPs are related to its shape and size [105]. The biological effectiveness of NPs can be increased due to their increased surface area, and this can be attributed to increase in surface energy [108]. Martínez-Castánn et al. [109] suggest that the improved chemical activity of metallic NPs is due to their crystallographic surface structure and higher surface area which allows large number of atoms to interact with cells. The next section gives an account on the antibacterial efficiency exhibited by the phytosynthesized AgNPs.

10.4.1 Antibacterial Activity of Phytosynthesized AgNPs: Case Studies

The antibacterial activity of silver ions and silver nanoparticles in particular is of significant interest as the activity appears to be independent of the strain of bacteria. The use of phytosynthesized AgNPs as antimicrobial agent has been exploited by a number of researchers in the past and is still continuing. Phytosynthesized NPs exhibited a higher antibacterial activity than the physiochemically synthesized NPs. Bacterial susceptibility to antimicrobial agents depends on the cell wall structure. Case studies span the antibacterial activity observed for AgNPs against both gram-positive and gram-negative bacteria.

The study by Prabakar et al. [110] observed the phytofabrication of AgNPs using the leaf extract of *Mukia scabrella* (Musumuskkai) [110]. The AgNPs exhibited significant antimicrobial activity against multidrug-resistant gram-negative bacteria (MDR-GNB) nosocomial pathogens such as *Acinetobacter sp., Klebsiella pneumoniae, and Pseudomonas aeruginosa.* They stated that the phytosynthesized AgNPs exhibited stronger antibacterial activity than the chemically synthesized NPs and leaf extract alone. In our previous study, we were able to successfully synthesize AgNPs and ZnO NPs using the leaf extract of *Ziziphous oenoplia*, and a higher antibacterial activity was observed in zone of inhibition studies for the phytosynthesized silver and zinc oxide NPs compared to the chemically synthesized ones. This needs to be further validated using *in vitro* cytotoxicity and cell viability studies.

The phytosynthesized AgNPs using *Acalypha indica* [111] against waterborne pathogens such as *Escherichia coli* and *Vibrio cholera* gave a minimum inhibitory concentration (MIC) value for both the pathogens at 10 μg/mL, whereas MIC of AgNO$_3$ was 20 μg/mL. The lower value of MIC for AgNPs is due to the smaller size of the NPs [112]. In the study, the authors measured the changes in membrane permeability and respiratory action of the bacterial cell, in order to understand the bactericidal nature of the AgNPs. The membrane permeability test results showed an increase in the permeability which is due to the serious damage caused to the cell membrane structure by the AgNPs (in TEM micrographs) and reduction in respiration rate of AgNPs-treated bacterial cells decreased, which might be due to the inhibitory activity of AgNPs on the respiratory enzymes (cytochrome oxidases, malate dehydrogenase, and succinate dehydrogenase). Table 10.1 lists summary of a few case studies done in the past decade toward the antimicrobial activity of phytosynthesized AgNPs.

Table 10.1 Lists summary of a few case studies done in the past decade toward the antimicrobial activity of phytosynthesized AgNPs

Study	Plant Used	Organism Under Study	Controls	Outcomes
[65]	*Ocimum sanctum*	*E. coli* and *S. aureus*	Silver nitrate Ciprofloxacin	MIC gives higher and stronger activity than both controls
[70]	*Garcinia mangostana*	*E. coli* and *S. aureus*	Amikacin, Cefalotaxin Penicillin, Methoxazole	The zone obtained was comparable to the antibiotics
[113]	*Cornus mas*	*E. coli, S. aureus, Enterocoocus faecalis*	–	Well diffusion assay. Strong activity was observed against *S. aureus.*
[114]	*Rhinacanthus nasutus*	*Bacillus Subtilis, S. aureus, P. aeruginosa, K. pneumonia, E. coli, A. niger, A. flavus*	Leaf extract, Ciprofloxacin	Disc diffusion activity was higher than to extract and comparable with Ab
[115]	*Lantana camara*	*E. coli, P. aeruginosa, Bacillus* and *Staphylococcus*	–	Disc diffusion method used, Bacillus and Pseudomonas showed a higher activity
[116]	*Ocimum tenuiflorum, Solanum tricobactum, Syzygium cumini, Centella asiatica, Citrus sinensis*	*S. aureus, P. aeruginosa, K. pneumonia, E. coli*	–	Well diffusion. Solanum and Ocimum showed highest activity against *S. aureus* and *E. coli.* different volumes of AgNPs used, as volume increased, activity increased

[21]	*Aegle marmelos*	*E. coli*	Leaf extract	Well diffusion, significant activity was obtained but was insensitive to AuNPs
[117]	*Desmodium trifolium*	*E. coli, S. aureus, Bacillus subtillis*	Gentamiycin	Disc diffusion was done. MIC value of $57\mu g/mL$ against *S. aureus* and *E. coli* was seen. Above $100\mu g/mL$, 100% killing was observed for all three.
[118]	*Mentha piperita*	*E. coli* and *S. aureus*	Leaf extract	Disc diffusion, higher activity against *E. coli* than *S. aureus*
[119]	*Ocimum tenuiflorum*	*E. coli, Corney bacterium, B. subtilis*	Leaf extract	Well diffusion, activity observed, but no change as the in activity as volume was increased
[120]	*Olive leaf extract*	*E. coli* and *S. aureus*	Leaf extract	Well diffusion, *E. coli* less sensitive than *S. aureus*
[121]	*Aloe vera*	*E. coli* and *S. aureus*	Leaf extract	Higher activity was observed against *E. coli* than *S. aureus*
[122]	*Piper nigrum*	*E. coli* and *S. aureus*	Ciprofloxacin	Antibacterial activity was dependent on the size of Ag particles. As size increases, activity decreases
[123]	*Iresine herbestii*	*S. aureus, P. aeruginosa, K. pneumonia, E. coli,* and *E. Faecalis*	Kanamycin, Norfloxacin, Ciprofloxacin	MIC < for AgNP than antibiotic, Klebsiella more sensitive to AgNPs at lower concentration
[124]	*Eucalyptus chapmaniana*	*S. aureus, K. pneumonia, E. coli, P. aeruginosa, Candida albicans, Proteus vulgaris,*	Leaf extract, silver nitrate	Antimicrobial effect was dose-dependent activity against, yeast, both gram-positive and negative organisms
[110]	*Mukia scabrella*	*P. aeruginosa, K. pneumonia, Acetobacter*	Leaf extract, commercial AgNPs	Phytosynthesized AgNPs have higher activity than the controls

(Continued)

Table 10.1   Continued

Study	Plant Used	Organism Under Study	Controls	Outcomes
[125]	*Anigozanthos manglesii*	*Deinococcus, E. coli, Staphylococcus Epidermis*	Leaf extract, commercial AgNPs	Disc diffusion activity against *Deinococcus* only, to others resistant
[126]	*Argemone Mexicana*	*E. coli, P. syringae, Aspergillus flavus*	–	Disc diffusion, toxic against the microorganisms, and a higher ZOI observed for treated than control
[127]	*Psidium guajava*	*E. coli*	Chemically synthesized AgNPs	Green synthesized showed higher activity than chemically synthesized, not much difference in morphology

10.5 Mechanism of Action of AgNPs

The need to understand the bioactivity/toxicity mechanism of NPs is highly relevant due to the fact that they should be simultaneously toxic against the pathogens and safe for humans and environment by nature. AgNPs with at least one dimension in range of 1–100 nm are able to physically interact with cell wall of different bacteria. AgNPs have better physical, chemical, or biological properties than their bulk counterparts due to small size and higher surface to volume ratio. Numerous studies were done in gram-negative bacteria where researchers have observed adhesion and accumulation of AgNPs to bacterial surface. The effect on gram positive is different due to the presence of a thick peptidoglycan layer.

10.5.1 Different Postulates of Mechanism of AgNPs Toxicity to Bacteria

Researchers in the past have tried to evaluate the different functional groups interacting with bacterial cell membrane, which might be one reason for their bacterial toxicity. One of the earlier studies by Kim et al. [103] demonstrated that the silver ions released from $AgNO_3$ resulted in bacterial death due to interaction of Ag^+ with thiol groups in cell wall of bacteria. Alongside, they also observed slight morphological changes in *S. aureus* compared to *E. coli* which explains a defense mechanism by gram positive against inhibitory effects of silver ions. Few researchers have demonstrated that action of NPs is more effective than their precursor ionic salts (like $AgNO_3$). Holt et al. [128] analyzed the interaction of silver ions with the respiratory chain of *E. coli*. They attribute the death of bacterial cells to the interaction of Ag^+ with the membrane-bound proteins found in the cytoplasmic membrane such as transport or respiratory chain enzymes. Scanning electrochemical microscopy (SECM) shows Ag^+ uptake by immobilized cells, of that about 60% of silver are seen transported into the cell and remaining 40% binds to the outside of the cell. According to Shrivastava et al. [129], the major mechanism of action of nanoparticles on bacteria might be by anchoring and penetrating the bacterial cell wall and then modulating cellular signaling by dephosphorylating peptide substrates on tyrosine residues.

Antibacterial activity of AgNPs relies on factors such as shape, size of the particles, surface condition, synthesis method, accumulation of nanoparticles near the cytoplasm, and electrostatic forces [130]. In addition, hydrophobocity and concentration of NPs are other major factors. Daima and Bansal [131] and Samberg et al. [132] were successful in establishing the relation of antibacterial activity of AgNPs to size, surface condition, and synthesis method. The authors stated that the activity increased with decreasing particle size [109]. Along with that, it was noted unmodified AgNPs synthesized by base catalyzed reduction method have greater antibacterial activity than carbon-modified AgNPs synthesized by carbon-stabilized laser ablation method which gave an insight into effect of surface modification of the particles on biological activity. Agnihotri et al. [22] explained the bacteriostatic and bactericidal effect of AgNPs to be size- and dose-dependent and an enhanced effect was observed with particles less than 10nm in size. In another study, the same authors [133] illustrated an enhanced contact killing nature of immobilized AgNPs than the colloidal AgNPs. There is a positive correlation of size of NPs and its effect on bacterial cells. Sotiriou and Pratsinis [134] explained that when particles are smaller, the antibacterial activity is dominated by the large number of Ag^+ ions released. However, when particles of size >10 nm are employed, low Ag^+ ions are released and the particles themselves influence antibacterial activity. Also, the shape of formed NPs has some influence on the damage caused due to the interaction of NP and bacteria [135]. Ivask et al. [136] state that spherical NPs with a diameter smaller than 20 nm are effective than other morphologically different NPs and the bulk counterparts. In general, smaller the NP size, higher the surface area for interaction and higher the ionic release.

Many studies have reported that AgNPs cause cell membrane damage which leads to structural changes in membrane leading to the death of bacterial cells. For instance, Sondi and Salpoek-Sondi [137] observed "pits" in cell wall of *E. coli* cells treated with AgNPs and also observed accumulated particles on the cell membrane from TEM images. Prabha et al. [69] assumed that the mechanism of action of silver ions on bacterial cells involves shrinkage of the cytoplasm membrane and its detachment from the cell wall and they made similar observations as that of Morones et al. [112]. The cellular uptake of NPs generally occurs through interaction of NP surface with the lipopolysacchaaride of gram-negative bacteria as well as with the cell wall or cytoplasmic membrane of other bacteria [135]. Usually, the NP internalization is promoted due to the electrostatic interaction between NP surface and cell membrane of the bacteria. The observations by Sportelli et al. [138] account for the mechanism of action of AgNPs and ZnO NPs as they can cross the cell membrane by endocytosis, diffusion, and perforation.

In the study of Agrobacterium and AgNPs interaction using TEM, Sarmast and Salehi [139] observe that when the bacteria are treated with 10µg/mL AgNPs for 20 h, the bacterial growth was completely suppressed. In TEM, they observed that NPs in clusters not only anchored around the cell wall but they are completely accumulated in the cell as well. This suggests that the cellular balance is compromised due to the effect of NPs on cell membrane, thus causing hindrance to the metabolic process inside the cell as well. Also, they could visualize very small NPs in the range of 3–50 nm inside the cell.

In our previous work [66], we have observed some similar changes in *E. coli* treated with phytosynthesized AgNPs as noted by Sarmast and Salehi [139]. Figure 10.4(a) shows the transition electron micrographs of untreated *E. coli* displayed the characteristic bacilli shape, (b) explains the *E. coli* cells treated with AgNPs for 30 min was seen with ruptured cell wall, condensed cytoplasm, cells with shrinkage, and detached flagella. (c) The images showed that on exposure, the AgNPs were present on the cell membrane, and (d) they appear to be attached to the lipopolysacchaaride layer in cell wall of gram-negative *E. coli* as reported in previous studies. Figure 10.4(e–g) shows the changes in *E. coli* cell at 24 h treatment with AgNPs.

A few authors in the past have looked at the changes in growth phase of bacteria on treatment with AgNPs. Kim et al. [103] observed growth inhibitory effects of AgNPs against yeast, *S. aureus* and *E. coli*. A close observation of growth kinetics study of AgNP–*E. coli* interaction suggested that the lag phase of treated bacteria was prolonged than the untreated control

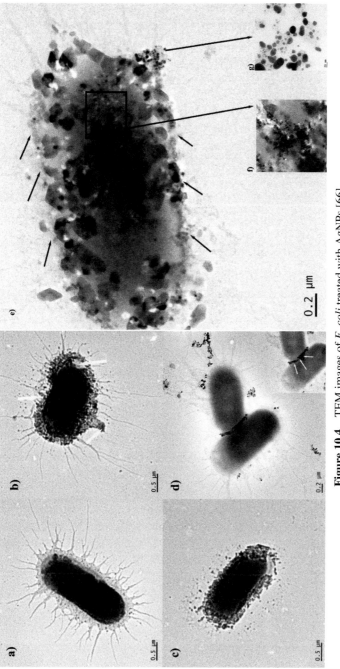

Figure 10.4 TEM images of *E. coli* treated with AgNPs [66].

and a higher number of live bacteria were observed in stationary phase than the treated bacterial samples [140]. The study by Lara et al. describe the role of AgNPs as a broad-spectrum antibacterial agent [141] and a bactericidal agent [142] rather than bacteriostatic agent. These studies suggest the role of AgNPs as bactericidal agent and the effect it has on growth kinetics of bacteria.

In addition to the changes in the membrane of bacteria and binding of ions to the different groups of functional groups in cell membrane by AgNPs, [143] explained the alteration in protein synthesis due to AgNPs treatment and proteomic data have shown an accumulation of immature precursors of membrane proteins which results in destabilization of outer membrane composition. Several studies also consider the release of reactive oxygen species by AgNP induction that forms free radicals having powerful bactericidal action.

10.6 Antibacterial Applications of AgNPs

AgNPs synthesized by the various methods have been used in different *in vitro* diagnostic applications [144–147]. Sudaramoorthi et al. [148] could describe wound-healing application of biosynthesized AgNPs, synthesized using fungus *Aspergillus niger.* The efficiency was established by experimenting in rat model and similarly [150] tested AgNPs produced using *Indigofera* for their wound-healing applications following excision in animal models. They supported both the antimicrobial effects of Ag and the ability of AgNPs to modulate cytokines involved in wound healing. The antibacterial activity of AgNPs was exploited by He et al. [149], when it was used in the preparation of clinical ultrasound gel. The gel was used in an ultrasound probe, and along with that, bactericidal activity and instrument sterility were evaluated.

Samberg et al. [151] have developed an LIDC-activated prophylactic surface system which provides controlled release of antibacterial silver ions, which is a major way to prevent transmission of infections from the surface of medical devices (scalpel handles, endoscopes, forceps), surgical sites, stethoscopes, hospital beds rails, door handles, and elevator handles. The system was very effective in completely inhibiting the growth of both *E. coli* and *S. aureus* after an exposure of 1.5 h. In one of the recent studies by Nasser and Wu (WRP 1) [143] synthesized AgNPs coated on central venous catheters (CVC) showed higher antibacterial activity and they exhibited improved biocompatibility and evident sterilization effect. The cytocompatibility analysis of catheters using WST-1 assay illustrates it to be safe for human use. Similar results of antimicrobial activity of plastic catheters coated with AgNPs prevented biofilm formation of *E. coli, S. aureus, Candida albicans,* and *P. aeruginosa* [152].

Several reports in the last decade have exploited the use of AgNPs as a component in water filters [153] and as air filter [154]. Hasan et al. and Kelly et al. [155, 156] account for the use of silver-based coatings in medical implants due to the release of silver ions from the surface which are quite bactericidal against gram-negative and gram-positive bacteria. The observed bactericidal activity of silver ions is due to the latter's binding or interaction with thiol groups on bacteria. In addition, silver ions also inactivate respiratory chain and TCA enzymes and induce hydroxyl radical formation causing change in the cellular DNA [157]. Table 10.2 accounts for the list of ongoing clinical trials for AgNPs [158].

Table 10.2 List of clinical trials of AgNPs [158]

Clinical Trials.gov Identifier	Status	Year	Study
NCT00341354	Completed	2006	Coated Endotracheal Tube and Mucus Shaver to Prevent Hospital-Acquired Infections.
NCT00337714	Completed	2006	Comparison of Central Venous Catheters With Silver Nanoparticles Versus Conventional Catheters (NanoAgCVC).
NCT00659204	Unknown	2008	Efficacy of AgNP Gel Versus a Common Antibacterial Hand Gel.
NCT00965198	Completed	2009	Comparison of Infection Rates Among Patients Using Two Catheter Access Devices.
NCT01258270	Completed	2010	Efficacy and Patient Satisfaction With AQUACEL® Ag Surgical Dressing Compared to Standard Surgical Dressing.
NCT01598480	Completed	2012	To Study the Healing Effect of Silver Impregnated Activated Carbon Fiber Wound Dressing on Superficial Dermal Burn.
NCT01598493	Completed	2012	To Study the Healing Effect of Silver Impregnated Activated Carbon Fiber Wound Dressing on Deep Dermal Burn.
NCT02213237	Recruiting	2013	The Application of SERS and Metabolomics in Sepsis.
NCT02241005	Recruiting	2014	Theraworx Bath Wipes Versus Standard Bath Wipes in the Reduction of Vancomycin Resistant Enterococci.
NCT02761525	Completed	2016	Topical Application of Silver Nanoparticles and Oral Pathogens in Ill Patients.

10.7 Conclusion

The results discussed in the review are very much encouraging which supports the application of green chemistry, i.e., synthesis from plant extracts that can produce NPs of desired shape by controlling the synthesis parameters. The widespread use of AgNPs in consumer products in the areas such as electronics, filtration, purification, neutralization, sanitization, personal care and cosmetics, household products, textiles, and shoes indicates that human exposure to these NPs is indispensable. The exact data on the concentrations of AgNPs used up in the products, its shape and size, and the form in which it is present in the products and materials are hardly available. The NPs synthesized using physicochemical methods are used in the synthesis of NPs often and is relatively toxic. Therefore, the switch over to fabrication of NPs using plants, microorganisms, fungi, and algae have become promising and can potentially diminish the toxicity issues arising from the earlier methods. In addition, there is also a need to understand the biological corona which surrounds the nanoparticles synthesized using biological ways that not only dominate the functionality of material but also its ultimate performance. Very few studies were done in the past analyzing the same except a few have used techniques such as SERS and IRRAS to look at surface chemistry of Ag/CM and Pd/CM [43].

In order to predict the pharmacokinetic parameters which are important determinant of effective dosage for understanding the efficacy of antimicrobial activity of nanocompounds, studies in cell line and animal models are the need of the hour. Some initial studies using the AgNPs are there in the literature. But in one of the earliest studies [102], Shimanovich and Gedanken describe the absence of toxicity among silver-based nanocomplexes developed as nano-antibiotic system. In addition, issues relating to the biomedical application including the distribution profile, excretion, and clearance of particles in *in vivo* trial need to be addressed. The investigations into the biocompatibility and bioavailability of nanoparticles are still at initial stages and considerable research needs to be done in this direction. As phytosynthesis has evolved to produce NPs using environment friendly methods, its widespread application in medical field exploiting its antimicrobial application has seen tremendous growth in the last decade. The question of being nontoxic to humans is of importance, and few studies have come up regarding the same. The AgNPs produced by tea leaf extracts by Moulton et al. [159] were found to be nontoxic to humans. Apart from *in vitro* and *in vivo* tests, NPs should be evaluated using a rational characterization strategy involving adsorption, distribution, metabolism, and excretion (ADME) tests [160].

In order to be considered as an alternative for chemically synthesized nanoparticles, phytosynthesized nanoparticles need to be compatible with chemically synthesized particles; i.e., the yield should be higher which solely depends on the source considered for production of nanoparticles. And the choice depends on biomass type considered. And this remains as a major challenge to scalable production and applicability [161] of phytosynthesized AgNPs.

References

[1] Rai, M., Gade, A., and Yadav, A. (2011). "Biogenic Nanoparticles: an Introduction to What they are, How they are synthesized and their applications," in *Metal Nanoparticles in Microbiology,* eds M. Rai and N. Duran (Berlin: Springer), 1–14.

[2] Daniel, M.-C., and Astruc, D. (2004). Gold nanoparticles: assembly, supramolecular chemistry, quantum-size-related properties, and applications toward biology, catalysis, and nanotechnology. *Chem. Rev.* 104, 293–346.

[3] Bogunia-kubik, K., and Sugisaka, M. (2002). From molecular biology to nanotechnology and nanomedicine. *Biosystems* 65, 123–138.

[4] Zharov, V. P., Kim, J., Curiel, D. T., and Everts, M. (2005). Self-assembling nanoclusters in living systems: application for integrated photothermal nanodiagnostics and nanotherapy. *Nanomedicine* 1, 326–345. doi: 10.1016/j.nano.2005.10.006

[5] Lee, H., Li, Z., Chen, K., Hsu, A. R., Xu, C., Xie, J., et al. (2008). PET/MRI dual-modality tumor imaging using radiolabeled iron oxide nanoparticles. *J. Nucl. Med.* 49, 1371–1380. doi: 10.2967/jnumed.108.051243

[6] Panigrahi, S., Kundu, S., Ghosh, S. K., Nath, S., and Pal, T. (2004). General method of synthesis for metal nanoparticles. *J. Nanoparticle Res.* 6, 411–414.

[7] Pissuwan, D., Valenzuela, S. M., and Cortie, M. B. (2007). Therapeutic possibilities of plasmonically heated gold nanoparticles. *Trends Biotechnol.* 24, 62–67. doi: 10.1016/j.tibtech.2005.12.004

[8] Tan, M., Wang, G., Ye, Z., and Yuan, J. (2006). Synthesis and characterization of titania-based monodisperse fluorescent europium nanoparticles for biolabeling. *J. Lumin.* 117, 20–28. doi: 10.1016/j.jlumin.2005.04.004

[9] Singh, P., Kim, Y., Zhang, D., and Yang, D. (2016). Biological synthesis of nanoparticles from plants and microorganisms. *Trends Biotechnol.* 34, 588–599. doi: 10.1016/j.tibtech.2016.02.006

[10] Borase, H. P., and Salunke, B. K. (2014). Plant extract: a promising biomatrix for ecofriendly, controlled synthesis of silver nanoparticles. *Appl. Biochem. Biotechnol.* 173, 1–29. doi: 10.1007/s12010-014-0831-4

[11] Gardea-torresdey, J. L., Gomez, E., Peralta-videa, J. R., Parsons, J. G., Troiani, H., and Jose-yacaman, M. (2003). Alfalfa sprouts: a natural source for the synthesis of silver nanoparticles. *Langmuir* 19, 1357–1361.

[12] Kumar, V., Yadav, S. C., and Yadav, S. K. (2010). *Syzygium cumini* leaf and seed extract mediated biosynthesis of silver nanoparticles and their characterization. *J. Chem. Technol. Biotechnol.* 85, 1301–1309. doi: 10.1002/jctb.2427

[13] Iravani, S. (2011). Green chemistry green synthesis of metal nanoparticles using plants. *Green Chem.* 13, 2638–2650. doi: 10.1039/c1g c15386b

[14] Haverkamp, R. G., and Marshall, A. T. (2009). The mechanism of metal nanoparticle formation in plants: limits on accumulation, *J. Nanoparticle Res.* 11, 1453–1463. doi: 10.1007/s11051-008-9533-6

[15] Narayanan, K. B., and Sakthivel, N. (2011). Green synthesis of biogenic metal nanoparticles by terrestrial and aquatic phototrophic and heterotrophic eukaryotes and biocompatible agents. *Adv. Colloid Interface Sci.* 169, 59–79. doi: 10.1016/j.cis.2011.08.004

[16] Ahmad, N., Sharma, S., Alam, K., Singh, V. N., Shamsi, S. F., Mehta, B. R., et al. (2010). Rapid synthesis of silver nanoparticles using dried medicinal plant of basil. *Colloids Surf. B Biointerfaces* 81, 81–86. doi: 10.1016/j.colsurfb.2010.06.029

[17] Kannan, N., Mukunthan, K. S., and Balaji, S. (2011). A comparative study of morphology, reactivity and stability of synthesized silver nanoparticles using Bacillus subtilis and *Catharanthus roseus* (L.) G. Don. *Colloids Surf. B Biointerfaces* 86, 378–383. doi: 10.1016/j.colsurfb.2011.04.024

[18] Monopoli, M. P., Åberg, C., Salvati1, A., and Dawson, K. A. (2012). Biomolecular coronas provide the biological identity of nanosized materials. *Nat. Nanotechnol.* 7, 779–786.

[19] Makarov, V. V., Love, A. J., Sinitsyna, O. V., Makarova, S. S., and Yaminsky, I. V. (2014). *"Green" nanotechnologies: synthesis of metal nanoparticles using plants. Acta Nat.* 6, 35–44.

[20] Baker, S., Rakshith, D., Kavitha, K. S., and Santosh, P. (2013). Plants: emerging as nanofactories towards facile route in synthesis of nanoparticles. *BioImpacts* 3, 111–117. doi: 10.5681/bi.2013.012

[21] Rao, K. J., and Paria, S. (2015). *Aegle marmelos* leaf extract and plant surfactants mediated green synthesis of Au and Ag nanoparticles by optimizing process parameters using taguchi method, ACS *Sustain. Chem. Eng.* 3, 483–491. doi: 10.1021/acssuschemeng.5b00022

[22] Agnihotri, S., Mukherji, S., and Mukherji, S. (2014). Size-controlled silver nanoparticles synthesized over the range 5–100 nm using the same protocol and their antibacterial effica. *RSC Adv.* 4, 3974–3983. doi: 10.1039/c3ra44507k

[23] Noruzi, M. (2015). Biosynthesis of gold nanoparticles using plant extracts. *Bioprocess Biosyst. Eng.* 38, 1–14. doi: 10.1007/s00449-014-1251-0

[24] Dahl, J. A., Maddux, B. L. S., and Hutchison, J. E. (2007). Toward greener nanosynthesis. *Chem. Rev.* 107, 2228–2269. doi: 10.1021/cr050943k

[25] Sharma, V. K., Yngard, R. A., and Lin, Y. (2009). Silver nanoparticles: Green synthesis and their antimicrobial activities. *Adv. Colloid Interface Sci.* 145, 83–96. doi: 10.1016/j.cis.2008.09.002

[26] Thakkar, K. N., Mhatre, S. S., and Parikh, R. Y. (2010). Biological synthesis of metallic nanoparticles. *Nanomedicine* 6, 257–62. doi: 10.1016/j.nano.2009.07.002

[27] Song, J. Y., Jang, H., and Kim, B. S. (2009). Biological synthesis of gold nanoparticles using *Magnolia kobus* and *Diospyros kaki* leaf extracts. *Process Biochem.* 44, 1133–1138. doi: 10.1016/j.procbio.2009.06.005

[28] Shankar, S. S., Ahmad, A., and Sastry, M. (2003). Geranium leaf assisted biosynthesis of silver nanoparticles. *Biotechnol. Prog.* 19, 1627–31. doi: 10.1021/bp034070w

[29] Shankar, S. S., Rai, A., Ahmad, A., and Sastry, M. (2004). Rapid synthesis of Au, Ag, and bimetallic Au core – Ag shell nanoparticles using Neem (*Azadirachta indica*) leaf broth. *J. Colloid Interface Sci.* 275, 496–502. doi: 10.1016/j.jcis.2004.03.003

[30] Chandran, S. P., Chaudhary, M., Pasricha, R., Ahmad, A., and Sastry, M. (2006). Synthesis of gold nanotriangles and silver nanoparticles using *Aloe vera* plant extract. *Biotechnol. Prog.* 22, 577–583.

[31] Li, S., Shen, Y., Xie, A., Yu, X., Qiu, L., Zhang, L., et al. (2007). Green synthesis of silver nanoparticles using *Capsicum annuum* L. extract. *Green Chem.* 9, 852–858. doi: 10.1039/b615357g

[32] Huang, J., Lin, L., Li, Q., Sun, D., Wang, Y., Lu, Y., et al. (2008). Continuous-Flow Biosynthesis of Silver Nanoparticles by Lixivium of Sundried Cinnamomum camphora Leaf in Tubular Microreactors. *Ind. Eng. Chem. Res.* 47, 6081–6090.

[33] Kasthuri, J., Veerapandian, S., and Rajendiran, N. (2009). Biological synthesis of silver and gold nanoparticles using apiin as reducing agent. *Colloids Surf. B Biointerfaces* 68, 55–60. doi:10.1016/j.colsurfb.2008.09.021

[34] Tripathy, A., Raichur, A. M., Chandrasekaran, N., Prathna, T. C., and Mukherjee, A. (2010). Process variables in biomimetic synthesis of silver nanoparticles by aqueous extract of *Azadirachta indica* (Neem) leaves. *J. Nanoparticle Res.* 12, 237–246. doi:10.1007/s11051-009-9602-5

[35] Christensen, L. (2011). Biosynthesis of silver nanoparticles using *Murraya koenigii* (Curry leaf): an investigation on the effect of broth concentration in reduction mechanism and particle size. *Adv. Mater. Lett.* 2, 429–434. doi:10.5185/amlett.2011.4256

[36] Annamalai, A., Babu, S. T., Jose, N.A., Sudha, D., and Lyza, C. V. (2011). Biosynthesis and characterization of silver and gold nanoparticles using aqueous leaf extraction of *Phyllanthus amarus* Schum. & Thonn. *World Appl. Sci. J.* 8, 1833–1840.

[37] Zayed, M. F., Eisa, W. H., and Shabaka, A. A. (2012). *Malva parviflora* extract assisted green synthesis of silver nanoparticles. *Spectrochim. Acta Part A Mol. Biomol. Spectrosc.* 98, 423–428. doi:10.1016/j.saa.2012.08.072

[38] Philip, D. (2011). *Mangifera indica* leaf-assisted biosynthesis of well-dispersed silver nanoparticles. *Spectrochim. Acta Part A Mol. Biomol. Spectrosc.* 78, 327–331. doi:10.1016/j.saa.2010.10.015

[39] Philip, D., and Unni, C. (2011). Extracellular biosynthesis of gold and silver nanoparticles using Krishna tulsi (*Ocimum sanctum*) leaf. *Phys. E Low Dimens Syst. Nanostruct.* 43, 1318–1322. doi:10.1016/j.physe.2010.10.006

[40] Philip, D., Unni, C., Aromal, S. A., and Vidhu, V. K. (2011). *Murraya koenigii* leaf-assisted rapid green synthesis of silver and gold nanoparticles. *Spectrochim. Acta Part A Mol. Biomol. Spectrosc.* 78, 899–904. doi:10.1016/j.saa.2010.12.060

[41] Prathna, T. C., Chandrasekaran, N., Raichur, A. M., and Mukherjee, A. (2011). Biomimetic synthesis of silver nanoparticles by *Citrus limon* (lemon) aqueous extract and theoretical prediction of particle size. *Colloids Surf. B Biointerfaces* 82, 152–159. doi:10.1016/j.colsurfb.2010.08.036

[42] Yilmaz, M., Turkdemir, H., Kilic, M. A., Bayram, E., Cicek, A., Mete, A., et al. (2011). Biosynthesis of silver nanoparticles using leaves of *Stevia rebaudiana*. *Mater. Chem. Phys.* 130, 1195–1202. doi:10.1016/j.matchemphys.2011.08.068

[43] Metz, K. M., Sanders, S. E., Pender, J. P., Dix, M. R., Hinds, D. T., et al. (2015). Green synthesis of metal nanoparticles via natural extracts: the biogenic nanoparticle corona and its effects on reactivity. *ACS Sustain. Chem. Eng.* 3, 1610–1617. doi:10.1021/acssuschemeng.5b00304

[44] Nadagouda, M. N., Iyanna, N., Lalley, J., Han, Dionysiou, D. D and Varma R. S. (2014). Synthesis of silver and gold nanoparticles using antioxidants from Blackberry, Blueberry, Pomegranate, and Turmeric Extracts. *ACS Sustain. Chem. Eng.* 2, 1717–1723.

[45] Tahir, M. N., Tremel, W., Al-warthan, A., and Rafiq, M. (2013). Green synthesis of silver nanoparticles mediated by *Pulicaria glutinosa* extract. *Int. J. Nanomed.* 8, 1507–1516.

[46] Murugan, K., Senthilkumar, B., Senbagam, D., and Al-sohaibani, S. (2014). Biosynthesis of silver nanoparticles using *Acacia leucophloea* extract and their antibacterial activity. *Int. J. Nanomed.* 9, 2431–2438.

[47] Sathishkumar, M., Sneha, K., Won, S. W., Cho, C., Kim, S., and Yun, Y. (2009). *Cinnamon zeylanicum* bark extract and powder mediated green synthesis of nano-crystalline silver particles and its bactericidal activity. *Colloids Surf. B Biointerfaces* 73, 332–338. doi:10.1016/j.colsurfb.2009.06.005

[48] Bar, H., Bhui, D. K., Sahoo, G. P., Sarkar, P., De, S.P., and Misra, A. (2009). Green synthesis of silver nanoparticles using latex of *Jatropha curcas*. *Colloids Surfaces A Physicochem. Eng. Asp.* 339, 134–139. doi:10.1016/j.colsurfa.2009.02.008

[49] Vidhu, V. K., Aromal, S. A., and Philip, D. (2011). Green synthesis of silver nanoparticles using *Macrotyloma uniflorum*. *Spectrochim.*

Acta Part A Mol. Biomol. Spectrosc. 83, 392–397. doi:10.1016/j.saa. 2011.08.051

[50] Kaviya, S., Santhanalakshmi, J., Viswanathan, B., Muthumary, J., Srinivasan, K. (2011). Biosynthesis of silver nanoparticles using citrus sinensis peel extract and its antibacterial activity. *Spectrochim. Acta Part A Mol. Biomol. Spectrosc.* 79, 594–598. doi:10.1016/j.saa. 2011.03.040

[51] Mariselvam, R., Ranjitsingh, A. J. A., Raja, A. U., Kalirajan, K., Padmalatha, C., and Selvakumar, P. M. (2014). Green synthesis of silver nanoparticles from the extract of the inflorescence of *Cocos nucifera* (Family: Arecaceae) for enhanced antibacterial activity. *Spectrochim. Acta Part A Mol. Biomol. Spectrosc.* 129, 537–541. doi:10.1016/j.saa.2014.03.066

[52] Velmurugan, P., Park, J., Lee, S., Jang, J., Lee, K., Han, S., et al. (2015). Synthesis and characterization of nanosilver with antibacterial properties using *Pinus densiflora* young cone extract. *J. Photochem. Photobiol. B Biol.* 147, 63–68. doi:10.1016/j.jphotobiol.2015.03.008

[53] Ghosh, S., Patil, S., Ahire, M., Kitture, R., Kale, S., Pardes, K., et al. (2012). Synthesis of silver nanoparticles using *Dioscorea bulbifera* tuber extract and evaluation of its synergistic potential in combination with antimicrobial agents. *Int. J. Nanomed.* 7, 483–496.

[54] Babu, S. A., and Prabu, H. G. (2011). Synthesis of AgNPs using the extract of *Calotropis procera* flower at room temperature. *Mater. Lett.* 65, 1675–1677.

[55] Nagajyothi, P. C., and Lee, K. D. (2011). Synthesis of plant-mediated silver nanoparticles using *Dioscorea batatas* rhizome extract and evaluation of their antimicrobial activities. *J. Nanomater.* 2011:573429. doi:10.1155/2011/573429

[56] Jyothi, A., Sashidhar, R. B., Arunachalam, J., (2010). Gum kondagogu (*Cochlospermum gossypium*): a template for the green synthesis and stabilization of silver nanoparticles with antibacterial application. *Carbohydr. Polym.* 82, 670–679. doi:10.1016/j.carbpol.2010.05.034

[57] Njagi, E. C., Huang, H., Stafford, L., Genuino, H., Galindo, H. M., Collins, J. B., et al. (2011). Biosynthesis of iron and silver nanoparticles at room temperature using aqueous sorghum bran extracts. *Langmuir* 27, 264–271. doi:10.1021/la103190n

[58] Darroudi, M., Bin Ahmad, M., Zamiri, R., Zak, A. K., Abdullah, A. H., and Ibrahim, N. A. (2011). Time-dependent effect in green

synthesis of silver nanoparticles. *Int. J. Nanomed.* 6, 677–681. doi: 10. 2147/IJN.S17669

[59] Vigneshwaran, N., R. Nachane, P., Balasubramanya, R. H., and Varadarajan, P. V. (2012). A novel one-pot "green" synthesis of stable silver nanoparticles using soluble starch. *Carbohydr. Res.* 341, 2012–2018. doi:10.1016/j.carres.2006.04.042

[60] Singh, P., Kim, Y. J., Wang, C., Mathiyalagan, R., Farh, M. E.-A., and Yang, D. C. (2016). Biogenic silver and gold nanoparticles synthesized using red Ginseng root extract, and their applications. *Artif. Cells Nanomed. Biotechnol.* 44, 811–816.

[61] Okafor, F., Janen, A., Kukhtareva, T., and Edwards, V. (2013). Green synthesis of silver nanoparticles, their characterization, application and antibacterial activity. *Int. J. Environ. Research Public Heal.* 10, 5221–5238. doi: 10.3390/ijerph10105221

[62] Choi, K. C., and Kim, K. H. (2014). Rapid green synthesis of silver nanoparticles from *Chrysanthemum indicum* L and its antibacterial and cytotoxic effects: an *in vitro* study. *Int. J. Nanomed.* 9, 379–388.

[63] Mohanpuria, P., Rana, N. K., and Yadav, S. K. (2008). Biosynthesis of nanoparticles: technological concepts and future applications. *J. Nanopart. Res.* 10, 507–517. doi: 10.1007/s11051-007-9275-x

[64] Patil, C. D., and Patil, S. V. (2012). Larvicidal activity of silver nanoparticles synthesized using *Plumeria rubra* plant latex against *Aedes aegypti* and *Anopheles stephensi. Parasitol. Res.* 110, 1815–1822. doi: 10.1007/s00436-011-2704-x

[65] Singhal, G., and Bhavesh, R. (2011). Biosynthesis of silver nanoparticles using *Ocimum sanctum* (Tulsi) leaf extract and screening its antimicrobial activity. *J. Nanopart. Res.* 13, 2981–2988. doi: 10.1007/s11051-010-0193-y

[66] Soman, S., and Ray, J. G. (2016). Silver nanoparticles synthesized using aqueous leaf extract of *Ziziphus oenoplia* (L .) Mill: characterization and assessment of antibacterial activity. *J. Photochem. Photobiol. B Biol.* 163, 391–402. doi: 10.1016/j.jphotobiol.2016.08.033

[67] Kathiresan, K., Manivannan, S., Nabeel, M. A., and Dhivya, B. (2009). Studies on silver nanoparticles synthesized by a marine fungus, *Penicillium fellutanum* isolated from coastal mangrove sediment. *Colloids Surf. B Biointerfaces* 71, 133–137. doi: 10.1016/j.colsurfb. 2009.01.016

[68] Mock, J. J., Barbic, M., Smith, D. R., Schultz, D. A., and Schultz, S. (2015). Shape effects in plasmon resonance of indivi-

dual colloidal silver nanoparticles. *J. Chem. Phys.* 6755, 10–15. doi: 10.1063/1.1462610

[69] Prabha, S., Lahtinen, M., and Sillanpää, M. (2010). Tansy fruit mediated greener synthesis of silver and gold nanoparticles. *Process Biochem.* 45, 1065–1071. doi: 10.1016/j.procbio.2010.03.024

[70] Veerasamy, R., Xin, T. Z., Gunasagaran, S., Xiang, T. F. W., Yang, E. F. C., Jeyakumar, N., et al. (2011). Biosynthesis of silver nanoparticles using Mangosteen leaf extract and evaluation of their antimicrobial activities. *J. Saudi Chem. Soc.* 15, 113–120. doi: 10.1016/j.jscs.2010.06.004

[71] Ghodake, G. S., Deshpande, N. G., Lee, Y. P., and Jin, E. S. (2010). Pear fruit extract-assisted room-temperature biosynthesis of gold nanoplates. *Colloids Surf. B Biointerfaces* 75, 584–589. doi: 10.1016/j.colsurfb.2009.09.040

[72] Sathishkumar, M., Sneha, K., and Yun, Y. (2010). Immobilization of silver nanoparticles synthesized using *Curcuma longa* tuber powder and extract on cotton cloth for bactericidal activity. *Bioresour. Technol.* 101, 7958–7965. doi: 10.1016/j.biortech.2010.05.051

[73] Prabha, S., Lahtinen, M., Sillanpää, M. (2010). Green synthesis and characterizations of silver and gold nanoparticles using leaf extract of *Rosa rugosa. Colloids Surf. A Physicochem. Eng. Asp.* 364, 34–41. doi: 10.1016/j.colsurfa.2010.04.023

[74] Dubey, M., Bhadauria, S., and Kushwah, B. S. (2009). Green synthesis of nanosilver particles from extract of *Eucalyptus hybrida* (safeda) leaf. *Dig. J. Nanomat. Biostruct.* 4, 537–543.

[75] Andreescu, D., Eastman, C., Balantrapu, K., and Goia, D. V. (2007). A simple route for manufacturing highly dispersed silver nanoparticles. *J. Mater. Res.* 22, 1–9. doi: 10.1557/JMR.2007.0308

[76] Dwivedi, A. D. and Gopal, K. (2010). Biosynthesis of silver and gold nanoparticles using *Chenopodium album* leaf extract. *Colloids Surf. A Physicochem. Eng. Asp.* 369, 27–33. doi: 10.1016/j.colsurfa.2010.07.020

[77] Cheong, S., Watt, J. D., and Tilley, R. D. (2010). Shape control of platinum and palladium nanoparticles for catalysis. *Nanoscale* 2, 2045–2053. doi: 10.1039/c0nr00276c

[78] Shaligram, N. S., Bule, M., Bhambure, R., Singhal, R. S., Kumar, S., Szakacs, G., et al. (2009). Biosynthesis of silver nanoparticles using aqueous extract from the compact in producing fungal strain. *Process Biochem.* 44, 939–943. doi: 10.1016/j.procbio.2009.04.009

[79] Fayaz, A. M., Balaji, K., Kalaichelvan, P. T., and Venkatesan, R. (2009). Fungal based synthesis of silver nanoparticles – An effect of temperature on the size of particles. *Colloids Surf. B Biointerfaces* 74, 123–126. doi: 10.1016/j.colsurfb.2009.07.002

[80] Philip, D. (2009). Biosynthesis of Au, Ag and Au – Ag nanoparticles using edible mushroom extract. *Spectrochim. Acta A Mol. Biomol. Spectrosc.* 73, 374–381. doi: 10.1016/j.saa.2009.02.037

[81] Lengke, M. F., Fleet, M. E., Southam, G. (2007). Biosynthesis of Silver Nanoparticles by Filamentous Cyanobacteria from a Silver (I) Nitrate Complex, Langmuir. 23, 2694–2699.

[82] Ghoreishi, S. M., Behpour, M., and Khayatkashani, M. (2011). Green synthesis of silver and gold nanoparticles using *Rosa damascena* and its primary application in electrochemistry, *Phys. E Low-Dimensional Syst. Nanostructures.* 44 97–104. doi: 10.1016/j.physe.2011.07.008

[83] Vijayaraghavan. K, Nalini. S P, (2010). Biotemplates in the green synthesis of silver nanoparticles, *Biotechnolo J.* 5 1098–1110.

[84] Thanh, N. T. K., Maclean, N., and Mahiddine, S. (2014). Mechanisms of Nucleation and Growth of Nanoparticles in Solution, *Chem. Commun.* (Camb). 3 7610–7630.

[85] Watzky, M. A., and Finke, R. G. (1997). Nanocluster Size-Control and "Magic Number" Investigations. Experimental Tests of the "Living-Metal Polymer" Concept and of Mechanism-Based Size-Control Predictions Leading to the Syntheses of Iridium (0) Nanoclusters Centering about Four Sequentia, *Chem. Mater.* 4756 3083–3095.

[86] Sagui, C., and Grant, M. (1999). Theory of nucleation and growth during phase separation, Phys. Rev. E. 59 4175–4187.

[87] Shields, S. P., Richards, V. N., and Buhro, W. E. (2010). Nucleation Control of Size and Dispersity in Aggregative Nanoparticle Growth. A Study of the Coarsening Kinetics of, *Chem. Mater.* 22 3212–3225. doi: 10.1021/cm100458b

[88] Lukman, A. I., Gong, B., Marjo, C. E., Roessner, U., and Harris, A. T. (2011). Facile synthesis, stabilization, and anti-bacterial performance of discrete Ag nanoparticles using *Medicago sativa* seed exudates, *J. Colloid Interface Sci.* 353 433–444. doi: 10.1016/j.jcis.2010.09.088

[89] Gan, P. P., and Li, S. F. Y. (2012). Potential of plant as a biological factory to synthesize gold and silver nanoparticles and their applications, *Rev. Environ. Sci. Biotechnol.* 11 169–206. doi: 10.1007/s11157-012-9278-7

[90] Kumar, A., Chisti, Y., and Chand, U. (2013). Synthesis of metallic nanoparticles using plant extracts, *Biotechnol. Adv.* 31 346–356. doi: 10.1016/j.biotechadv.2013.01.003

[91] Bali, R., and Harris, A. T. (2010). Biogenic Synthesis of Au Nanoparticles Using Vascular Plants, *Ind. Eng. Chem. Res.* 49 12762–12772.

[92] Rizzoa, L., Manaiab, C., Merlinc, C., Schwartzd, T., Dagote, C., Ployf, M. C., Michaelg, I., and Fatta-Kassinosg, D. (2013). Urban wastewater treatment plants as hotspots for antibiotic resistant bacteria and genes spread into the environment: A review, *Sci. Total Environ.* 447 345–360.

[93] Hooper, L. V., Littman, D. R., and Macpherson, A. J. (2015). Program, Interactions between the microbiota and the immune system, *Science.* 336 1268–1273. doi: 10.1126/science.1223490.Interactions

[94] Sandegren, L., Hughes, D., Dan, I., Gullberg, E., Cao, S., Berg, O. G., et al. (2011). Selection of resistant bacteria at very low antibiotic concentrations. *PLoS ONE Pathog.* 7, e1002158. doi:10.1371/journal.ppat.1002158.

[95] Cira, N. J., Ho, J. Y., Dueck, M. E., and Weibel, D. B. (2012). Lab on a Chip Miniaturisation for chemistry, physics, biology, materials science and bioengineering. *RSC Adv.* 12, 4239–4468. doi: 10.1039/c2lc20887c

[96] Soo, V. W. C., Hanson-Manful, P., and Patrick, W. M. (2010). Artificial gene amplification reveals an abundance of promiscuous resistance determinants in *Escherichia coli. Proc. Natl. Acad. Sci. U.S.A.* 108, 1484–1489.

[97] Colomer-lluch, M., Jofre, J., and Muniesa, M. (2011). Antibiotic resistance genes in the bacteriophage DNA fraction of environmental samples. *PLoS ONE* 6, 1–11. doi: 10.1371/journal.pone.0017549

[98] Van Meervenne, E., De Weirdt, R., Van Coillie, E., Devlieghere, F., Herman, L., and Boon, N. (2014). Biofilm models for the food industry: hot spots for plasmid transfer? *Pathog. Dis.* 70, 332–338. doi: 10.1111/2049-632X.12134

[99] Fernández, L., and Hancock, R. E. W. (2012). Adaptive and mutational resistance: role of porins and efflux pumps in drug resistance. *Clin. Microbiol. Rev.* 25, 661–681. doi: 10.1128/CMR.00043-12

[100] Arias, C. A., and Murray, B. E. (2013). The rise of the Enterococcus: beyond vancomycin resistance. *Nat. Rev. Microbiol.* 10, 266–278. doi: 10.1038/nrmicro2761

[101] Spellberg, B., Bartlett, J. G., and Gilbert, D. N. (2013). The future of antibiotics and resistance. *N. Engl. J. Med.* 368, 299–302. doi: 10.1056/NEJMp1215226

[102] Shimanovich, U., and Gedanken, A. (2016). Nanotechnology solutions to restore antibiotic activity. *J. Mater. Chem. B.* 4, 824–833. doi: 10.1039/C5TB01527H

[103] Kim, J. S., Kuk, E., Yu, N., Kim, J., Park, S. J., Lee, J., et al. (2007). Antimicrobial effects of silver nanoparticles. *Nanomedicine* 3, 95–101. doi: 10.1016/j.nano.2006.12.001

[104] Burrell, R. E., Apte, P. S., Gill, K. S., Precht, R. J., and Morris, L. R. (1999). Process for producing anti-microbial effect with complex silver ions. U.S. Patent No. 5985308. Washington, DC: U.S. Patent and Trademark Office.

[105] Cho, M. R. A. T., Hu, P., Lee, S. J., Deming, C. P., Sweeney, S. W., Saltikov, C., et al. (2015). Enhanced antimicrobial activity with faceted silver nanostructures. *J. Mater. Sci.* 50, 2849–2858. doi: 10.1007/s10853-015-8847-x

[106] Beyth, N., Houri-haddad, Y., Domb, A., Khan, W., and Hazan, R. (2015). Alternative antimicrobial approach: nano-antimicrobial materials. *Evid. Based Complement. Altern. Med.* 2015, 1–16.

[107] Ann, M., and Irving, S. (2007). "Nanotechnology and life cycle assessment," in *Proceedings of A systems approach to Nanotechnology and the environment Synthesis of Results Obtained at a Workshop*, Washington, DC.

[108] Song, J. Y., and Kim, B. S. (2009). Rapid biological synthesis of silver nanoparticles using plant leaf extracts. *Bio* 32, 79–84. doi: 10.1007/s00449-008-0224-6

[109] Martínez-Castañn, G. A., Martínez, N. N., Martínez-Gutierrez, F., Mendoza, M., and Ruiz, F. (2008). Synthesis and antibacterial activity of silver nanoparticles with different sizes. *J. Nanoparticle Res.* 10, 1343–1348. doi: 10.1007/s11051-008-9428-6

[110] Prabakar, K., Sivalingam, P., Ibrahim, S., Rabeek, M., Muthuselvam, M., Devarajan, N., et al. (2013). Evaluation of antibacterial efficacy of phyto fabricated silver nanoparticles using *Mukia scabrella* (Musumusukkai) against drug resistance nosocomial gram negative bacterial pathogens. *Colloids Surf. B Biointerfaces* 104, 282–288. doi: 10.1016/j.colsurfb.2012.11.041

[111] Krishnaraj, C., Jagan, E. G., Rajasekar, S., Selvakumar, P., Kalaichelvan, P. T., and Mohan, N. (2010). Synthesis of silver nanoparticles

using *Acalypha indica* leaf extracts and its antibacterial activity against water borne pathogens. *Colloids Surf. B Biointerfaces* 76, 50–56. doi: 10.1016/j.colsurfb.2009.10.008

[112] Morones, J. R., Elechiguerra, J. L., Camacho, A., Holt, K., Kouri, J. B., Ram, J. T., et al. (2005). The bactericidal effect of silver nanoparticles. *Nanotechnology.* 16, 2346–2353. doi: 10.1088/0957-4484/16/10/059

[113] Barbinta-patrascu, M. E., Ungureanu, C., and Iordache, M. (2014). Green silver nanobioarchitectures with amplified antioxidant and antimicrobial properties. *J. Mater. Chem. B.* 2, 3221–3231. doi: 10.1039/c4tb00262h

[114] Shiekh, R. A., and Balam, S. K. (2013). Biogenic silver nanoparticles using *Rhinacanthus nasutus* leaf extract: synthesis, spectral analysis, and antimicrobial studies. *Int. J. Nanomed.* 8, 3355–3364.

[115] Ajitha, B., Kumar, Y. A., Shameer, S., Rajesh, K. M., Suneetha, Y., and Reddy, P. S. (2015). *Lantana camara* leaf extract mediated silver nanoparticles: antibacterial, green catalyst. *J. Photochem. Photobiol. B Biol.* 149, 84–92. doi: 10.1016/j.jphotobiol.2015.05.020

[116] Logeswari, P., Silambarasan, S., and Abraham, J. (2015). Synthesis of silver nanoparticles using plants extract and analysis of their antimicrobial property. *J. Saudi Chem. Soc.* 19, 311–317. doi: 10.1016/j.jscs.2012.04.007

[117] Ahmad, N., Sharma, S., Singh, V. N., Shamsi, S. F., Fatma, A., and Mehta, B. R. (2011). Biosynthesis of silver nanoparticles from *Desmodium triflorum*: a novel approach towards weed utilization. *Biotechnol. Res. Int.* 2011, 1–8. doi: 10.4061/2011/454090

[118] Mubarakali, D., Thajuddin, N., Jeganathan, K., and Gunasekaran, M. (2011). Plant extract mediated synthesis of silver and gold nanoparticles and its antibacterial activity against clinically isolated pathogens. *Colloids Surf. B Biointerfaces* 85, 360–365. doi: 10.1016/j.colsurfb.2011.03.009

[119] Patil, S. V., and Borase, H. P. (2012). Biosynthesis of silver nanoparticles using latex from few euphorbian plants and their antimicrobial potential. *Appl. Biochem. Biotechnol.* 167, 776–790. doi: 10.1007/s12010-012-9710-z

[120] Khalil, M. M. H. (2014). Green synthesis of silver nanoparticles using olive leaf extract and its antibacterial activity. *Arab. J. Chem.* 7, 1131–1139. doi: 10.1016/j.arabjc.2013.04.007

[121] Zhang, Y., Cheng, X., Zhang, Y., Xue, X., and Fu, Y. (2013). Biosynthesis of silver nanoparticles at room temperature using aqueous aloe

leaf extract and antibacterial properties. *Colloids Surf. A Physicochem. Eng. Asp.* 423, 63–68. doi: 10.1016/j.colsurfa.2013.01.059

[122] Augustine, R., Kalarikkal, N., and Thomas, S. (2014). A facile and rapid method for the black pepper leaf mediated green synthesis of silver nanoparticles and the antimicrobial study. *Appl. Nanosci.* 4, 809–818. doi: 10.1007/s13204-013-0260-7

[123] Dipankar, C., and Murugan, S. (2012). The green synthesis, characterization and evaluation of the biological activities of silver nanoparticles synthesized from *Iresine herbstii* leaf aqueous extracts. *Colloids Surf. B Biointerfaces* 98, 112–119. doi: 10.1016/j.colsurfb.2012.04.006

[124] Sulaiman, G. M., Mohammed, W. H., Marzoog, T. R., Al-Amiery, A. A. A., Kadhum, A. A. H., and Mohamad, A. B. (2013). Green synthesis, antimicrobial and cytotoxic effects of silver nanoparticles using *Eucalyptus chapmaniana* leaves extract. *Asian Pac. J. Trop. Biomed.* 3, 58–63.

[125] Shah, M., Eddy, G., Poinern, J., and Fawcett, D. (2016). Biogenic synthesis of silver nanoparticles via indigenous *Anigozanthos manglesii* (red and green kangaroo paw) leaf extract and its potential antibacterial activity. *Int. J. Res. Med. Sci.* 4, 3427–3432.

[126] Singh, A., Jain, D., Upadhyay, M. K., and Khandelwal, N. (2010). Green synthesis of silver nanoparticles using *Argemone mexicana* extract and evaluation of their antimicrobial activities. *Dig. J. Nanomat Biostruct.* 5, 483–489.

[127] Bose, D., and Chatterjee, S. (2016). Biogenic synthesis of silver nanoparticles using guava (*Psidium guajava*) leaf extract and its antibacterial activity against *Pseudomonas aeruginosa*. *Appl. Nanosci.* 6, 895–901. doi:10.1007/s13204-015-0496-5

[128] Holt, K. B., Bard, A. J., May, R. V., Re, V., Recei, M., and July, V. (2005). Interaction of Silver (I) ions with the respiratory chain of *Escherichia coli*: an electrochemical and scanning electrochemical microscopy study of the antimicrobial mechanism of micromolar $Ag^{\dagger}$. *Biochemistry* 44, 13214–13223.

[129] Shrivastava, S., Bera, T., Roy, A., Singh, G., Ramachandrarao, P., and Dash, D. (2007). Characterization of enhanced antibacterial effects of novel silver nanoparticles. *Nanotechnology* 18, 1–9. doi:10.1088/0957-4484/18/22/225103.

[130] Hajipour M. J., Fromm K. M., Ashkarran A. A., Aberasturi D. J. De, Larramendi I. R. De, Rojo T., Serpooshan V., Parak W. J.,

Mahmoud M., and Mahmoudi M. (2012). Antibacterial properties of nanoparticles. *Trends Biotechnol.* doi:10.1016/j.tibtech.2012.06.004

[131] Daima, H. K., and Bansal, V. (2015). Influence of physicochemical properties of nanomaterials on their antibacterial applications, in: nanotechnol. diagnosis. *Treat. Prophyl. Infect. Dis.* 151–166.

[132] Samberg, M. E., Orndorff, P. E., and Monteiro-Riviere, N. A. (2011). Antibacterial efficacy of silver nanoparticles of different sizes, surface conditions and synthesis methods. *Nanotoxixology* 5, 244–253.

[133] Agnihotri, S., Mukherji, S., and Mukherji, S. (2013). Immobilized silver nanoparticles enhance contact killing and show highest efficacy: elucidation of the mechanism of bactericidal action of silver. *Nanoscale* 5, 7328–7340. doi:10.1039/c3nr00024a

[134] Sotiriou, G. A., and Pratsinis, S. E. (2010). Antibacterial activity of nanosilver ions and particles. *Environ. Sci. Technol.* 44, 5649–5654.

[135] Saleh, N. B., Chambers, B., Aich, N., Plazas-Tuttle, J., Phung-Ngoc, H. N., and Kirisits, M. J. (2015). Mechanistic lessons learned from studies of planktonic bacteria with metallic nanomaterials: implications for interactions between nanomaterials and biofilm bacteria. *Front. Microbiol.* 6:677. doi:10.3389/fmicb.2015.00677

[136] Ivask, A., Kurvet, I., Kasemets, K., Blinova, I., Aruoja, V., Suppi, S., et al. (2014). Size-Dependent toxicity of silver nanoparticles to bacteria, yeast, algae crustaceans and mammalian cells *in vitro. PLoS ONE* 9:e102108 doi:10.1371/journal.pone.0102108.

[137] Sondi, I., and Salopek-sondi, B. (2004). Silver nanoparticles as antimicrobial agent: a case study on *E. coli* as a model for Gram-negative bacteria. *J. Colloid Interface Sci.* 275, 177–182. doi:10.1016/j.jcis.2004.02.012

[138] Sportelli, M. C., Picca, R. A., and Cioffi, N. (2016). Recent advances in the synthesis and characterization of nano-antimicrobials. *Trends Anal. Chem.* 84, 131–138. doi:10.1016/j.trac.2016.05.002

[139] Sarmast, M. K., and Salehi, H. (2016). Silver nanoparticles: an influential element in plant nanobiotechnology. *Mol. Biotechnol.* 58, 441–449. doi:10.1007/s12033-016-9943-0

[140] Dror-Ehre, A., Mamane, H., Belenkova, T., Markovich, G., and Adin, A. (2009). Silver nanoparticle – *E. coli* colloidal interaction in water and effect on *E. coli* survival. *J. Colloid Interface Sci.* 339, 521–526. doi:10.1016/j.jcis.2009.07.052

[141] Lara, H. H., Ayala-Núũez, N. V., Ixtepan, C., and Rodrí, C. (2010). Bactericidal effect of silver nanoparticles against multidrug-resistant bacteria. *World J. Microbiol. Biotechnol.* 26, 615–621. doi:10.1007/s11274-009-0211-3

[142] Ayala-Núũez, N., Vanesa, H. H. L., and Turrent, L. D. C. (2009). Silver nanoparticles toxicity and bactericidal effect against methicillin-resistant *Staphylococcus aureus*: nanoscale does matter. *NanoBiotechnol.* 5, 2–9.

[143] Nasser, H., and Wu, H. (2015). Proteomics analysis of the mode of antibacterial action of nanoparticles and their interactions with proteins. *Trends Anal. Chem.* 65, 30–46. doi:10.1016/j.trac.2014.09.010

[144] Veigas, B., Giestas, L., Almeida, C., Baptista, P. V., De Lisboa, C. De Caparica, M., et al. (2012). Noble metal nanoparticles for biosensing applications. *Sensors* 12, 1657–1687. doi:10.3390/s120201657

[145] Chen, X. J., Sanchez-Gaytan, B. L., Qian, Z., and Park, S. J. (2012). Noble metal nanoparticles in DNA detection and delivery. *Wiley Interdiscip. Rev. Nanomed. Nanobiotechnol.* 4, 273–290.

[146] Fortina, P., Kricka, L. J., Graves, D. J., Park, J., Hyslop, T., Tam, F., et al. (2007). Applications of nanoparticles to diagnostics and therapeutics in colorectal cancer. *Trends Biotechnol.* 25, 145–152. doi:10.1016/j.tibtech.2007.02.005

[147] Thanh, N. T. K., and Green, L. A. W. (2010). Functionalisation of nanoparticles for biomedical applications. *Nano Today* 5, 213–230. doi:10.1016/j.nantod.2010.05.003

[148] Sundaramoorthi, C., Kalaivani, M., Mathews, D. M., and Palanisamy, S. (2009). Biosynthesis of silver nanoparticles from *Aspergillus niger* and evaluation of its wound healing activity in experimental rat model. *Int. J. Pharma Tech Res.* 1, 1523–1529.

[149] He, Y., Du, Z., Lv, H., Jia, Q., Tang, Z., Zheng, X., et al. (2013). Green synthesis of silver nanoparticles by *Chrysanthemum morifolium* Ramat. extract and their application in clinical ultrasound gel. *Int. J. Nanomedicine.* 8, 1809–1815.

[150] Arunachalam, K. D., Annamalai, S. K., Arunachalam, A. M., and Kennedy, S. (2013). Green synthesis of crystalline silver nanoparticles using Indigofera aspalathoides-medicinal plant extract for wound healing applications. *Asian J. Chem.* 25, S311–S314.

[151] Samberg, M. E., Tan, Z., Paul, R., and Rohan, E.O. (2013). Biocompatibility analysis of an electrically-activated silver-based antibacterial

surface system for medical device applications. *J. Mater. Sci. Mater. Med.* 755–760. doi:10.1007/s10856-012-4838-5

[152] Roe, D., Karandikar, B., Bonn-savage, N., Gibbins, B., and Roullet, J. (2008). Antimicrobial surface functionalization of plastic catheters by silver nanoparticles. *J. Antimicrob. Chemother.* 61, 869–876. doi:10.1093/jac/dkn034

[153] Mangala, S., Ahmad, P., and Aris, Z. (2015). Application of low-cost materials coated with silver nanoparticle as water filter in *Escherichia coli* removal. *Water Qual. Expo. Health* 7, 617–625. doi:10.1007/s12403-015-0167-5

[154] Yoon, K. Y., Byeon, J. H., Park, C. W., and Hwang, J. (2008). Antimicrobial effect of silver particles on bacterial contamination of activated carbon fibers. *Environ. Sci. Technol.* 42, 1251–1255.

[155] Kelly, P. J., Li, H., Whitehead, K. A., Verran, J., Arnell, R. D., and Iordanova, I. (2009). A study of the antimicrobial and tribological properties of TiN/Ag nanocomposite coatings. *Surf. Coat. Technol.* 204, 1137–1140. doi:10.1016/j.surfcoat.2009.05.012

[156] Hasan, J., Crawford, R. J., and Ivanova, E. P.(2013). Antibacterial surfaces: the quest for a new generation of biomaterials. *Trends Biotechnol.* 31, 295–304. doi:10.1016/j.tibtech.2013.01.017

[157] Gordon, O., Slenters, V., Brunetto, P. S., Villaruz, A. E., Sturdevant, D. E., Otto, M. et al. (2010). Silver Coordination Polymers for Prevention of Implant Infection: Thiol Interaction Impact on Respiratory Chain Enzymes and Hydroxyl Radical Induction?. *Antimicrob. Agents Chemother.* 54, 4208–4218. doi:10.1128/AAC.01830-09

[158] Franci, G., Falanga, A., Galdiero, S., Palomba, L., Rai, M., Morelli, G., et al. (2015). Silver nanoparticles as potential antibacterial agents. Molecules 20, 8856–8874. doi:10.3390/molecules20058856

[159] Moulton, M. C., Braydich-stolle, L. K., Nadagouda, M. N., Kunzelman, S., Hussain, S. M., and Varma, R. S. (2010). Synthesis, characterization and biocompatibility of "'green'" synthesized silver nanoparticles using tea polyphenols. *Nanoscale* 2, 763–770. doi:10.1039/c0nr00046a

[160] Sanvicens, N., and Marco, M. P. (2008). Multifunctional nanoparticles–properties and prospects for their use in human medicine. *Trends Biotechnol.* 26, 425–33. doi:10.1016/j.tibtech.2008.04.005

[161] Sintubin, L., Verstraete, W., and Boon, N. (2012). Biologically produced nanosilver: current state and future perspectives. *Biotechnol. Bioeng.* 109, 2422–2436. doi:10.1002/bit.24570

11

Recreation of Turmeric Matrix with Enhanced Curcuminoids—Enhances the Bioavailability and Bioefficacy

Augustine Amalraj and Sreeraj Gopi

R&D Centre, Aurea Biolabs Pvt. Ltd, Kolenchery, Cochin, India

Abstract

Health benefits of curcuminoid are highly limited due to their poor aqueous solubility, very low systemic bioavailability, fast metabolic alterations, and rapid elimination. In this study, a novel bioavailable curcuminoid formulation Cureit^{TM} was prepared by using polar–nonpolar–sandwich (PNS) technology with complete natural turmeric matrix (CNTM). The synthesized bioavailable curcuminoid formulation Cureit^{TM} was characterized by nuclear magnetic resonance spectroscopy (NMR), scanning electron microscopy (SEM), X-ray diffraction (XRD), Fourier transform infrared (IR), current–voltage (I–V) study, quadrupole time-of-flight mass spectrometry (Q-TOF), and thermogravimetric analysis (TGA). The metabolic profile of Cureit^{TM} was analyzed and confirmed the presence of curcuminoids (curcumin, demethoxycurcumin, and bismethoxycurcumin); lactones; sesquiterpenes; and their derivatives derived from polar layer, aromatic turmerone, dihydroturmerone, turmeronol, curdione, and bisacurone derived from nonpolar layer by Q-TOF. This study was also conducted to assess the bioavailability of natural turmeric formulation (Cureit^{TM}) with standard curcumin in healthy human adult male subjects under fasting conditions. Each form of curcumin was administrated orally as a single dose of 500 mg in capsule form, and blood samples were analyzed by LC-MS/MS at different time intervals up to 24 h. The extent of absorption of total curcuminoids in the blood for the Cureit^{TM} was 5.5 times greater than the standard curcumin. The rates of absorption of curcuminoids in the blood for the Cureit^{TM} formulation were higher than the standard curcumin (10 times).

11.1 Introduction

Natural products have been used in traditional medicines for thousands of years and have shown promise as a source of components for the development of new drugs [1, 2]. Turmeric (*Curcuma longa Linn*) is a member of the *Zingiberaceae* family and is cultivated in tropical and subtropical regions around the world, and it originates from India, Southeast Asia, and Indonesia [3]. Turmeric powder is used extensively as a coloring and flavoring agent in curries and mustards. Turmeric has been used in India to maintain oral hygiene [4]. It has traditionally been used for medical purposes for many centuries in countries such as India and China for the treatment of jaundice and other liver ailments [5, 6]. Turmeric is one of the most popular medicinal herbs, with a wide range of pharmacological activities such as antioxidant [7], anti-protozoal [8], anti-venom activities [9], antimicrobial [10], anti-malarial [11], anti-inflammatory [12], anti-proliferative [13], anti-angiogenic [14], anti-tumor [15], and anti-aging [16] properties. It has also been used to treat ulcers, parasitic infections, various skin diseases, and anti-immune diseases, and curing the symptoms of colds and flus [17]. The pharmacological activity of turmeric has been attributed mainly to curcuminoids consists of curcumin (CUR) and two related compounds demethoxycurcumin (DMC) and bisdemethoxycurcumin (BDMC) [3]. CUR itself appears as a crystalline compound with a bright orange-yellow color. Curcuminoids are commonly used as coloring agent as well as food additives. World Health Organization (WHO) stated the acceptable daily intake of curcuminoids as a food additive in the range of 0–3 mg/kg. Curcuminoids and turmeric products have been characterized as safe by the Food and Drug Administration (FDA) in USA. The average intake of turmeric in the Indian diet is approximately 2–2.5 g for a 60-kg individual which corresponds to a daily intake of approximately 60–100 mg of CUR [18]. Curcuminoids have achieved the potential therapeutic interest to cure immune-related, metabolic diseases and cancer due to a vast number of biological targets and virtually no side effects [17, 18].

11.2 Discovery of Curcumin

Curcumin is the active ingredient of the dietary spice turmeric and is extracted from the rhizomes of *Curcuma longa*, a plant in the *Zingiberaceae* family. It was first discovered about two centuries ago when Vogel and Pelletier reported the isolation of a "yellow coloring matter" from rhizomes of

Curcuma longa and named it curcumin [19]. It is characterized by Milobedeska et al. [20] and first synthesized by Lampe et al. [21].

11.3 Isolation of Curcumin

Curcumin is insoluble in water; an organic solvent has been used for its isolation. Anderson et al. [22] developed a technique for isolating CUR from ground turmeric. They magnetically stirred the ground turmeric in dichloromethane and heated at reflux for 1 h. The mixture was suction-filtered, and the filtrate was concentrated in a hot-water bath maintaining at 50°C. The reddish-yellow oil residue was triturated with hexane, and the resulting solid was collected by suction filtration. Further thin-layer chromatography (TLC) analysis (3% methanol and 97% dichloromethane) showed the presence of all three components [22]. Bagchi [23] explained the extraction of CUR from turmeric powder with the use of a solvent consisting of a mixture of ethanol and acetone. Chemical analyses have shown that turmeric contains carbohydrates (69.4%), moisture (13.1%), protein (6.3%), fat (5.1%), and minerals (3.5%). The essential oil (5.8%) obtained by steam distillation of the rhizomes contains α-phellandrene (1%), sabinene (0.6%), cineol (1%), borneol (0.5%), zingiberene (25%), and sesquiterpines (53%), and curcumin (3–6%) is responsible for the yellow color.

11.4 Physical, Chemical, and Molecular Properties of Curcuminoids

Two active components of turmeric are the curcuminoids and volatile oil, and both are present in oleoresin extracted from the turmeric root. The chemical structures of curcuminoids make them much less soluble in water at acidic and neutral pH, but soluble in methanol, ethanol, dimethyl sulfoxide, and acetone. The curcuminoids give a yellow–orange coloration to turmeric powder due to the wide electronic delocalization inside the molecules that exhibit strong absorption between 420 and 430 nm in an organic solvent. The curcuminoids are a mixture of curcumin, chemically a diferuloyl-methane [1,7-bis(4-hydroxy-3-methoxy-phenyl)-hepta-1,6-diene-3,5-dione] mixed with its two derivatives, demethoxy curcumin [4-hydroxycinnamoyl-(4-hydroxy-3-methoxycinnamoyl) methane] and bis-demethoxy curcumin [bis-(4-hydroxy cinnamoyl) methane], defining the chemical formulae as $C_{21}H_{20}O_6$, $C_{20}H_{18}O_5$, and $C_{19}H_{16}O_4$, respectively [17]. The essential oils are composed mainly of sesquiterpenes, many of which are specific for the Curcuma genus. The aroma of this spice is principally derived from α- and β-turmerones and aromatic turmerone (Ar-turmerone) [24]. The chemical structures of important constituents present in turmeric are given in Figure 11.1.

Curcuminoids share the same structure with two benzenemethoxy rings, joined by an unsaturated chain. It has three important functions: an aromatic methoxy phenolic group; α,β-unsaturated β-diketo linker, and keto-enol tautomerism. All these compounds exist in the trans-trans keto-enol form. The aromatic groups provide hydrophobicity, and the linker gives flexibility. The tautomeric structures also influence the hydorphobicity and polarity. The hydrophobicity of curcuminoids makes them poorly soluble in water. Three acidity constants (pKa) were measured for curcuminoids as follows: pKa1 = 8.38 ± 0.04, pKa2 = 9.88 ± 0.02, and pKa3 = 10.51 ± 0.01 [25]. Typical curcuminoids' composition of popular Indian varieties was found to be in the range of CUR 52–63%, DMC 19–27%, and BDMC 18–28% [17].

In recent decades, curcuminoid draw great attentions for its broad spectrum of therapeutic actions, including antioxidant, anti-inflammatory,

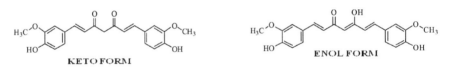

Figure 11.1   Chemical structures of important constituents present in turmeric.

anti-cancer, antimicrobial, wound-healing, hepatoprotective, and potential ability in preventing neurodegenerative diseases [26–29]. Besides, it has a superior safety profile determined by clinical study that as high as 8 g/day of dosage would not induce any observable adverse effects [30]. This safe profile has been reflected by the continuous increase of preparations based on curcuminoid marketed as a food ingredient or constituent of dietary supplements. However, the functional applications of curcuminoid have been seriously limited by its very low systemic bioavailability, attributable to poor absorption, fast metabolic alterations, and rapid elimination [31, 32]. Available evidence indicates that only minute amounts of curcuminoid reach the circulation after high-dose oral administration in animals and humans. The majority of the orally administered curcuminoid is excreted in the feces and the urine, with very little being detected in the blood plasma [33]. Curcuminoid has very low solubility in aqueous media due to inter- and intra hydrogen bonding [34]. Higher solubility was observed in alkaline solution, when dissolved curcuminoid was quickly degraded into vanillin, ferulic acid, and feruloyl methane [35]. Other environmental factors such as UV irradiation and heating also contribute to the decomposition of this yellowish polyphenolic compound [36–38]. These dramatically affect the absorption and bioavailability of this active molecule with consequent unsatisfactory pharmacokinetic profile and reduced efficacy.

To increase its water solubility, stability, bioavailability, and potential applications, different methods have been proposed and investigated. Several strategies such as nanoparticles, liposomes, solid dispersions, solid lipid nanoparticles, microemulsions, and complexation with phospholipids and cyclodextrins have been developed to improve the bioavailability of curcumin/curcuminoid [39–43]. Phospholipid complexes (phytosomes) increased the area under the blood concentration–time curve (AUC) of curcumin after oral administration in rats by 5 times [44]; association with cyclodextrin, with a ten-fold increase in curcumin AUC [45]; BCM-95 extract, whose bioavailability in rats was 7.8 times higher than unformulated curcumin [46]; and mixing of curcumin with an essential oil obtained from standardized turmeric where AUC was 7–8 times higher than unformulated curcumin [47]. Most of the best performing curcumin formulations produced so far have provided no more than a ten-fold increase bioavailability compared to unformulated curcumin [48].

Nevertheless, it is notable that most of the nanodelivery systems are not readily suitable for food, drug, and related applications due to their inherent demerits. While each of these novel delivery strategies offers significant promise, there are still limitations to their potential use in food/medicine. In addition, most of these technologies are not able to accommodate high loading of curcumin/curcuminoid, thus limiting the bioactivity of the finished products. Most of the delivery systems have limited application for use as a powder formulation as their stability will be affected when converted to powder. Moreover, micelle, microemulsion, and liposome complexes might be degraded in the stomach before reaching their targeted sites, hence compromising the bioavailability of the active ingredient. In this regard, natural matrix-based formulations without sophisticated fabrication and chemical modification have been investigated for delivering curcumin/curcuminoid.

Our research group (Aurea Biolabs Pvt. Ltd, Cochin, India) developed the bioavailable curcuminoid—"Cureit^{TM}" based on the recreation of the complete natural turmeric matrix (CNTM) with active curcuminoids ($\sim$50%) by a method known as polar–nonpolar–sandwich (PNS) technology, a patent pending formulation. The PNS technology involves the polar fractions of aqueous extract from turmeric are mixed with 95% curcuminoids ad nonpolar fractions of turmeric essential oil to form a uniform mixture, and the mixture is micronized using bead mills. The PNS technology can be used to preserve functional properties, improve the stability of compounds, enhance health benefits, control the release of bioactive compounds at desired time and specific target, and increase the bioavailability of bioactive compounds. The PNS technology is one of the most promising techniques among various techniques used to improve the dissolution of poorly soluble curcuminoid. This is because it is simple, cost-effective, and commercially attractive for industrial production. The characterizations of Cureit^{TM} are furthermore analyzed by nuclear magnetic resonance spectroscopy (NMR), scanning electron microscopy (SEM), X-ray diffraction (XRD), Fourier transform infrared (IR), current–voltage (I–V) study, quadrupole time-of-flight mass spectrometry (Q-TOF), and thermogravimetric analysis (TGA). This chapter also discussed about the bioavailability of Cureit^{TM} and its potential applications in different biological activities.

11.5 Experimental Part

11.5.1 Preparation Method of Cureit^{TM}

The design of the product, Cureit^{TM}, is a very smart design to develop an efficacious and potential bioavailable curcuminoid through the application of the natural physicochemical properties of curcuminoid toward the molecular physiology of the intestinal absorption of a xenobiotic. An orally consumed molecule has to land up at the inner intestinal wall and should pass through the barriers of the cell membrane, before it is to be available in the blood stream for the purported biological action. The molecule needs to be soluble inside the gastrointestinal tract for reaching the inner intestinal walls and permeable through the lipid bilayer of cell membrane. As curcuminoid is a hydrophobic molecule, it cannot dissolve easily in the intestinal tract, and also, it cannot easily pass through the cellular membrane due its larger structure. The unique product Cureit™—the bioavailable curcuminoid developed by Aurea Biolabs Pvt. Ltd, Cochin, India, was designed to retain curcuminoid as a single free molecule, inside a turmeric matrix created *ex situ*. The turmeric matrix was recreated by extracting three different entities: curcuminoid, turmeric essential oil, and water extract of turmeric. Curcuminoid with 95% purity was extracted from dried turmeric rhizomes, using food-grade solvent-ethanol and the obtained oleoresin crystallized to get curcuminoid powder. Essential oil was separated by steam distillation. The powdered turmeric was extracted with water to get the carbohydrates ($\sim$40%), dietary fiber ($\sim$5%), and turmerin protein ($\sim$2%). The water-soluble protein turmerin is more efficient to cross over the lipid bilayer. These three components are combined together through a unique process of PNS technology, and the curcuminoid is well protected as a single molecule inside this matrix. The bioactive molecules present in the Cureit™ other than curcuminoid play an important role in the bioavailability of curcuminoid rather than the curcuminoid itself. The bisabolanes and sesquiterpenes present in the Cureit™ help the curcuminoid to make a nonpolar sandwiched matrix, while the water-soluble proteins and the carbohydrates make the polar matrix. The Cureit™ also retains the advantages of traditional modified systems such as enhanced physical stability, protection of drug molecules from degradation in the body, controlled drug release, biocompatibility, and laboratory to industrial scalability. The composition of the product has been standardized, and a continuous quality control program is initiated to monitor and maintain the quality standards for the product. The PNS technology allows curcuminoid to be delivered to the intestinal walls and pass through the cell membrane by simple diffusion through enhanced solubility and absorption. The schematic representation of the PNS technology design is depicted in Figure 11.2.

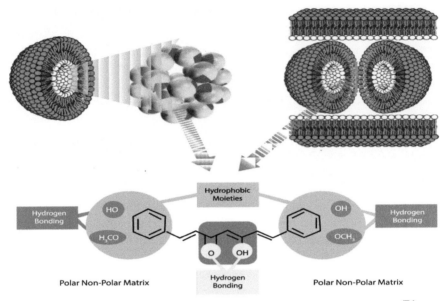

Figure 11.2   Schematic representation of the PNS technology design of Cureit™.

11.5.2 Analytical Method for Analysis of Plasma Curcumin Level in Cureit™ and Standard Curcumin

Analysis for Curcumin has carried out in UHPLC (Shimadzu, NEXERA, model LC30 AD). The column used was 2×150 mm, shim pack, XR-ODS III. An isocratic elution was used with mobile phase of composition 40% THF and 60% water containing 1% citric acid at a flow rate of 0.5 ml/min. Under identical conditions, a six-point linearity curve was plotted using standard Curcumin of 97% purity, in a range of 2 ng/ml to 500 ng/ml.

11.5.3 Statistical Analysis

Pharmacokinetic parameters including C_{max}, AUC_{0-t}, and $AUC_{0-\infty}$ and T_{max} for both standard curcumin and Cureit™ groups were generated using WinNonlin Version 5.0.1. The pharmacokinetic parameters C_{max}, AUC_{0-t}, and $AUC_{0-\infty}$ were analyzed using the GLM ANOVA model with the main effect of treatment. Statistical analyses were performed using a SAS® package (SAS Institute Inc., USA). Values with $p < 0.05$ were considered statistically significant.

11.6 Results and Discussion

11.6.1 Chemical Analysis of Cureit™

The chemical analysis of Cureit™ with Q-TOF was done by Xevo G2-S Q-TOF (Waters Corporation, Milford, USA) via the direct infusion method to obtain the metabolic profile of Cureit™. The identity of the presence of curcuminoid viz. curcumin, demethoxycurcumin, and bisdemethoxycurcumin were confirmed by mass spectra (Figure 11.3(a)). The mass spectra of noncurcuminoid fraction, present in the Cureit™, has shown in Figure 11.3(b). The spectra distinctly demonstrate the presence of sesquiterpenes and their derivatives, and lactones in polar layer. The presence of aromatic turmerone, dihydroturmerone, turmeronol, curdione, and bisacurone in the spectra indicated the occurrence of these compounds in nonpolar layer of Cureit™. This analysis clearly confirmed that the Cureit™ is completely designed by the polar–nonpolar fractions of turmeric with curcuminoid.

11.6.2 Characterization of Cureit™

11.6.2.1 NMR studies of cureit™

^{1}H and ^{13}C NMR spectra of both curcuminoid and Cureit™ were recorded by Bruker Avance III 400 MHz, Switzerland, using DMSO-d_6 as a solvent. The NMR study was investigated to clarify the PNS technology of the Cureit™ (Figure 11.4). For curcuminoids (Figure 11.5(a)) ^{1}H NMR (400 MHz, DMSO-d_6): δ 10.09 (s, 1H), 9.69 (2H, OH, s), 7.57 (2H, d, J = 15.6 Hz), 7.32 (2H, s), 7.15 (2H, d, J = 8.4 Hz), 6.84 (2H, d, J = 8.4 Hz), 6.74 (2H, d, J = 16.0 Hz), 6.05 (1H, s), 3.83 (OCH$_3$, 6H). ^{13}C NMR (400 MHz, DMSO-d_6): δ 183.17, 149.32, 147.97, 140.68, 126.34, 123.05, 121.07, 115.91, 111.34, 100.89, 55.65 (Figure 11.5(b)). For Cureit™ (Figure 11.6(a)), ^{1}H NMR (400 MHz, DMSO-d_6): δ 11.05 (s, 1H), 7.54 (2H, d, J = 15.6 Hz), 7.31 (2H, s), 7.15 (2H, d, J = 8.0 Hz), 6.83 (2H, d, J = 8.0 Hz), 6.75 (2H, d, J = 16.0 Hz), 6.07 (1H, s), 3.84 (OCH$_3$, 6H). ^{13}C NMR (400 MHz, DMSO-d6): δ 183.15, 149.50, 148.01, 140.67, 126.21, 123.10, 120.99, 115.83, 111.31, 100.82, 55.66 (Figure 11.6(b)).

^{1}H NMR spectrum of curcuminoid (Figure 11.4(a)) shows a sharp singlet peak at 9.69 ppm, indicating the presence of hydroxyl groups, while a small singlet peak at 10.09 ppm indicates the presence of intramolecular H-bonding in curcuminoids [49]. In Figure 11.4(b), a new signal at 11.05 ppm was observed, due to the shifting of the peaks at 9.69 and 10.09 downfield. This is attributed to the presence of strong interaction between the –OH group

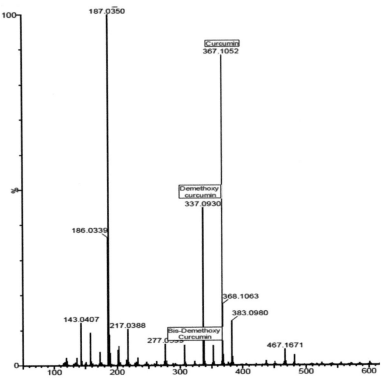

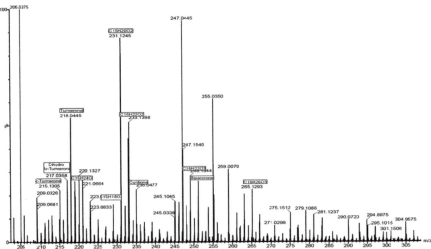

Figure 11.3   Q-TOF of (a) curcuminoid and (b) noncurcuminoid fraction present in Cureit™.

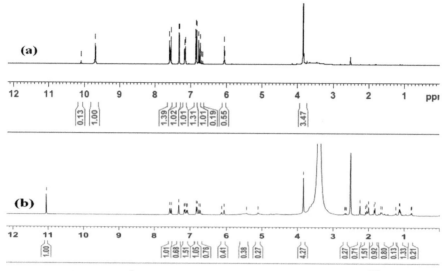

Figure 11.4 ^{1}H NMR spectra of (a) curcuminoid and (b) Cureit^{TM}.

of curcuminoids viz., phenolic –OH and intramolecular/enol form of –OH, and polar and nonpolar environment. Moreover, the interaction of polar and nonpolar entities with the hydrogen of –OH group deshielded the electron density in the hydrogen. This NMR data are clear evidence for the presence of hydrogen bonding interactions with curcuminoids, polar, and nonpolar compounds. The presence of other peaks in Figure 11.4(b) from 2.67 to 1.12 ppm indicates the existence of other polar and nonpolar compounds in the PNS technology.

11.6.2.2 FT-IR studies of cureit^{TM}

Fourier transform infrared (FT-IR) spectra of curcuminoid and Cureit^{TM} were recorded by JASCO FT/IR-460 plus instrument in the range of 4000 to 400 cm^{-1} with 32 scans per samples. The FI-IR spectra of curcumin, demethoxycurcumin (DMC), bisdemethoxycurcumin (BDMC), curcuminoid, and Cureit^{TM} are shown in Figure 11.7 (a–e), respectively. The bands at 1510 and 1428 cm^{-1} are due to the stretching vibrations of C–C of the benzene ring and olefin bending vibration of the C=C group bound to the benzene ring, respectively [50]. The peak at 1629 cm^{-1} is attributed to the carbonyl (C=O) stretching of the conjugated ketone. The sharp absorption peaks at 1602 cm^{-1} correspond to the benzene ring stretching [51], while the peak at 1281 cm^{-1} corresponds to aromatic C–O stretching vibrations.

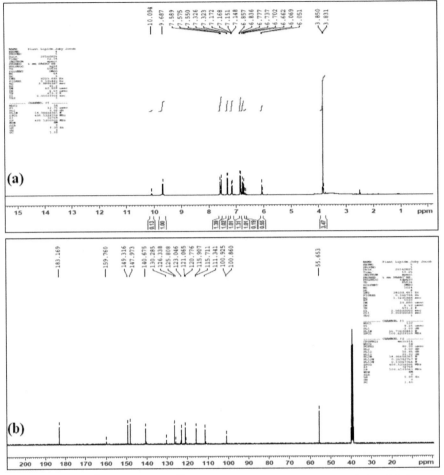

Figure 11.5 ^{1}H NMR (a) and ^{13}C NMR (b) spectra of curcuminoid in DMSO-d$_6$.

In addition, the peak at 811 cm^{-1} is attributed to the stretching vibration of C–O in –C–OCH$_3$, while the peak at 1027 cm^{-1} corresponds to the C–O–C stretching of ether [52, 53]. The peak for the phenyl ring was also observed at 857 cm^{-1}, while the peak at 714 cm^{-1} corresponds to aromatic in plane bending of curcuminoids [54].

Furthermore, a characteristic absorption bands was observed in all the spectra in the range between 3390 and 3500 cm^{-1}. This is attributed to the phenolic O–H stretching vibration [55]. However, in the Cureit^{TM}

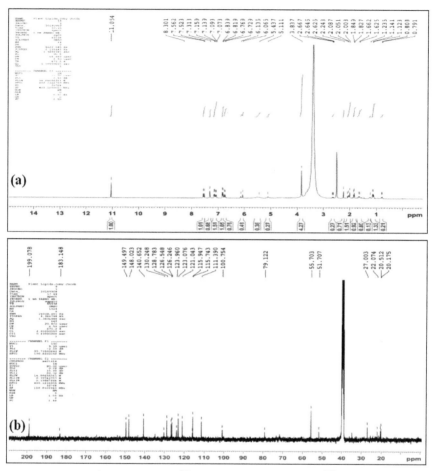

Figure 11.6 ^{1}H NMR (a) and ^{13}C NMR (b) spectra of Cureit^{TM} in DMSO-d$_6$.

(Figure 11.7(e)) this band became broad. This broadness clearly indicated that the band does not due only to the presence of phenolic O–H, but also due to the hydrogen bonding of curcuminoids between the polar and nonpolar layer. Furthermore, the prepared Cureit^{TM} showed characteristic peaks which were very close to the bands of curcumin analogs confirming the presence of curcuminoid in the PNS technology with weak hydrogen bonding. These data are in very good agreement with NMR analysis and further confirmed the presence of hydrogen bonding interactions with curcuminoids, polar, and nonpolar compounds.

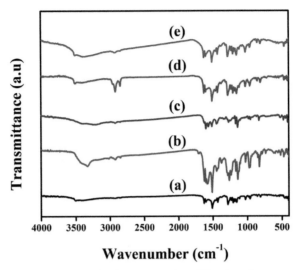

Figure 11.7 FT-IR spectra of (a) curcumin, (b) demethoxycurcumin (DMC), (c) bisdemethoxy-curcumin (BDMC), (d) curcuminoid, and (e) Cureit[TM]

11.6.2.3 XRD studies of cureit[TM]

The crystalline natures of the curcuminoid and Cureit[TM] were determined using X-ray diffraction (XRD) (Xpert-Pro). The XRD pattern of curcuminoid and Cureit[TM] is shown in Figure 11.8. In the XRD pattern of both

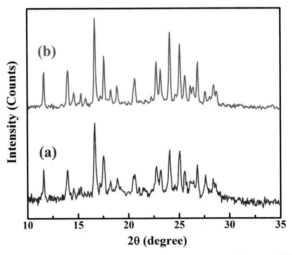

Figure 11.8   XRD pattern of (a) curcuminoid and (b) Cureit[TM].

curcuminoid and Cureit™, a number of peaks are seen in the region of 5–30°C, without any significant difference between the two samples. This also confirmed the presence of curcuminoid in the polar–nonpolar–sandwich design without any modifications.

11.6.2.4 TGA/DTA studies of cureit™

TGA curves of curcuminoid and Cureit™ are shown in Figure 11.9. Curcuminoids (Figure 11.9(a)) do not show any stage of water loss up to 200°C due to its high hydrophobicity. However, it shows a gradual weight loss with first weight loss at 279.52°C and second at 373.54°C [43]. In Figure 11.9(b), the first degradation step due to elimination of moisture from the polar matrix occurs at 67.79°C, while the second degradation step at 168.47°C corresponds to the breakage of protein chain [43, 56] present in the polar layer in the Cureit™. The third weight loss is recorded at 283.78°C and corresponds to the degradation of available curcuminoid in the Cureit™ [43]. Curcuminoid and Cureit™ show almost similar weight loss pattern in the range between 279 and 284°C. This also confirmed that curcuminoid encapsulated PNS technology does not alter thermal degradation pattern.

11.6.2.5 SEM analysis of cureit™

Samples of curcuminoid and Cureit™ were fabricated on scanning electron microscopy (SEM) aluminum stubs with double-side carbon tape and sputter-coated with gold. Surface morphology images were captured with SEM (Vega3Tescan, Germany). SEM analysis was conducted to investigate the morphology of Cureit™. It was evident from SEM image that the

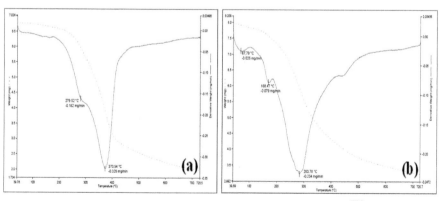

Figure 11.9   TGA curve of (a) curcuminoid and (b) Cureit™.

formulation was almost spherical and well dispersed with rough morphology (Figure 11.10(a) insert). The mean diameter is in the range of 400–600 μm. Figure 11.10(c) clearly indicated the presence of three different layers with slight morphological modifications. It might be the characteristic morphology of PNS technology.

11.6.2.6 I–V studies of cureit™

Current–voltage curve (I–V curve) is a relationship represented as a graph between the electric current through a material and the corresponding voltage difference across it. The direct current (DC) electrical characteristics study (the I–V study) shows that the conductivity of the Cureit™ is being increased by 1:62 ratio than the curcuminoid (Figure 11.11(a)) due to the presence of

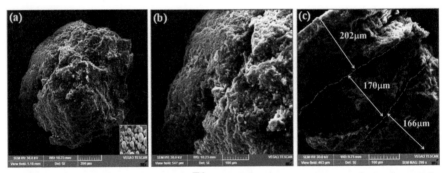

Figure 11.10 SEM images of Cureit™ with different cross sections (a) single granule, (b) cross section of semilayer form, and (c) three layers of PNS form.

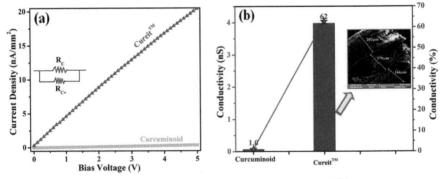

Figure 11.11 Current–voltage study of (a) curcuminoid and Cureit™ [insert: equivalent current] (b) conductivity and percentage of increase in conductivity of curcuminoid and Cureit™ [insert: SEM micrograph of Cureit™].

layered structure (Figure 11.11(b)). Current–voltage studies on different layers of Cureit™ showed that it is highly positive and that Cureit™ conducts electricity. This is attributed to the incorporation of polar and nonpolar matrix, which is an evidence of the PNS technology, whereas curcuminoid is very poor conducting.

11.7 Bioavailability and Bioefficacy Studies of Cureit™

11.7.1 Comparative Oral Bioavailability Study of Cureit™ with Standard Curcumin

A pilot crossover study was carried out to analyze the bioavailability of Cureit™. Twelve human healthy male adults were selected for this study in an age-group of 18–45 years old. They were medically healthy, with clinically acceptable laboratory profiles, ECG, and chest X-ray, and they did not have any major disease during screening medical history and whose physical examination was performed within 21 days prior to the period one of testing. They were not habitual users of tobacco or alcohol.

This study was a two period, two sequences, and two treatment study. A single oral dose of the test product (Cureit™ capsules 500 mg) and reference product (Curcumin capsules 500 mg) were administered on 2 different days for the subjects. The first and second dosages were separated by a washout period of 14 days. The duration of the study from the check-in of period one to the last blood draw of period two was 15 days. Subjects were reported to the clinical facility from at least 11 h prior to the drug administration such that there was at least 8 h supervised fasting before the scheduled dose of each treatment period and subjects remained housed in the clinical facility until the 8 h post-dose blood sample has drawn in each of the study periods.

The volunteers were divided into 2 groups of 6 members. After a supervised overnight fasting for at least eight hours, subjects of first group were administered a single oral dose of the test product (Cureit™ capsules 500 mg) and the subjects of second group were administrated the reference product (Curcumin capsules 500 mg). After 14 days of washout period, the first group of subjects consumed the reference product, and the second group of subjects consumed the test product. Subjects remained seated upright for first four hours after dosing.

Thereafter, they were allowed to engage in normal activities avoiding severe physical activities. The pre-dose sample was collected within 1 h prior to drug administration. The post-dose blood samples were collected at 1, 2, 3, 4, 5, 6, and 8 h in each study period. The blood samples were collected for each volunteer in pre-specified vacuum tubes containing anti-coagulant. The plasma was separated from the blood samples and analyzed.

A method was validated with curcumin standard, and the curcumin content in blood plasma was quantified. Figure 11.12 shows the time profile of the plasma concentration of Cureit^{TM} with the standard curcumin. The absorption of curcumin is gradually increasing in the successive hours after the consumption of samples. The plasma curcumin level is attained maximum at 4th hours for Cureit^{TM}, 3rd and 4th hour for standard curcumin, and then decreased gradually.

Cureit^{TM} has been shown tremendous increase in the plasma curcumin level, and it can be the best promising formulation to enhance the bioavailability. Curcumin in Cureit^{TM} and standard curcumin were tested, and the C_{max} values were 434.3 ng/mL and 43.1 ng/L, respectively (Table 11.1). The area under the blood concentration–time curve (AUC) for Cureit^{TM} was 904.0 ng/mL, and for standard curcumin, it was 165.7 ng/mL, respectively (Table 11.1). From the results, Cureit^{TM} possesses a good bioavailability, comparing to standard curcumin, and it proves that the bioavailability as measured by the C_{max} as well as AUC of Cureit^{TM} was ten-folds more, compared to standard curcumin.

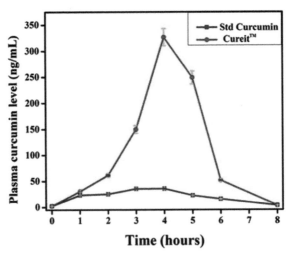

Figure 11.12   Mean curcumin concentration in plasma after treatment with Cureit^{TM} and standard curcumin at different time intervals.

Table 11.1 The average pharmacokinetic variables from plasma curcumin levels after administration of the Cureit[TM] and standard curcumin at different time intervals

Parameters	Parmacokinetic Values (Mean $\pm$ SD)	
	Cureit[TM]	Standard Curcumin
C_{max} (ng/mL)	434.25 $\pm$ 256.61	43.10 $\pm$ 16.73
AUC_{0-t} (ng $\times$ h/mL)	904.00 $\pm$ 459.73	165.70 $\pm$ 55.76
AUC_{0-inf} (ng $\times$ h/mL)	980.00 $\pm$ 508.10	192.80 $\pm$ 63.89
T_{max} (h)	4.2 $\pm$ 0.6	3.6 $\pm$ 0.7

11.7.2 Recent Studies on the Cureit[TM]

An anticancer study was carried out by our research group on the cytotoxicity demonstrated by a spectrochemical study using MTT assay on the effect of Cureit[TM] on cell proliferation such as growth kinetics of MCF-7, LnCAP, and HEK 293 T cells. The study clearly revealed that Cureit[TM] could serve as a good anticancer medication [57]. A cell culture study of the elastase inhibition activity of Cureit[TM] was measured. The study showed that the Cureit[TM] showed potent elastase inhibiting activity at 100 mg/L concentration. It was observed from the co-culture study of melanocytes and keratinocytes that the cell integrity, morphology, and functionality were intact by the Cureit[TM]. Its elastic inhibiting activity in human cell lines inferred that Cureit[TM] can inhibit elastases activity at high concentration [58]. Another study was conducted on the antioxidant property of Cureit[TM] by DPPH free radical scavenging activity method. The antioxidant potential of Cureit[TM] was equivalent to that of ascorbic acid; as a result, it could be a good source of natural antioxidant [59].

Hyluronic acid is a naturally occurring mucopolysaccharide in all living organisms. It is an important extracellular matrix in the various tissues such as skin, lungs, and ligaments. Inhibitors of hyaluronidase could serve as therapeutic targets in alleviating the age-related compromised functions associated with imbalance of hyaluronic acid synthesis and lyaluronidases. The effect of Cureit[TM] on anti-aging study was exhibited by our earlier study with hyaluronidase inhibition through cell culture study. Cureit[TM] was inhibiting hyaluronidases up to 42%. This confirmed that the Cureit[TM] could be a useful anti-aging medication [60].

11.8 Conclusions

A novel bioavailable curcuminoid formulation, Cureit[TM], was prepared by using PNS technology with complete natural turmeric matrix. The PNS technology of Cureit[TM] was confirmed by various instrumental techniques.

SEM images clearly indicated that Cureit™ was almost spherical and well dispersed with rough morphology and separated with three layers of PNS formulation. The metabolic profile of Cureit™ was analyzed by Q-TOF and confirmed the presence of curcuminoids, lactones, sesquiterpenes, and their derivatives in polar layer, aromatic turmerone, dihydroturmerone, turmeronol, curdione, and bisacurone in nonpolar layer. NMR data clearly confirmed the presence of hydrogen bonding interactions with curcuminoids, polar, and nonpolar compounds in the PNS technology. IR, XRD, and TGA further confirmed the presence of curcuminoids with high stability in the PNS formulation. All the analyses confirm the presence of curcuminoid inside the PNS design without any alterations. This novel turmeric matrix formula Cureit™ greatly facilitates the absorption and bioavailability of curcumin as compared to 95% curcumin. The solubility and bioavailability were enhanced by PNS curcuminoids in highly polar (proteins, fibers, and polysaccharides) and nonpolar (curcumin essential oil) entities, resulting in a natural complex matrix. The earlier studies of Cureit™ ensured that the enhancement of bioavailability of curcuminoid, and the PNS technology can be a promising curcuminoid delivery system for various biological activities without any side effect.

References

[1] Newman, D. J., Cragg, G. M., and Snader, K. M. (2003). Natural products as sources of new drugs over the period. *J. Nat. Prod.* 66, 1022–1037.

[2] Pandit, S., Kim, H., Kim, J., and Jeon J. (2011). Separation of an effective fraction from turmeric against Streptococcus mutans biofilms by the comparison of curcuminoid content and anti-acidogenic activity. *Food Chem.* 126, 1565–1570.

[3] Paramasivam, M., Poi, R., Banerjee, H., and Bandyopadhyay, A. (2009). High performance thin layer chromatographic method for quantitative determination of curcuminoids in Curcuma longa germplasm. *Food Chem.* 113, 640–644.

[4] Chaturvedi, T. P. (2009). Uses of turmeric in dentistry: an update. *Indian J. Dent. Res.* 20, 107–109.

[5] Mukerjee, A., and Vishwanatha, J. K. (2009). Formulation, characterization and evaluation of curcumin-loaded PLGA nanospheres for cancer therapy. *Anticancer Res.* 29, 3867–3875.

[6] Perko, T., Ravber, M., Knez, Z., and Skerget M. (2015). Isolation, characterization and formulation of curcuminoids and in vitro release study of the encapsulated particles. *J. Supercrit Fluids* 103, 48–54.

[7] Kalpravidh, R. W., Siritanaratkul, N., Insain, P., Charoensakdi, R., Panichkul, N., Hatairaktham, S., et al. (2010). Improvement in oxidative stress and antioxidant parameters in β-thalassemia/Hb E patients treated with curcuminoids. *Clin Biochem.* 43, 424–429.

[8] Changtam, C., de Koning, H. P., Ibrahim, H., Sajid, M. S., Gould, M. K., and Suksamrarn, A. (2010). Curcuminoid analogs with potent activity against *Trypanosoma* and *Leishmania* species. *Eur. J. Med. Chem.* 45, 941–956.

[9] Lim, H. S., Park, S. H., Ghafoor, K., Hwang, S. Y., and Park, J. (2011). Quality and antioxidant properties of bread containing turmeric (*Curcuma longa* L.) cultivated in South Korea. *Food Chem.* 124, 1577–1582.

[10] Peret-Almeida, L., Cherubino, A. P. F., Alves, R. J., Dufossé, L., and Glória, M. B. A. (2005). Separation and determination of the physico-chemical characteristics of curcumin, demethoxycurcumin and bisdemethoxycurcumin. *Food Res. Int.* 38, 1039–1044.

[11] Aditya, N. P., Chimote, G., Gunalan, K., Banerjee, R., Patankar, S., and Madhusudhan B. (2012). Curcuminoids-loaded liposomes in combination with arteether protects against *Plasmodium berghei* infection in mice. *Exp. Parasitol.* 131, 292–299.

[12] Khan, M. A., El-Khatib, R., Rainsford, K. D., Whitehouse, M. W. (2012). Synthesis and anti-inflammatory properties of some aromatic and heterocyclic aromatic curcuminoids. *Bioorg. Chem.* 40, 30–38.

[13] Yue, G. G. L., Chan, B. C. L., Hon, P., Kennelly, E. J., Yeung, S. K., Cassileth, B. R., Fung, K., Leung, P., Lau, C. B. S. (2010a). Immunostimulatory activities of polysaccharide extract isolated from *Curcuma longa Int. J. Biol. Macromol.* 47, 342–347.

[14] Tapal, A., and Tiku, P. K. (2012). Complexation of curcumin with soy protein isolate and its implications on solubility and stability of curcumin. *Food Chem.* 130, 960–965.

[15] Panahi, Y., Saadat, A., Beiraghdar, F., Nouzari, S. M. H., Jalalian, H. R., Sahebkar, A. (2014). Antioxidant effects of bioavailability-enhanced curcuminoids in patients with solid tumors: A randomized double-blind placebo-controlled trial. *J. Funct. Foods* 6, 615–622.

[16] Zhan, P. Y., Zeng, X. H., Zhang, H. M., and Li, H. H. (2011). High-efficient column chromatographic extraction of curcumin from *Curcuma longa*. *Food Chem.* 129, 700–703.

[17] Siviero, A., Gallo, E., Maggini, V., Gori, L., Mugelli, A., Firenzuoli, F., and Vannacci, A. (2015). Curcumin, a golden spice with a low bioavailability. *J. Herb. Med.* 5, 57–70.

[18] Mahmood, K., Zia, K. M., Zuber, M., Salman, M., and Anjum, M. N. (2015). Recent developments in curcumin and curcumin based polymeric materials for biomedical applications: a review. *Int. J. Biol. Macromol.* 81, 877–890.

[19] Vogel, H., and Pelletier, J. (2006). Curcumin-biological and medicinal properties. *J. Pharma* 2:50.

[20] Milobedeska, J., Kostanecki, S., and Lampe, V. (1910). Structure of curcumin. *Ber. Dtsch. Chem. Ges.* 43, 2163–2170.

[21] Lampe, V., and Milobedeska, J. (1913). Studien über curcumin. *Ber. Dtsch. Chem. Ges.* 46, 2235–2240.

[22] Anderson, A. M., Mitchell, M. S., and Mohan, R. S. (2000). Isolation of curcumin from turmeric. *J. Chem. Educ.* 77, 359–360.

[23] Bagchi, A. (2012). Extraction of Curcumin. *IOSR J. Environ. Sci. Toxicol. Food Technol.* 1, 1–16.

[24] Ravindran, P. N., Nirmal Babu, K., and Sivaraman, K. (2007). *Turmeric: The genus Curcuma*. Boca Raton, FL: CRC Press, 235.

[25] Bernabé-Pineda, M., Ramírez-Silva, M. T., Romero-Romo, M., González-Vergara, E., and Rojas-Hernández, A. (2004). Determination of acidity constants of curcumin in aqueous solution and apparent rate constant of its decomposition. *Spectrochim. Acta A Mol. Biomol. Spectrosc.* 60, 1091–1097.

[26] Darvesh, A. S., Carroll, R. T., Bishayee, A., Novotny, N. A., Geldenhuys, W. J., and Van der Schyf, C. J. (2012). Curcumin and neurodegenerative diseases: a perspective. *Expert Opin. Investig. Drugs* 21, 1123–1140.

[27] Basnet, P, and Skalko-Basnet, N. (2011). Curcumin: an anti-inflammatory molecule from a curry spice on the path to cancer treatment. *Molecules* 16, 4567–4598.

[28] Basniwal, R. K., Buttar, H. S., Jain, V. K., and Jain, N. (2011). Curcumin nanoparticles: preparation, characterization, and antimicrobial study. *J. Agric. Food Chem.* 59, 2056–2061.

[29] Amalraj, A., Pius, A., Gopi, S., and Gopi, S. (2016). Biological activities of curcuminoids, other biomolecules from turmeric and their derivatives – A review. *J. Tradit. Complement. Med.* 7, 205–233. doi: 10.1016/j.jtcme.2016.05.005

[30] Cheng, A. L., Hsu, C. H., Lin, J. K., Hsu, M. M., Ho, Y. F., Shen, T. S., et al. (2001). Phase I clinical trial of curcumin, a chemopreventive agent, in patients with high-risk or pre-malignant lesions. *Anticancer Res.* 21, 2895–2900.

[31] Jantarat, C. (2013). Bioavailability enhancements techniques of herbal medicine: A case example of curcumin. *Int. J. Pharm. Pharm. Sci.* 5, 493–500.

[32] Prasad, S., Tyagi, A. K., and Aggarwal, B. B. (2014). Recent developments in delivery, bioavailability, absorption and metabolism of curcumin: The golden pigment from golden spice. *Cancer Res. Treat.* 46, 2–18.

[33] Pan, M. H., Huang, T. M., and Lin, J. K. (1999). Biotransformation of curcumin through reduction and glucuronidation in mice. *Drug Metab. Dispos.* 27, 486–494.

[34] Heger, M., van Golen, R. F., Broekgaarden, M., and Michel, M. C. (2014). The molecular basis for the pharmacokinetics and pharmacodynamics of curcumin and its metabolites in relation to cancers. *Pharmacol. Rev.* 66, 222–307.

[35] Tonnesen, H. H., and Karlsen, J. (1985). Studies on Curcumin and Curcuminoids 5. Alkaline-degradation of curcumin. *Z. Lebensm. Unters. For.* 180, 132–134.

[36] Tonnesen, H. H., Karlsen, J., and Vanhenegouwen, G. B. (1986). Studies on curcumin and curcuminoids 8. photochemical stability of curcumin. *Z. Lebensm. Unters. For.* 183, 116–122.

[37] Paramera, E. I., Konteles, S. J., and Karathanos, V. T. (2011). Stability and release properties of curcumin encapsulated in Saccharomyces cerevisiae, β-cyclodextrin and modified starch. *Food Chem.* 125, 913–922.

[38] Li, J., Shin, G. H., Lee, I. W., Chen, X., and Park, H. J. (2015). Soluble starch formulated nanocomposite increases water solubility and stability of curcumin. *Food Hydrocolloids* 56, 41–49.

[39] Bergonzi, M. C., Hamdouch, R., Mazzacuva, F., Isacchi, B., and Bilia, A. R. (2014). Optimization, characterization and in vitro evaluation of curcumin microemulsions. *LWT-Food Sci. Technol.* 59, 148–155.

[40] Chaurasia, S., Patel, R. R., Chaubey, P., Kumar, N., Khan, G., and Mishra, B. (2015). Lipopolysaccharide based oral nanocarrier for the improvement of bioavailability and anticancer efficacy of curcumin. *Carbohydr. Polym.* 130, 9–17.

[41] Sanoj Rejinold, N., Sreerekha, P. R., Chennazhi, K. P., Nair, S. V., and Jayakumar, R. (2011). Biocompatible, biodegradable and thermosensitive chitosan-g-poly (N-isopropylacrylamide) nanocarrier for curcumin drug delivery. *Int. J. Biol. Macromol.* 49, 161–172.

[42] Righeschi, C., Bergonzi, M. C., Isacchi, B., Bazzicalupi, C., Gratteri, P., and Bilia, A. R. (2016). Enhanced curcumin permeability by SLN formulation: the PAMPA approach. *LWT – Food Sci. Technol.* 66, 475–483.

[43] Sarika, P. R., James, N. R., Anil kumar, P. R., and Raj, D. K. (2016). Preparation, characterization and biological evaluation of curcumin loaded alginate aldehyde – gelatin nanogels. *Mater. Sci. Eng. C.* 68, 251–257.

[44] Appendino, G., Belcaro, G, Cornelli, U., Luzzi, R., Togni, S., Dugall, M., Cesarone, M., Feragalli, B., Ippolito, E., and Errichi, B. (2011). Potential role of curcumin phytosome (Meriva) in controlling the evolution of diabetic microangiopathy – a pilot study. *Panminerva. Med.* 53, 43–49.

[45] Mohan, P. K., Sreelakshmi, G., Muraleedharan, C., and Joseph, R. (2012). Water soluble complexes of curcumin with cyclodextrins: Characterization by FT-Raman spectroscopy. *Vib. Spectrosc.* 62, 77–84.

[46] Antony, B., Merina, B., Iyer, V., Judy, N., Lennertz, K., and Joyal, S. (2008). A pilot cross-over study to evaluate human oral bioavailability of BCM-95® CG (Biocurcumax™), a novel bioenhanced preparation of curcumin. *Indian J. Pharm. Sci.* 70, 445–449.

[47] Jia, Q., Ivanov, I., Zlatev, Z. Z., Alaniz, R. C., Weeks, B. R., Callaway, E. S., et al. (2011). Dietary fish oil and curcumin combine to modulate colonic cytokinetics and gene expression in dextran sodium sulphate-treated mice. *Br. J. Nutr.* 106, 519–529.

[48] Zhongfa, L., Chiu, M., Wang, J., Chen, W., Yen, W., Fan-Havard, P., et al. Enhancement of curcumin oral absorption and pharmacokinetics of curcuminoids and curcumin metabolites in mice. *Cancer Chemother. Pharmacol.* 69, 679–689.

[49] Goren, A. C., Cikrikci, S., Cergel, M., and Bilsel, G. (2009). Rapid quantitation of curcumin in turmeric via NMR and LC-tandem mass spectrometry. *Food Chem.* 113, 1239–1242.

[50] Kolev, T. M., Velcheva, E. A., Stamboliyska, B. A., and Spiteller, M. (2005). DFT and experimental studies of the structure and vibrational spectra of curcumin. *Int. J. Quantum Chem.* 102, 1069–1079.

[51] Anitha, A., Sreeranganathan, M., Chennazhi, K. P., Lakshmanan, V. K., and Jayakumar, R. (2014). In vitro combinatorial anticancer effects of 5-fluorouracil and curcumin loaded N, O-carboxymethyl chitosan nanoparticles toward colon cancer and in vivo pharmacokinetic studies. *Eur. J. Pharm. Biopharm.* 88, 238–251.

[52] Renuka, K., Ramakant, M., Ashwni, V., Pankaj, D., Vivek, K., Varsha, G., et al. (2013). Colon-specific delivery of curcumin by exploiting Eudragit-decorated chitosan nanoparticles in vitro and in vivo. *J. Nanopart. Res.* 15:1893.

[53] Akl, M. A., Kartal-Hodzic, A., Oksanen, T., Ismael, H. R., Afouna, M. M., Yliperttula, M., et al. (2016). Factorial design formulation optimization and *in vitro* characterization of curcumin-loaded PLGA nanoparticles for colon delivery. *J. Drug Deliv. Sci. Technol.* 32, 10–20.

[54] Bourbon, A. I., Cerqueira, M. A., and Vicente, A. A. (2016). Encapsulation and controlled release of bioactive compounds in lactoferrin-glycomacropeptide nanohydrogels: curcumin and caffeine as model compounds. *J. Food Eng.* 180, 110–119.

[55] Benassi, R., Ferrari, E., Lazzari, S., Spagnolo, F., and Saladini, M. (2008). Theoretical study on Curcumin: A comparison of calculated spectroscopic properties with NMR, UV–vis and IR experimental data. *J. Mol. Struct.* 892, 168–176.

[56] Cai, Z., and Kim, J. (2010). Preparation and characterization of novel bacterial cellulose/gelatin scaffold for tissue regeneration using bacterial cellulose hydrogel. *J. Nanotechnol. Eng. Med.* 1:021002.

[57] Gopi, S., George, R., Jude, S., and Sriraam, V. T. (2014). Cell culture study on the cytotoxic effects of "Cureit" – a novel bio available curcumin-anti cancer effects. *J. Chem. Pharm. Res.* 6, 96–100.

[58] Gopi, S., George, R., and Sriraam, V. T. (2014). Cell culture study on the effect of bioavailable curcumin – "Cureit" on elastase inhibition activity. *Br. Biomed. Bull.* 2, 545–549.

[59] Gopi, S., George, R., and Sriraam, V. T. (2014). Anti oxidant potential of "Cureit" – A novel bioavailable curcumin formulation. *Asian J. Pharm. Tech. Innov.* 2, 123–127.

[60] Gopi, S., George, R., and Sriraam, V. T. (2014). Cell culture study on the effects of "cureit" hyaluronidase inhibition – Anti aging effects. *Int. J. Curr. Res.* 6, 8473–8474.

12

The Good Tooth, The Bad Influence of Aciduric Germs and The Ugly Stench of Decay

T. Jesse Joel[1] and P. W. Ramteke[2]

[1]Department of Biotechnology, School of Agriculture and Biosciences, Karunya Institute of Technology and Sciences, Tamil Nadu, India
[2]Head, Department of Biological Sciences, Sam Higginbottom University of Agriculture, Technology and Sciences, Allahabad, India

Abstract

Dental caries is a localized, transmissible, pathological infectious process that ends up in the destroying hard dentin of the teeth. In effect, the disease initiated by the organism under study in the mouth, is evidentially, a case of evolutionary incongruence, a lifestyle enjoyed only by those in developed countries. The fact that there are millions of microorganisms in the air we breathe, every second of our lives, is a miracle that we don't die instantly. The reason, however, is defended, if not answered, by various people with cognitive reasoning. We term such reasoning as scientific.

It is a growing industry milking revenue at a very definitive pace, paving ways to novel inventions in technology and employment as far as the art industry for flawless jaw lines and pearly white teeth. The human mouth or the oral cavity is comprised of many surfaces, each coated with a plethora of bacteria, the proverbial bacterial biofilm. Surprisingly, little is known about the various microflora of the healthy oral cavity. Saliva is a complex protein-rich medium that delivers nutrients, increased saliva flow prevents changes in oral pH, because the buffer bicarbonate is present in the saliva and acts as an acid sink at a time when acidic products are being introduced into the oral cavity [18].

12.1 The Good, The Bad, and The Ugly Microbe –
Streptococcus Mutans

Ever since the dental profession has been around, dentists have voiced their importance by innovative, improved, and efficient treatments. People are forewarned for the prevention of cavities. All of these interactions and microbial coexistence are brought about in an extraordinary medium which buffers any pH drop in the plaque resulting from fermentation of dietary carbohydrates resulting in an acidic pH. There are a lot of mineral ions which are removed from the enamel and thus cause demineralization. It is soon restored by a process called re-mineralization. Saliva is the medium through which re-mineralization occurs. Caries are formed when the rate of decay of the teeth caused by the lactic acid produced by anaerobic bacteria exceeds the rate of repair initiated by the phosphate and calcium ions in saliva. Thus, it is important to understand its intricate make up and its composition.

Significance of Walter Loesche's Work [54]

Walter's work was very pivotal in the understanding of Human dental decay and Streptococcus mutans as a pathogen. The etiologies of prominent dental maladies such as caries and periodontitis have been hypothesized even before the promotion of the germ theory of disease. Walter J. Loesche has been at the forefront, researching both the biological basis of disease processes as well as translating the resulting knowledge to effectively adapt clinical approaches that target the infectious nature of dental diseases. He and his collaborators have tackled an array of oral health issues. They investigated the taxonomy of the oral flora, virulence properties of the oral pathogens, and transmission of oral pathogens. Walter developed the landmark "Specific Plaque Hypothesis," which held that dental caries was a specific bacterial infection most prominently involving Mutans streptococci and possibly Lactobacilli. This hypothesis, which was perhaps the most provocative and highly debated in dental research history, served as a basis for the initiation of hypothesis-based diagnostic and treatment paradigms for clinical management of dental caries. This concept may well be the most revolutionary idea in oral microbiology and infectious disease. After this he expanded the hypothesis to include the role of anaerobic bacteria in periodontal disease. Walter's legacy that is not very well known is that he was willing to follow novel, far-sighted ideas based on his scientific intuition. Bibliography.

Summary

The hypothesis that has brought about this work is that individual isolates of *Streptococcus mutans* can be qualitatively different, exhibiting virulence characteristics different from each other and that some strains would have better transmissibility between individuals than are others. Dental caries disease causes a great deal of continuous discomfort through impaired function and aesthetics as well as inconvenient treatment. Dental caries may even lead to life-threatening infections, and the costs for operative dental treatment are significant both for individuals and society. Novel treatments derived from an ecological or evolutionary viewpoint may be instrumental in treating persistent dental diseases that are especially prevalent in low-income populations. The lifestyles of bountiful enjoyed by developed countries and the livelihood of paucity not much enjoyed by developing or underdeveloped countries throw the disease into complete uncertainty of a decent and proper research for preventive recommendations whatsoever. When this organism is pathogenic, it is as unbridled as ever, and if unattended or undertreated, it could originate into a serious ordeal. The present work has defined a study to ascertain common if not scientific facts for a better cliché to life.

12.1.1 Introduction

12.1.1.1 Streptococcus mutans: Isolation and identification

In 1924, Clark discovered the organism *Streptococcus mutans* he isolated them from conspicuous lesions in the teeth and the perception that dental infection is bacterial, offers no justification for any prophylactic measure. It was then proved to be the principle etiological agent for human dental decay [18]. Dental caries has been recognized as one of the most common types of bacterial infections in humans, often a recurring infection that most individuals have to contend with throughout the duration of their life [28]. *Streptococcus mutans* is pleomorphic, microaerophilic organism that is associated with dental caries and plaque. Various schemes for the identification of Viridans have been developed [17, 20], and their early work was modified and presented [38]. Clinical oral microbiology laboratories employ one or a combination of these methods, depending on the pathogens to be identified. Rarely does one detection method prove optimal for all situations. *Streptococcus mutans* is unique in its colony morphology and can be identified, but there is always a question of false-positive and false-negative colorations on the media. Earlier studies record such a problem [14]. A selective medium-color indicator test was developed to use in large-scale epidemiological surveys for the presence or absence of *Streptococcus mutans* [60]. The MSB medium is highly selective, and, except for occasional contamination of Enterococci, which are easily recognized, growth of extraneous organisms is not a problem. It is known that this medium does not support the full growth of all *Streptococcus mutans* strains, applying for mostly a & b serotypes, which are generally not found in humans [19, 45]. Simplified laboratory methods for detection and quantification of *Streptococcus mutans* have been proposed after it was proved as the etiology. In a study on similar intentions, a micromethod was developed for quantitative estimation for *Streptococcus mutans* and lactobacilli in Saliva. Here, they used a semiautomatic pipette and 25 microliter of diluted saliva was spotted on the surface of an agar plate containing a selective medium using it. This volume gave a spot with a diameter of about 10 mm in which separate colonies could be counted [76]. In another study, a wooden spatula moistened with saliva in the mouth is pressed directly onto the surface of a medium selective for *Streptococcus mutans* [39]. A group of researchers formulated a test for semiquantitative determinations of *Streptococcus mutans* based on the unique ability of this organism to develop adherent colonies which form on the walls of culture vessels [47]. A slide-scoring method for estimating *Streptococcus mutans* populations in saliva was described in which paraffin-stimulated saliva is poured onto a special slide coated with Mitis-salivarius-sucrose agar [5]. Discs impregnated with bacitracin are placed on the inoculated slides; growth density of *Streptococcus mutans* colonies which develop within inhibition zones is scored after incubation. A Cariescreen SM kit based on the previously described bacteriological growth medium, MSB agar [11], was devised. The system is suitable for use in dental offices and other nonlaboratory settings and can be utilized by clinical personnel not specially trained in microbiology. This Cariescreen SM system describes a significant simplification of cultural detection procedures by eliminating the need for media preparation, sample dilution, and plating and colony counting [32]. *Streptococcus mutans* is called so because on gram-stain, they were more oval than round and thus appeared to be a mutant form of the microorganism. *Streptococcus mutans* was more or less like a short rod (or) sphere depending on its environment [65]. *Streptococcus mutans* is capable of utilizing a broad range of carbohydrates, and it is thought that this versatility is one of the characteristics contributing to its survival in plaque and involvement in the caries process. Among the substrates utilized are b-glucosides such as aesculin, arbutin, cellobiose, and salicin. All these substrates contain a glucose moiety that can enter glycolysis, coupled to glucose or some other nonmetabolizable component. Different species of oral streptococci vary in their capacity to attack these substrates, and this variation has been of value in the development of identification schemes [77]. The oral cavity is comprised of many surfaces, each coated with a plethora of bacteria, the proverbial bacterial biofilm. Surprisingly, little is known about the microflora of the healthy oral cavity. Previously using culture-independent molecular methods, over 500 species or phylotypes in the subgingival plaque of healthy subjects and subjects with periodontal disease were detected [55]. In view of the inadequacy at hand, an effort was made to define the normal flora of the oral cavity. More than 700 bacterial species or phylotypes, of which over 50% were not cultivated, have been detected in the oral cavity. After utilizing culture-independent molecular techniques to determine the site and subject specificity of bacterial colonization, and further determining species identity by sequencing the 16S rRNA genes from the sample DNA, species specifically associated with periodontitis and dental caries was not detected. However, the oral cavity harbored distinctively predominant bacterial flora that was highly diverse and both site and subject specific [4]. In a study conducted to both isolate and identify nonmutans streptococci organisms from dental plaque, mitis-salivarius bacitracin agar (MSB) plates were used. The bacteria growing on the MSB were identified with biochemical as well as 16S rRNA sequencing techniques from 63 human subjects. Out of 21 typical colonies, 12 were identified as nonmutans colonies. This result strongly implicates and encourages the formulation of a new selective medium for the reliable isolation and identification of Mutans streptococci [27]. In another study, three pairs of primers that were specific for each serotype and a multiplex PCR assay readily identified serotypes of *Streptococcus mutans* strains were developed. Using the same cross-sectional samples in this study, the relationship was identified between the results of serotyping PCR assays and caries experience in preschool children. Clearly, a caries development study should be done to identify the clinical usefulness of the assay. It is also reasonable to identify caries-susceptible individuals from the correlation between the presence of mutans streptococci and caries development. Using the aforementioned mutans streptococci typing PCR assays (*Streptococcus sobrinus* and *Streptococcus mutans*, including serotypes c, e, and f), it might be possible to identify caries-susceptible individuals more clearly [58]. In a comparative study, the Mutans streptococci (MS) in stimulated saliva from 27 subjects were detected by MSB (Mitis-Salivarius Bacitracin)-modified MSB agar medium and commercial kits to detect and evaluate MS conveniently. The ratios of MS in detected bacteria were compared by ELISA. The scores using an mMSB kit on the basis of modified MSB agar medium were tabulated. Saliva samples showed different levels of MS between culture methods and the commercial kit. Some samples which were full of MS were not detected by the commercial kit. The detection of MS by modified MSB agar medium and mMSB kit was significantly higher when compared with MSB agar medium. The sensitivity for the detection of MS was found to be higher for modified MSB agar medium when compared with MSB agar medium. It could be concluded that the mMSB kit can be used simply for fast processing, and could be an important contributor for the evaluation of MS in dental caries [54].

12.1.1.2 Habitat and nature of source

The carcinogenicity of *Streptococcus mutans* in laboratory animals and its repeated associations with coronal dental caries in humans and frequent isolation from established root lesions implicate its etiology [30]. In another study affirming the above findings, numerically significant and predominant SM were found within lesions [64]. In contradiction to earlier studies, several earlier established root lesions did not yield SM. In a 3-year prospective investigation of bacterial flora on the supra-gingival root surfaces, three approaches for enumeration and isolation were used during the 16-month study. Total streptococcal and *Streptococcus mutans* populations were found to be much lower than in previous reports [18, 19, 45, 63]. The microbe under study *Streptococcus mutans* is mostly found between adjacent teeth or in the deep crevices on occlusal (the biting surface) of teeth. It is as mentioned before the main cavity and halitosis causing bacteria in the mouth (although there have been hundreds of other bacteria as well as other microorganisms identified). The microorganism under study is known to adhere to the tooth depending on the concentration and period of contact. The tooth is made of the strongest material in the body, the enamel. This surface selectively absorbs various glycoprotein known as mucin from the saliva, leading to the formation of the acquired enamel pellicle (AEP) [21, 44, 53, 56]. The AEP is the precursor for the caries to occur. The Mutans streptococci are not particularly good at colonizing teeth [68], but this theory proved ineffective as they do have a great likeness to adhere to the plaque that is formed and synthesize glucans from the sucrose diet of the individual and survive in the microenvironment. Previous reports have shown that the oral cavity of an individual can be colonized by one or by multiple clonal types of mutans streptococci [8]. Walter Loesche, sarcastically hinted at the taste for sugar inherited from our forefathers to consume maximum calories and to indulge us into snacking in-between meals has churned out a multibillion dollar industry, and he called *"the slow-release device for sucrose known as a candy."* The instinctive defense mechanism is disturbed when the plaque is formed to facilitate only the facultative and adhesive organisms like the Mutans streptococci dominate it. The very presence of *Streptococcus mutans* would have been no harm at all and might even have benefited humans by preventing the colonization of harmful bacteria [18].

12.1.1.3 Taxonomy

Over the past 20 years, the serological studies reveal something new each time. Strains of *Streptococcus mutans* representing 6 different serotypes were cultured in pour plates containing a glucose agar medium under an atmosphere of 95% nitrogen and 5% CO_2. Three colonial types were obtained: Type 1 colonies were rough spheroid shaped, with no white haze around any colony; type 2 were rough spheroid shaped with a white haze around each colony; and type 3 were smooth lens shaped with white haze. On the other hand, all strains tested grew as rough spheroid-shaped colonies in a sucrose agar medium plates. When these colonies were compared with groupings based on serological and genetic characterization, they revealed that the strains belonging to the colony type I were of serotype c or e and genetic group I. Strains belonging to the colonial type III were of serotype a, d, or g, genetic group III or IV respectively [29]. The cultural identification of this organism in plaque is not likely to be used routinely in a clinical or epidemiological situation and this suggests that the fluorescent labelled-antibody technique is more sensitive than cultural methods for the detection of *Streptococcus mutans* in dental plaque [24].

12.1.2 Dangerous Etiology or a Farce

12.1.2.1 Anaerobiosis

The isolation of this particular bacterium has shown that the main product of its anaerobic fermentation is lactic acid, a very harmful substance to the outer surfaces of teeth. When fermentation has run its course, lactic acid is secreted onto the surface of the hard outer shell surface of the tooth, the enamel it begins breaking down the mineral content leaving holes in the outer surface, then progressing into the inner surface of the tooth, the dentin. Once the dentin has been penetrated, defense is futile and further damage can progress, resulting in caries. Jar systems constitute the most popular anaerobic culture methodology in the clinical laboratory. The efficiency of available anaerobic culture systems has been studied for medical bacteria [23]. Microbial diagnosis of oral infections is performed by using culture, direct microscopic examination, immunoserological identification, and nucleic acid-based methods [66]. Clinical oral microbiology laboratories employ one or a combination of these methods, depending on the pathogens to be identified. Rarely does one detection.

12.1.2.2 Pathogenecity

Cavities initially were attributed to an increased colonization of oral bacteria, termed the "nonspecific plaque hypothesis" which was not approved and compared the bacterial makeup of caries-active and caries-inactive hamsters and found large proportions of a group of bacteria termed the "Mutans streptococci" in the cariogenic hamsters. He further showed that hamsters without caries did not develop them until exposed to caries-active hamsters or their feces [36]. Once the bacteria were introduced into blood stream by trauma, they may colonize hear tissue and cause subacute bacterial endocarditis. This is mediated by the adherence factors of Streptococci to platelet and fibrin vegetations that are formed at the site of damage of cardiac epithelium. These patients are believed to be suffering from a systemic infection. In another report, adherence of the organism to host fibronectin is a major factor for causing endocarditis. Plaque associations, epidemiological surveys, and correlations collectively support the view that *Streptococcus mutans* is an important dental pathogen which can serve as a useful indicator of the microbiological risk factor in dental caries. These organisms also exhibit a positive association with the caries activity and rank very high among the oral microbiota with regard to their ability to produce acid as well as to survive at acidic pH values [18, 69]. Therefore, it is very evident that the role of bacterial acidogenesis at a low pH is an apt ecological determinant [72]. *Streptococcus mutans* can actually continue to lower or maintain the oral pH at an unnaturally acidic value, leading to conditions favorable for its own metabolism and unfavorable for other species it once coexisted with. It is at this lowered pH that results in demineralization of the teeth and a cavity results. Both of which increase with increased rates of SM. Under acidic conditions, SM succeeds in creating a cycle that is favorable for it and unfavorable for others involved in the oral ecology—becoming, in effect, a pathogen. The only organisms for which a positive relationship has been established between their population levels in plaque or saliva and plaques's pH lowering potential are the Mutans streptococci [73]. A secret "window" might occur between ages 6 and 12, as permanent dentition emerges [15]. Dental caries is a chronic, cummulative disease, and the caries status of an individual is through a lifetime. Caries occurs at any age or stage in life. Longitudinal studies with the objective to identify and describe trajectory caries experience in the permanent dentitions to age 32, and 955 participants in a longstanding birth cohort study were analyzed. All were linear, although the higher trajectories were more effect disappeared following adjustment for the number of unaffected surfaces suggesting that, among individuals following a similar caries trajectory, caries is constant across time [9]. Unlike most infectious diseases, in which classic virulence factors, such as a toxin, play a clear role in the damage elicited by the organism, the pathology of dental caries is associated almost exclusively with bacterial metabolism. Catabolism of the nutrients in saliva and the host's diet creates stressors in the form of acids, reactive oxygen species, and other agents that damage biomolecules. Thus, stress tolerance by the bacteria is intimately intertwined with virulence.

12.1.2.3 Virulence

The greatest virulence factor and determinant of caries susceptibility is extra-bacterial: the consumption of sugar-rich carbohydrates. Each influx of sugar into the mouth results in a sharp drop in pH, conditions which favor demineralization of the teeth and heightened *Streptococcus mutans* (SM) activity. Under the conditions in which the human–oral bacteria relationship evolved, dietary sugar levels were dramatically lower and humans ate a few large meals, rather than constantly introducing sugar into the mouth by snacking. In studies tracking oral pH and eating habits over time, those who ate three regular meals a day experienced the same post-meal drop in pH as those who snacked constantly, with the critical distinction that the time difference between each drop allowed the saliva to raise pH and rematerialize the teeth, undoing the damage caused by Streptococcus mutans metabolism at each meal time. In contrast, those who ate three meals as well as sucrose snacks experienced an overall drop in pH with no recovery ability, because the saliva did not have enough time to rise the pH before more sugar was consumed [18]. In studies in rats, the attributes thought to contribute to the virulence of *Streptococcus mutans* are its ability to elaborate antimicrobial or bacteriocin-like substances, which may provide a selective advantage for initial or sustained colonization in a milieu of densely packed competing organisms found in dental plaque [56]. One of the important virulence properties of these organisms is their ability to tenaciously colonize teeth in the presence of dietary sucrose solely dependent on the synthesis of water-soluble glucans by the glucosyltransferases (GTFs) elaborated by the *Streptococcus mutans* strains. Both the genetic and biochemical characterizations of the GTFs produced by the mutans group streptococci (*Streptococcus mutans* and Streptococcus sobrinus) have suggested two different classes of these enzymes: GTF-I, which catalyzes the formation of α-1, 3-linked water-insoluble glucans, and GTF-S, which synthesis water-soluble glucans containing primarily α-1, 6-glucose linkages [18]. The Streptococcus mutans is a potent initiator of caries because there is a variety of virulence factors unique to the bacterium that have been isolated that play an important role in caries formation. Firstly, SM is an anaerobic bacterium known to produce lactic acid as part of its metabolism. Secondly, there is the ability of SM to bind to tooth surfaces in the presence of sucrose by the formation of water-insoluble glucans, a polysaccharide that aids in binding the bacterium to the tooth. Mutant strains developed to produce water-soluble glucans instead have extremely diminished cariogenicity, especially on the smooth surfaces of the teeth which require greater tenacity for binding to occur [18]. The arrangement of that particular tooth surface is quint essential for the adherence phenomenon. *Streptococcus mutans* has been implicated most of all as the initiator of dental caries. In the matter of adhesion, *Streptococcus mutans* stood next only to the enterococci, *Streptococcus faecalis*. *Streptococcus mutans* has the capability to hydrolyze aesculin and ferment amygdalin, arbutin, and salicin but it has been reported that many isolates of S. mutans do not possess one or more of these properties [5]. While such variation can confuse identification schemes, variation within the species may also reflect variation in the competitiveness of different strains, and possibly in their virulence. *Streptococcus mutans* as a species demonstrates variability in sugar utilization patterns; it seems probable that other important properties, for which the phenotype is not so readily observable, may also be subject to variability. Such variation clearly is of considerable relevance to our understanding of its virulence and in the selection of potential targets for preventive measures. However, little is known as yet about the extent of genomic variation or the mechanisms for generating diversity. The mechanisms underlying this ability to survive and proliferate at low pH are an area of that remains unscathed. In a novel study, five proteins identified as being associated with acid tolerance in *Streptococcus mutans* have been identified which is set to provide new information on targets for mutagenic studies that will allow future assessment of their physiological significance in the survival and proliferation in low pH environment. Additionally, proteomic analysis of Streptococcus mutans grown in continuous culture under neutral and acidic conditions identified 30 different proteins with altered levels of expression at pH 5.0 relative to pH 7.0, representing cellular and extracellular gene products associated with stress response pathways. More notably, five of these proteins (Ssb, GreA, PnpA, ClpL, and PepD) were isolated based on the fact that they have never been associated with acid tolerance and/or have never been studied in oral streptococci in detail [15]. The early development of the oral biofilm shows colonies from the mitis-group (*Streptococcus mitis* and *Streptococcus sanguis*), and over the years, little or no information is available on the role of carbohydrate in the formation of these early settlers [3, 4, 6]. In a study that compared the early and late settlers, a simulated mouth model was used. The development of biofilms in the presence and absence of sucrose was experimented in three different nutrient conditions for both early and late settlers. A contained sterile saliva, B contained 1% brain heart infusion (BHI) which was added three times a day 6-hourly, and C contained 1% BHI with glucose to show the effect of glucose in the presence of sucrose on the biofims. The presence of both glucose and sucrose enhanced the biofilm production in both the settlers, and overall, the growth of late settlers (*Streptococcus mutans*) was found to be significantly high in the presence of sucrose as when compared to glucose. *Streptococcus mutans* has been shown to produce three types of GTFs (GTFB, GTFC, and GTFD), whose cooperative action is essential for cellular adhesion, with the highest level of sucrose-dependent cellular adhesion found at the ratio of 5 GTFB: 0.25 GTFC: 1 GTFD. These organisms also produce multiple glucan-binding proteins (Gbp proteins), which presumably promote the adhesion of the organisms. Three Gbp proteins (GbpA protein, GbpB protein, and GbpC protein; [24], respectively. These have been identified and characterized; however, the mechanisms of attachment between glucans and bacterial cell surfaces are still unknown. In several reports, *Streptococcus mutans* cell-associated proteins (FruA, WapA, and GbpC) have been shown to play an important role in the pathogenesis of dental caries [22]. Sortase, a membrane-localized transpeptidase mediates cell-wall anchoring of the GbpC protein and dextran-dependent aggregation of this organism. The mechanism of sucrose-dependent adhesion and aggregation may be multifactorial and complex. Cell surface-associated GbpC is related to the PAc family of streptococcal proteins and participates in dextran-dependent aggregation (Banas et al. 2003). Since the sucrose-dependent adhesion of Streptococcus mutans to tooth surfaces is mediated by GTF glucans, GbpC protein appears to contribute to its virulence. The study at present was to examine the role of GbpC protein in the cellular adhesion of the organism. GTF- or GbpC-deficient mutants were used, but however, there were no changes in the level of sucrose-dependent adhesion [49]. The *Streptococcus mutans* UA159 genome sequence has subsequently identified genes potentially involved in β-glucoside metabolism, on the basis of homology to genes from other bacteria. The genome of *Streptococcus mutans* UA159 contains two phospho-β-glucosidase genes, bglA and celA, which occur in operon-like arrangements along with genes for components of phosphotransferase transport systems and a third phospho-β-glucosidase encoded by the arb gene, which does not have its own associated transport system but relies on uptake by the bgl or cel systems. Targeted inactivation of each of the phospho-β-glucosidase genes revealed that bglA is involved in aesculin hydrolysis, celA is essential for utilization of cellobiose, amygdalin, gentobiose, and salicin, and arb is required for utilization of arbutin. Inactivation of genes for the phosphotransferase systems revealed an overlap of specificity for transport of β-glucosides and also indicated that further, unidentified transport systems existed. The cel and arb genes are subject to catabolite repression by glucose, but the regM gene is not essential for catabolite repression. Screening a collection of isolates of Streptococcus mutans revealed strains with deletions affecting the msm, bgl, or cel operons. The phenotypes of these strains could largely be explained on the basis of the results obtained from the knockout mutants of *Streptococcus mutans* UA159 but also indicated the existence of other pathways apparently absent from UA159. The extensive genetic and phenotypic variation found in β-glucoside metabolism indicates that there may be extensive heterogeneity in the species [55]. In the dental pathogen *Streptococcus mutans*, Glucan plays a central role in sucrose-dependent biofilm formation. Several proteins capable of binding glucan are synthesized by this oragnism. These are divided into the glucosyltransferases (Gtfs) that catalyze the synthesis of glucan and the non-Gtf glucan-binding proteins (Gbps). The biological significance of the Gbps has not been thoroughly defined, but studies suggest these proteins influence virulence and play a role in maintaining biofilm architecture by linking bacteria and extracellular molecules of glucan. Here, the researchers engineered a panel of Gbp mutants, targeting GbpA, GbpC, and GbpD, in which each gene encoding a Gbp was deleted individually and in combination. These strains were then analyzed by confocal microscopy and the biofilm properties quantified by the biofilm quantification software COMSTAT. The study revealed that all biofilms produced by mutant strains lost significant depth, but the basis for the reduction in height depended on which particular Gbp was missing. The loss of the cell-bound GbpC appeared dominant as might be expected based on losing the principal receptor for glucan. The loss of an extracellular Gbp, either GbpA or GbpD, also profoundly changed the biofilm architecture, each in a unique manner. This study was indicative of the fact that the non-GTF GBPs influence virulence and in vitro phenotypes traditionally associated with sucrose-dependent adhesion and accumulation [46]. While the molecular mechanisms underlying the control of carbohydrate acquisition, acid production, and adaptation to low pH by *Streptococcus mutans* have been the focus of a substantial number of studies, the role of oxygen in fundamental aspects of gene regulation and physiologic homeostasis is poorly understood. An earlier study revealed that oxygen is required by many oral organisms for respiration and energy generation. Organisms that initially colonize the surfaces of the mouth are exposed to levels of oxygen approaching those found in air or air-saturated water [48]. Recently, it was demonstrated that the ability of *Streptococcus mutans* to form biofilms, an essential virulence attribute of this organism, was dramatically reduced when cells were cultivated in the presence of oxygen [6]. Oxygen profoundly affects the composition of oral biofilms. Recently, studies reveal that exposure of *Streptococcus mutans* to oxygen strongly inhibits biofilm formation and alters cell surface biogenesis. To begin to dissect the underlying mechanisms by which oxygen affects known virulence traits of *Streptococcus mutans*, transcription profiling was used to show that roughly 5% of the genes of this organism are differentially expressed in response to aeration. This ability of *Streptococcus mutans* to form biofilms was severely impaired by oxygen exposure, transcription of the gtfB gene, which encodes one of the primary enzymes involved in the production of water-insoluble, adhesive glucan exopolysaccharides, was down-regulated in cells growing aerobically. Further investigation also revealed that transcription of gtfB, but not gtfC, was responsive to oxygen and that aeration causes major changes in the amount and degree of cell association of the Gtf enzymes. Inactivation of the VicK sensor kinase affected the expression and localization of the GtfB and GtfC enzymes. Thus, this study provides a novel insight into the complex transcriptional and posttranscriptional regulatory networks used by *Streptococcus mutans* to modulate virulence gene expression and exopolysaccharide production in response to changes in oxygen availability [6]. Energy-dependent cytoplasmic proteases or ClpP complexes are proposed to be more central for stress tolerance and global regulation in streptococci than in other bacterial groups. In the ClpP proteolytic system, ClpP must associate with a Clp ATPase partner that possesses nucleotide-binding domains characteristic of the AAA+ superfamily of ATPases to form a functional complex (Frees et al. 2007). Biofilm formation by the ΔclpP and ΔclpX strains was impaired when grown in glucose but enhanced in sucrose. Searching the Streptococcus mutans genome, two putative spx genes, designated spxA and spxB were identified. The inactivation of either of these genes bypassed phenotypes of the clpP and clpX mutants. On the whole, this study revealed that the proteolysis of ClpL and ClpXP plays a role in the expression of key virulence traits of S. mutans and indicates that the underlying mechanisms by which ClpXP affect virulence traits are associated with the accumulation of two Spx orthologues [37]. Adaptive responses of bacteria to environmental changes, such as nutrient limitation, oxygen deprivation, antibiotic stress, and osmotic shock, are regulated by the so-called two-component signal transduction system (TCS) pathways [38–40]. TCSs typically consist of a membrane-bound sensor histidine kinase and a cytoplasmic response regulator, with a common biochemical mechanism involving phosphoryl-group transfer between two distinct protein components. These TCSs are critical for survival under adverse conditions, as well as for regulation of virulence-associated factors of this pathogen [9, 10]. LiaSR is a TCS that is believed to be a part of a complex regulatory network that monitors and responds to cell envelope stress in *Bacillus subtilis* [33–35]. Since its discovery in 1924 by J. Clarke, [10] *Streptococcus mutans* has been the focus of rigorous research efforts due to its involvement in caries initiation and progression. A major problem of bacteria that live in acidic environments is the potential of these surroundings to acidify the intracellular cytoplasm. Negative consequences of this include loss of glycolytic enzyme activity and structural damage to the cell membrane, proteins, and DNA. Conclusively, the above findings make it obvious that the ability of *Streptococcus mutans* to ferment a range of dietary carbohydrates can rapidly drop the external environmental pH, thereby making dental plaque inhabitable to many competing species, and can ultimately lead to tooth decay. Acid production by this oral pathogen would prove suicidal if not for its remarkable ability to withstand the acid onslaught by utilizing a wide variety of highly evolved acid-tolerance mechanisms [41, 73].

12.1.2.4 Transmissibility

It is vital for the researcher to have a clear understanding of the route of transmission of Mutans streptococci. It is the major tool to develop measures to prevent, delay, or lower colonization, thereby decreasing caries experience [2]. Since the earlier the infection of mouth with *Streptococcus mutans*, the greater is the caries risk of the deciduous dentition, and that since salivary transfer is required to spread the infection, nurturing habits such as cleaning a pacifier by putting in the mother's mouth before it is given to the child, kissing the child directly on the mouth, and pre-tasting food before it is given to the child have been studied. However, these studies have shown that a frequent transfer of saliva to the mouth of the baby from the mother is actually protective. Children with a high frequency of maternal salivary contact before tooth eruption had lower numbers of *Streptococcus mutans* and less dental caries than those with rare contact, possibly because the infant's exposure to cariogenic bacteria prior to tooth eruption might have increased the child's immunological resistance to the infection [1, 2]. *Streptococcus mutans* (SM) depends on transmission routes to proliferate like any other infectious pathogen. It harbors on hard, nonshedding surfaces for the establishment of permanent colonies, a fact which may lead one to presume that levels of SM were not detectable in toddlers until the eruption of the primary teeth. Current studies have revealed that SM can colonize on the furrows of the tongue in pre-dentate infants [11]. When the teeth erupt, typically between the ages of one and two, SM can establish active colonies on the teeth that eventually lead to cavities, most notably early childhood caries (ECC). Detection of SM in the furrows of the tongue reinforces the conclusion that the most common transmission route for the bacteria is vertical, from mother to child, most likely shortly after birth. Studies of the saliva samples of two- to five-year-old children and their mothers revealed a high fidelity in the genetic makeup of each host's SM population [42]. Similarly, experimenters concluded that plasmid DNA similarities correlate to different races, also implying primarily vertical transmission. As a result, mothers with high titers of the bacteria or who have suffered from many dental caries themselves are likely to pass the same virulence and associated problems on to their children. In fact, mothers whose salivary SM levels exceeded 105 colony-forming units were about nine times more likely to pass the bacteria on to their children; Streptococcus mutans also appears capable of horizontal transmission. Children in the same nursery school class often had identical strains of the bacteria in their saliva [11]. Children with no detectable SM until the age of five often shared strains with both mother and father when the bacteria were finally acquired. Frequent close contact with others seems to be enough to transmit SM, a fact which could impact prevention techniques in environments such as daycare centers, where children harboring more virulent strains could increase the caries-inducing potential of the bacteria in other caries-free children [42]. In another study whole saliva samples were collected from 70 healthy school children and 46 healthy dental students, ages being

11 and 22–32, respectively. A majority, i.e., 70%, of the test subjects received chlorhexidine rinses because of showing dental indications. However, when a background check was done, it became very evident that the *Streptococcus mutans* infection spreads within a family from mother to child [40]. A variety of unique traits of *Streptococcus mutans* (SM) contribute to its virulence and enable it to be so prevalent and so damaging in the human oral cavity. The time at which *Streptococcus mutans* is acquired in relation to other common oral bacteria also plays a role in its prevalence in the ecology of the mouth. The successive colonization of many different species leaves only a certain realized niche for SM, minimizing its impact through a process that has been termed bacterial succession. Some pioneer species, such as *Streptococcus oralis* and *Streptococcus mitis*, are detectable in infants only a few days old, while SM is virtually undetectable until around age two. Any alteration in this progression of colonization can lead to increased risk of dental caries. They also found that babies delivered via Caesarian section had detectable levels of SM approximately a year earlier than those delivered vaginally, presumably because they were not colonized by pioneer bacteria found on the perineum of their mothers that the babies born vaginally were exposed to this bacterium [16]. A new reliable genotyping method, mutlilocus sequence typing (MLST), was used to evaluate vertical transmission of the cariogenic pathogen *Streptococcus mutans*. A total of 136 strains were isolated from saliva samples of 20 Japanese mother–child pairs. Similarity for all of those sequences between strains from mother–child pair was regarded as indicating transmission, which was shown in 70% of the pairs. Interestingly, the rate of transmitted strains was significantly higher in the girls (90%) than that in the boys. The highest distribution percentage in each maternal saliva sample was found to be transferred to their children. Thus, these findings indicate that mothers are the main source from transmission of *Streptococcus mutans* to their children, while the present MLST method was useful for investigating bacterial transmission [45]. A study with the aim of evaluating, by means of a questionnaire, the knowledge and usual attitude of 640 parents and caretakers regarding the transmissibility of caries disease was. Generally, dental caries is a transmissible infectious disease in which mutans streptococci are generally considered to be the main etiological agents. Although the transmissibility of dental caries is relatively well established in the literature, little is known whether information regarding this issue is correctly provided to the population. Most interviewed adults did not know the concept of dental caries being an infectious and transmissible disease, and reported the habit of blowing and tasting food, sharing utensils and kissing the children on their mouth.

A total of 372 (58.1%) adults reported that their children had already been seen by a dentist, 264 (41.3%) answered that their children had never gone to a dentist, and 4 (0.6%) did not know. When the adults were asked whether their children had already had dental caries, 107 (16.7%) answered yes, 489 (76.4%) answered no, and 44 (6.9%) did not know. The revealed data reinforce the need to provide the population with some important information regarding the transmission of dental caries in order to facilitate a more comprehensive approach toward the prevention of the disease (Sakai et al. 2008). Early acquisitions of mutans streptococci has been shown in many studies to be a major risk factor for early childhood caries and future caries experience [7]. Vertical transmission from mother to child has been suggested as the main pathway for mutans streptococci acquisitions. Several studies reported the transmission among similar genotypes common to mother, father, and child [29]. It has not been established whether transmission of mutans streptococci occurs between unrelated children older than 4 years of age or whether it is horizontal or vertical. Therefore, an investigation was carried out to find out the possible transmission of mutans streptococci genotypes from child to child at kindergarten level. A total of 96 children (ages 5–6 years) in 3 San Francisco Bay Area public schools were investigated. Mutans streptococci colonies from each child were isolated from selective culture on Mitis-salivarius-sucrose Bacitracin agar. Arbitary primed PCR was used to determine the mutans streptococci genotypes. The findings were, however, quite fascinating; two children (not siblings) in each of the three schools (6%) shared an identical amplitype of *Streptococcus mutans*, unique to each pair. There were 19 Streptococcus sobrinus amplitypes found in 12 children, and all were unique to each child. The presence of matching genotypes of *Streptococcus mutans* demonstrated horizontal transmission of this species between unrelated children aged 5–6 years [21]. If the transmission of Mutans streptococci (MS) could be thwarted by any means, it could be a very good strategy for caries prevention in children. Based on the above theory, a measure to interfere with transmission was carried out. The study comprised of 107 pregnant women with heavy salivary MS. They were administered with Xylitol gum to check the effectiveness of early maternal exposure to Xylitol chewing gum on the growth of MS. They were randomized into two groups, xylitol gum group ($n = 56$) and no gum group ($n = 51$) or the control group. Maternal exposure to the gum was started at the sixth month of pregnancy and completed 13 months later. The results after the stipulated time confirmed the effectiveness of early exposure so efficiently that the control group with no gum acquired MS nearly 9 months earlier than those in the xylitol group. The findings were found to be similar and beneficial as they were found in Japan to those demonstrated earlier in Nordic countries [63].

12.1.2.5 Risk factors

Humans' relationship with both food and with *Streptococcus mutans* (SM) evolved in very different conditions than those of today. In effect, the disease initiated by SM in our mouths is a case of evolutionary incongruence with the lifestyle enjoyed by those in developed countries. Caries-risk predictors may be found among the microbiota (dental plaque), the diet (carbohydrates), and the host (teeth), all three of which are indispensable for caries development [43]. Saliva may be added in view of its potentially powerful influence on the caries process. Quantity of plaque and composition of the microbiota of plaque in vivo are highly variable, as are diet (composition, quantity, frequency of consumption), caries "resistance" of the teeth (e.g., fluoride exposure), and salivary factors. Moreover, many factors are interactive; for example, salivary flow rate as well as dietary carbohydrate intake can affect the plaque levels of Mutans streptococci [17]. Strong evidence indicates the association of Mutans streptotococci (MS) with caries development, and this in turn is linked directly to carbohydrate consumption which, in turn, is one of the indispensable factors in caries development. Levels of Lactobacilli and MS in plaque or saliva are very sensitive to dietary carbohydrate. Less extreme changes in dietary carbohydrate intake, even during periods of only a few months, can also cause significant changes in the levels of Lactobacilli and MS [81]. Early studies involving extreme oral exposure to dietary carbohydrate, such as a very low level of exposure during stomach-tube-feeding of animals or humans, or a very high level of exposure of children with "nursing bottle caries" is evidence enough for this relation [83]; Further evidence suggests that the linkage between the oral levels of Lactobacilli and Mutans streptococci (MS) and dietary carbohydrate consumption is associated with the pronounced acid tolerance, which is higher than that of many other plaque organisms [84]. Accordingly, a scenario may be proposed in which an increased caries-conducive carbohydrate intake can lvead to the selective emergence of Lactobacilli and MS in plaque due to the increased frequency of a high plaque acidity which provides these organisms with a growth advantage over other less-acid-tolerant plaque organisms [81]; this process would reverse itself during a decreased carbohydrate intake. The selective emergence Lactobacilli and MS in plaque would also be conducive to a higher probability of their spread to hither to uninfected nearby tooth surface areas; in the case of MS and an increased intake of sucrose specifically, this process could be augmented by a more effective initial attachment of MS cells to the tooth surface [86]. This may even lead to incipient caries lesion formation in the most caries-prone dentition sites, whereas the subsequent development of cavities may provide another suitable milieu (prolonged acidic pH) for the MS. The above suggests the potential use of Lactobacilli and MS not only as caries risk indicators, but also as indicators of another caries-risk factor, the frequent consumption of carbohydrates. In a study conducted by Mayhall among societies in which access to sugar was pathetically low or amusingly high the caries record evidently showed variations in occurrence. Culturally traditional (low-sucrose) diets had many fewer cavities than wealthier people who could afford more foods [58]. In a similar study, Norwegian children who grew up during World War II, whose access to sugary foods were minimal experienced fewer caries than children growing up shortly after, when sugary foods were common [79]. The assessment of risk indicators may involve predictors such as specific causally related risk factors; it may also involve predictors which are associated with caries but not causally related to it [7]. Dental caries is widely recognized as an infectious disease induced by diet. The main players in the etiology of the disease are (a) cariogenic bacteria, (b) fermentable carbohydrates, (c) a susceptible tooth and host, and (d) time. However, in young children, bacterial flora and host defence systems are in the process of being developed; tooth surfaces are newly erupted and may show hypoplastic defects; and their parents must negotiate the dietary transition through breast/bottle feeding, first solids, and childhood tastes. Thus, it is thought that there may be unique risk factors for caries in infants and young children [71].

12.2 Other Organisms Associated with Caries

Many microorganisms inhabit human body externally (on the surface) or indigenously. Most of the indigenous microbes of the human body are commensals. Soluble nutrients and abundance of moisture continuously present in the mouth cavity provide a suitable environment for the growth of bacteria. Several microbes inhabit in the mouth cavity; some common ones are *Staphylococcus aureus*, *Streptococcus mitis*, peptostreptococci, lactobacilli, Actinomyces, *Bacteroides oralis*, *Fusobacterium nucleatum*, *Candida albicans*, etc. Caries-associated plaques were characterized by higher levels of mutans streptpocci and of particular interest regarding the drastic change in pH levels; an observation was made that a very rapid pH drop and a very low pH could occur when significant levels of mutans streptococci were absent.

This also suggested that other acidogenic organisms capable of producing low pH counts contribute significantly to the high pH lowering capacity of many plaques, thus owing to the increase in caries formations. These nonmutans streptococci generally constitute a high proportion of the dental plaque flora [18, 81]). These streptococci may include *Streptococcus gordonii, Streptococcus oralis, Streptococcus mitis*, and *Streptococcus anginosus*, according to a recent taxonomic redefinition of the streptococci (Kilian et al. 1989). It should be noted that past studies of the association between plaque organisms and caries have involved often poorly defined groups of bacteria such as the lactobacilli, the actinomyces, "*Streptococcus sanguis*," or "*Streptococcus mitis*" rather than clearly defined individual species [11, 18, 81, 86]. Culture-difficult bacteria, including asaccharolytic anaerobic gram-negative coccobacilli (AAGNC), may constitute a predominant of the oral sites. Bacterial species other than *Streptococcus mutans*, e.g., species of the genera Veillonella, Lactobacillus, Bifidobacterium, and Propionibacterium, low-pH non-*Streptococcus mutans*, Actinomyces spp., and Atopobium spp., likely play important roles in caries progression. The study suggested that bacterial profiles change with progression of the disease and that they differ from the primary to the secondary dentition. The present findings supported the ecological plaque hypothesis in caries disease. Changes in ecologic factors required different bacterial qualities and stimulated alterations in the bacterial composition. Further studies of the potential etiologic roles of these diverse bacterial communities, including additional and novel species, are recommended for the future research. Identified cultivable and not-yet-cultivable organisms might provide additional targets for caries intervention [1, 3]. In an antibacterial study involving triclosan-based mouthrinses against 28 salivary *Staphylococcus aureus* organisms the results were not so conclusive as there seem to be a different maximum inhibitory dilution for each mouthrinse, this may be due to other active ingredients in the triclosan formulations. In addition, statistical analyses were performed to analyze the correlations among clinical parameters and the detected species. The most frequently detected species were *Capnocytophaga sputigena* (28.3%), followed by *Aggregatibacter actinomycetemcomitans* (20.9%) and *Campylobacter rectus* (18.2%). *Eikenella corrodens, Capnocytophaga ochracea*, and *Prevotella nigrescence* were detected in approximately 10% of the specimens, whereas *Treponema denticola, Tannerella forsythia*, and *Prevotella intermedia* were rarely found, and *Porphyromonas gingivalis* was not detected in any of the subjects. The detected species were positively correlated with the age of the subjects. There were 10 subjects with positive reactions for *Treponema denticola* or *Tannerella forsythia*, in whom the total number of bacterial species was significantly higher as compared to the other subjects. Furthermore, subjects possessing Campylobacter rectus showed significantly greater values for periodontal pocket depth, gingival index, and total number of species. In conclusion, it was found that approximately one-fourth of the present subjects with disabilities who possessed at least one of *Treponema denticola*, *Tannerella forsythia*, and *Campylobacter rectus* were at possible risk of periodontitis [59].

References

[1] Aaltonen, A. S. (1991). The frequency of mother-infant salivary close contacts and maternal caries activity affect caries occurrence in 4-year-old children. *Proc. Finn. Dent. Soc.* 87, 373–382.

[2] Aaltonen, A. S., and Tenovuo, J. (1994). Association between mother-infant salivary contacts and caries resistance in children: a cohort study. *Ped. Dent.* 16, 110–116.

[3] Aas, J. A., Griffen, A. L., Dardis, S. R., Lee, A. M., Olsen, I., Dewhirst, F. E., et al. (2008). Bacteria of dental caries in primary and permanent teeth in children and young adults. *J. Clin. Microbiol.* 46, 1407–1417.

[4] Aas, J. A., Paster, B. J., Stokes, L. N., Olsen, I., and Dewhirst, F. E. (2005). Defining the normal bacterial flora of the oral cavity. *J. Clin. Microbiol.* 43, 5721–5732.

[5] Alaluusua, S., J. Savolainen, H. Tuompo, and L. Gron-roos. (1984). Slide-scoring method for estimation of *Streptococcus mutans* levels in saliva. *Scand. J. Dent. Res.* 92:127–133.

[6] Ahn, S. J., Z. T. Wen, and R. A. Burne. (2007). Ahn. Effects of Oxygen on Virulence Traits of *Streptococcus mutans*. J. Bacteriol. 189(23):8519–8527.

[7] Beck, J. D., G. G. Koch, R. G. Rozier, and G. E. Tudor. (1990). Prevalence and risk indicators for periodontal attachment loss in a population of older community-dwelling blacks and whites. *J. Periodont.* Res. 61:521–528.

[8] Beighton, D., Hardie, J. M., and Whiley, R. A. (1991). A scheme for the identification of Viridans Streptococci. *J. Med. Microbiol.* 35, 367–372.

[9] Biswas, I., L. Drake, and S. Biswas. (2007). Regulation of *gbp*C Expression in *Streptococcus mutans*. J. Bacteriol. 189(18):6521–6531.

[10] Biswas, I., L. Drake, D. Erkina, and S. Biswas. (2008). Involvement of Sensor Kinases in the Stress Tolerance Response of *Streptococcus mutans*. J. Bacteriol. 190(1):68–77.

[11] Bowden, G. H., I. R. Hamilton. (1998). Survival of oral bacteria. Critical. Rev. *Oral. Biol Med.* 9:54–84.

[12] Broadbent, J. M., W. M. Thomson, and R. Poulton. (2008). Trajectory patterns of dental caries experience in the dentition to the fourth decade of life. *J. Dent. Res.* 87(1):69–72.

[13] Holt, J. G. (1994). *Bergey's Manual of Determinative Bacteriology*, 9th Edn. Baltimore, MD: Williams and Wilkins, 1047–1063.

[14] Berkowitz, R. J. (2006). Mutans streptococci: acquisition and transmission. *Pediatr. Dent.* 28, 106–109.

[15] Bratthall, D. (1970). Demonstration of five serological groups of streptococcal strains resembling *Streptococcus mutans*. *Odontol. Revy.* 21, 143–152.

[16] Bratthall, D. (1997). Discovery! A *Streptococcus mutans* safari! *J. Dent. Res.* 76, 1332–1336.

[17] Brown, L. R., S. Dreizen, S. Handler. (1976). Effects of selected caries preventive regimens on microbiol changes following irradiation-induced xerostomia in cancer patients. In: Proceedings, Microbial aspects of dental caries. Vol: 1. Stiles HM, Loesche. W. J., T. C. O'Brien, editors. Washington, DC: *Information Retrieval, Inc.,* pp. 275–290.

[18] Carlén, A., J. Olsson, and A. C. Borjesson. (1996). Saliva-mediated binding *in vitro* and prevalence *in vivo* of *Streptococcus mutans*. Arch. Oral. Biol. 41:35–39.

[19] Clarke, J. K. (1924). On the bacterial factor in the aetiology of dental caries. *Br. J. Exp. Pathol.* 5, 141–147.

[20] Coleman, G., and R. A. D. Williams. (1972). Taxonomy of Some Human Viridans Streptococci. In: Streptococci and Streptococcal Diseases, Wannamaker, L. W. and Matsen, J. M., Eds., New York: Academic Press, Inc., pp. 281–299.

[21] Domejean, S., L. Zhan, P. K. DenBesten, J. Stamper, W. T. Boyce, and J. D. Featherstone. (2010). Horizontal transmission of mutans streptococci in children. *J. Dent. Res.* 89(1):51–55.

[22] Duchin, S., and J. van Houte. (1978). Relationship of *Streptococcus mutans* and lactobacilli to incipient smooth surface dental caries in man. *Arch. Oral. Biol.* 23:779–786.

[23] Emilson, C. G., and D. Bratthall. (1976). Growth of *Streptococcus mutans* on various selective media. *J. Clin. Microbiol.* 4:95–98.

[24] Facklam, R. R. (1974). Characteristics of *Streptococcus mutans* Isolated from Human Dental Plaque and Blood. *Int. J. Syst. Bacteriol.* 24: 313–319.

[25] Gibbons, R. J., and S. S. Socransky. (1962). Intracellar polysaccharide storage by organisms in dental plaques. *Arch. Oral Biol.* 7: 73–80.

[26] Gold, O. G., Jordan, H. V., and van Houte, J. (1973). A selective medium for *Streptococcus mutans*. *Arch. Oral. Biol.* 20, 473–477.

[27] Goldstein, E. J. C., D. M. Citron, and R. J. Goldman. (1992). National hospital survey of anaerobic culture and susceptibility testing methods: results and recommendation for improvement. *J. Clin. Microbiol.* 30:1529–1534.

[28] Grenier, Ella M., W. C. Eveland, and W. J. Loesche, (1973). Identification of *Streptococcus mutans* serotypes in dental plaque by fluorescent antibody techniques. *Arch. Oral. Biol.* 18(6):707–715.

[29] Hamada, S., and Slade, H. D. (1990). Biology, immunology and cariogenicity of *Streptococcus mutans*. *Microbiol. Rev.* 44, 331–384.

[30] Hames-Kocabas, E. E., F. Ucar, N. Kocatas Ersin, A. Uzel, and A. R. Alpoz. (2008). Colonization and vertical transmission of *Streptococcus mutans* in Turkish children. *Microbiol. Res.* 163:168–172.

[31] Hamilton, I. R. and Martin, E. J. S. (1982). Evidence for the involvement of proton motive force in the transport of glucose by a mutant of *Streptococcus mutans* starin DR0001 defective in Glucose-Phosphoenolpyruvate Phosphotransferase Activity. *Infect. Immun.* 36, 567–575.

[32] Hardie, J. M., and Bowden, G. H. (1976). Physiological classification of oral *Viridans streptococci*. *J. Dent. Res.* 55, 166–178.

[33] Heloe, L. A., and O. Haugejorden. (1981). "The rise and fall" of dental caries: some global aspects of dental caries epidemiology. *Comm. Dent. Oral. Epidemiol.* 9:294–299.

[34] Ikeda, T., K. Ochiai, and T. Shiota. (1979). Taxonomy of the oral *Streptococcus mutans* based on colonial characteristics and serological, biochemical and genetic features. *Arch. Oral. Biol.* 24(10–11):863–865, 867.

[35] Jordan H. V. and Hammond B. F. (1972). Filamentous Bacteria Isolated from Human Root Surface Caries. *Arch Oral Biol.* 17:1333–1342.

[36] Jordan, H. V. (1986). Cultural method for identification and quantitation of *Streptococcus mutans* and Lactobacilli in oral samples. *Oral. Microbiol. Immunol.* 1:23–30.

[37] Jordan, H. V., R. Laraway, R. Snirch, and M. Marmel. (1987). A simplified diagnostic system for cultural detection and enumeration of Streptococcus mutans. *J. Dent. Res.* 66(1):57–61.

[38] Jordan, S., A. Junker, J. D. Helmann, and T. Mascher. (2006). Regulation of LiaRS- dependent gene expression in *Bacillus subtilis*: identification of inhibitor proteins, regulator binding sites, and target genes of a conserved cell envelope stress-sensing two-component system. *J. Bacteriol.* 188:5153–5166.

[39] Jordan, S., E. Rietkotter, M. A. Strauch, F. Kalamorz, B. G. Butcher, J. D. Helmann, and T. Mascher. (2007). LiaRS-dependent gene expression is embedded in transition state regulation in Bacillus subtilis. *J. Microbio.* 153:2530–2540.

[40] Jordan, S., M. I. Hutchings, and T. Mascher. (2008). Cell envelope stress response in Gram-positive bacteria. *FEMS. Microbiol. Rev.* 32: 107–146.

[41] Kajfasz, J. K., A. R. Martinez, I. Rivera-Ramos, J. Abranches, H. Koo, R. G. Quivey, Jr., and J. A. Lemos. (2009). Role of Clp Proteins in Expression of Virulence Properties of *Streptococcus mutans*. *J. Bacteriol.* 191(7):2060–2068.

[42] Keyes, P. H., (1960). The infectious and transmissible nature of experimental dental caries. Findings and Implications. *Arch. Oral. Biol.* 1:304–320.

[43] Keyes, P. H. and Jordan, H. V. (1963). Factors influencing initiation, transmission and inhibition of dental caries. In: Harris RJ, ed. Mechanisms of hard tissue destruction. *New York: Academic Press*, pp. 261–83. 10.

[44] Kilian, M., L. Mikkelsen, and J. Henrichsen. (1989). Taxonomic study of viridans streptococci: description of *Streptococcus gordonii* sp. nov. and emended descriptions of *Streptococcus sanguis* (White and Niven 1946), *Streptococcus oralis* (Bridge and Sneath 1982), and *Streptococcus mitis* (Andrewes and Horder 1906). *Int. J. Syst. Bacteriol.* 39471434.

[45] Kohler, B., and D. Bratthall. (1978). Intrafamiliar levels of *Streptococcus mutans* and some aspects of the bacterial transmission. Scand. J. Dent. Res. 86:35–42.

[46] Kohler, B., and I. Andreen. (1994). Influence of caries-preventive measures in mothers on cariogenic bacteria and caries experience in their children. *Arch. Oral. Biol.* 39:907–911.

[47] Lapirattanakul, J., K. Nakano, R. Nomura, S. Hamada, I. Nakagawa, and T. Ooshima. (2008). Demonstration of mother-to-child transmission

of *Streptococcus mutans* using mutlilocus sequence typing. *Caries. Res.* 42:466–474.

[48] Len, A. C. L., Harty, D. W. S., and Jacques, N. A. (2004). Proteome analysis of *Streptococcus mutans* metabolic phenotype during acid tolerance. *Microbiology.* 150, 1353–1366.

[49] Li, Y. and Caufield, P. W. (2005). Mode of delivery and other maternal factors influence the acquisition of *Streptococcus mutans* in infants. *J. Dent. Res.* 84, 806–811.

[50] Li, Y. H., Tang, N., Aspiras, M. B., Lau, P. C. Y., Lee, J. H., Ellen, R. P., et al. (2002). A quorum-sensing signaling system in *Streptococcus mutans* is involved in biofilm formation. *J. Bacteriol.* 184, 2699–2708.

[51] Lie, T., (1977). Early dental plaque morphogenesis. *J. Periodont. Res.* 12:73–89.

[52] Little, W. A., D. C. Korta, L. A. Thomson, and W. H. Bowden. (1977). Comparative recovery of *Streptococcus mutans* on ten isolation media. *J. Clin. Microbiol.* 5:578–583.

[53] Lynch, D. J., T. L. Fountain, J. E. Mazurkiewicz, and J. A. Banas. (2007). Glucan Binding Proteins are Essential for Shaping *Streptococcus mutans* Biofilm Architecture. FEMS. *Microbiol. Lett.* 268(2):158–165.

[54] Loesche, W. J. (1986). Role of *Streptococcus mutans* in human dental decay. *Microbiol. Rev.* 50, 353–380.

[55] Loveren, C. V. and J. F. Buijs. (2000). Similarity of Bacteriocin Activity Profiles of *Mutans streptococci* within the Family When the Children Acquire the Strains After the age of 5. *Caries. Res.* 34(6): 481–5.

[56] Marquis, R. E. (1995). Oxygen metabolism, oxidative stress and acid-base physiology of dental plaque biofilms. *J. Ind. Microbiol.* 15:198–207.

[57] Matsukubo, T., K. Ohta, Y. Maki, M. Takeuchi, and I. Takazoe. (1981). A semi-quantitative determination of *Streptococcus mutans* using its adherent ability in a selective medium. *Caries. Res.* 15:40–45.

[58] Mayhall, J. T., and A. A. Dahlberg. (1970). Dental Caries in the Eskimos of Wainwright, *Alaska. J. Dent. Res.* 49(4): 886.

[59] Naka, S., A. Yamana, K. Nakano, R. Okawa, K. Fujita, A. Kojima, H. Nemoto, R. Nomura, M. Matsumoto, and T. Ooshima. (2009). Distribution of periodontopathic bacterial species in Japanese children with developmental disabilities. *BMC. Oral. Health.* 9:24–33.

[60] Nakano, K., Nemoto, H., Nomura, R., Homma, H., Yoshioka, H. Y., Shudo, H., et al. (2007). Serotype distribution of *Streptococcus mutans*

a pathogen of dental caries in cardiovascular specimens from Japanese patients. *J. Med. Microbiol.* 56, 551–556.

[61] Nakano, K., Nomura, R., Nakagawa, I., Hamada, S., and Ooshima, T. (2004a). Demonstration of *Streptococcus mutans* with a cell wall polysaccharide specific to a new serotype k in the human oral cavity. *J. Clin. Microbiol.* 42, 198–202.

[62] Nakano, K., Nomura, R., Shimizu, N., Nakagawa, I., Hamada, S., and Ooshima, T. (2004b). Development of a PCR method for rapid identification of new *Streptococcus mutans* serotype k strains. *J. Clin. Microbiol.* 42, 4925–4930.

[63] Nakai, Y., C. Shinga-Ishihara, M. Kaji, K. Moriya, K. Murakami-Yamanaka, and M. Takimura. (2010). Xylitol gum and maternal transmission of mutans streptococci. *J. Dent. Res.* 89(1):56–60.

[64] Newman, B. M., P. White, S. B. Mohan, and J. A. Cole. (1980). Effect of dextran and ammonium sulfate on the reaction catalyzed by a glucosyltransferase complex from *Streptococcus mutans*. *J. Gen. Microbiol.* 118:353–357.

[65] Nishikawaraa, F., Y. Nomuraa, S. Imaib, A. Sendac, and N. Hanadab. (2007). Evaluation of Cariogenic Bacteria. *Eur. J. Dent.* 1:31–39.

[66] Old, L. A., S. Lowes, and R. R. B. Russell. 2006. Genomic variation in *Streptococcus mutans:* deletions affecting the mutilple pathways of β-glucoside metabolism. 21:21–27.

[67] Paster, B. J., S. K. Boches, J. L. Galvin, R. E. Ericson, C. N. Lau, V. A. Levanos, A. Sahasrabudhe, and F. E. Dewhirst. (2001). Bacterial diversity in human subgingival plaque. *J. Bacteriol.* 183:3770–3783.

[68] Rogers, A. H. (1976). Bacteriocinogeny and the properties of some bacteriocins of *Streptococcus mutans*. *Arch. Oral Biol.* 21:99–104.

[69] Sato, Y., Senpuku, H., Okamoto, K., Hanada, N., and Kizaki, H. (2002). *Streptococcus mutans* binding to solid phase dextran mediated by the glucan binding protein C. Oral. *Microbiol. Immunol.* 17, 252–256.

[70] Seki, M., Y. Yamashita, Y. Shibata, H. Torigoe, H. Tsuda, and M. Maeno. (2006). Effect of mixed mutans streptococci colonization in caries development. *Oral. Microbiol. Immunol.* 21:47–52.

[71] Seow, W. K. (1998). Biological mechanisms of early childhood caries. *Community Dent Oral Epidemiol.* 26(1-Suppl):8–27.

[72] Shklair, I. L., and T. Walter. (1976). Evaluation of a selective medium-colour test for *Streptococcus mutans*, *IADR Progr & Abst.* 55: No. 241.

[73] Slavkin, H. C. (1999). *Streptococcus mutans*, early childhood caries and new opportunities. *J. Am. Dent. Assoc.* 130, 1787–1792.

[74] Smith, D. J., Akita, H., King, W. F., and Taubman, M. A. (1994). Purification and antigenicity of a novel glucan-binding protein of *Streptococcus mutans*. *Infect. Immun.* 62, 2545–2552.

[75] Staat, R. H. (1976). Inhibition of several *Streptococcus mutans* strains by different Mitis-Salivarius agar preparations. *J. Clin. Microbiol.* 3:378–380.

[76] Syed, S. A., Loesche, W. J., Pape, H. L., and Jr, grenier, E. (1975). Predominant cultivable flora isolated from human root surface caries plaque. *Infect Immun.* 11(4):727–731.

[77] Tao, L., MacAlister, T. J., and Tanzer, J. M. (1993). Transformation efficiency of EMS-induced mutants of *Streptococcus mutans* of altered cell shape. *J. Dent. Res.* 72(6):1032–9.

[78] Ting, M., and J. Slots. (1997). Microbiological diagnostic in periodontics. *Compend. Contin. Ed. Dent.* 18:861–876.

[79] Toverud, G. (1957). The Influence of War and Post-War Conditions on the Teeth of Norwegian School Children. *The Milbank Memorial Fund Quarterly.* 35(2): 127–196.

[80] van Houte, J., and Green, D. B. (1974). Relationship between the concentration of bacteria in saliva and the colonization of teeth in humans. *Infect. Immun.* 9, 624–630.

[81] van Houte J. (1980). Bacterial specificity in the etiology of dental caries. *Int Dent J.* 30(4):305–26.

[82] van Houte J, Yanover L, Brecher S. (1981). Relationship of levels of the bacterium *Streptococcus mutans* in saliva of children and their parents. *Arch Oral Biol.* 26(5):381–386.

[83] van Houte, J., Gibbs, G., and Butera, C. (1982). Oral flora of children with "nursing bottle caries". *J. Dent. Res.* 61(2):382–5.

[84] van Houte, J., and Russo, J. (1986). Variable colonization by oral streptococci in molar fissures of monoinfected gnotobiotic rats. *Infect Immun.* 52(2):620–2.

[85] van Houte., Sansone, C., Joshipura, K., and Kent, R. (1991). Mutans streptococci and non-mutans streptococci acidogenic at low pH, and in vitro acidogenic potential of dental plaque in two different areas of the human dentition. *J. Dent. Res.* 70(12):1503–7.

[86] van Houte J. (1993). Microbiological predictors of caries risk. *Adv. Dent. Res.* 7(2):87–96.

[87] Wan, A. K. L., Seow, W. K., Walsh, L. J., and Bird, P. S. (2002). Comparison of Five selective media for the growth and enumeration of *Streptococcus mutans. Aust. Dental. J.* 47, 21–26.

[88] Westergren, G, and Krasse, B. (1978). Evaluation of a micromethod for determination of *Streptococcus mutans* and Lactobacillus infection. *J. Clin. Microbiol.* 7(1):82–3.

[89] Whiley, R. A., and D. Beighton. (1998). Current classification of the oral streptococci. *Oral. Microbiol. Immunol.* 13:195–216.

[90] Yoo, S. Y., Kim, P. S., Hwang, H. K., Lim, S. H., Kim, K. W., Choe, S. J. (2005). Identification of Non-mutans streptococci organisms in dental plaques recovering on Mitis-Salivarius Bacitracin agar medium. *J. Microbiol.* 43, 204–208.

13

Therapeutic Angiogenesis in Cardiovascular Diseases, Tissue Engineering, and Wound Healing

K. R. Rakhimol[1], Robin Augustine[2], Sabu Thomas[1] and Nandakumar Kalarikkal[1]

[1]International and Inter University Centre for Nanoscience and Nanotechnology, Mahatma Gandhi University, Kottayam, Kerala 686560, India
[2]School of Nano Science and Technology, National Institute of Technology Calicut, Kozhikode, Kerala 673601, India

Abstract

Angiogenesis is the formation of new blood vessels from the existing ones in any part of the body which is in need of oxygen-rich blood. Therapeutic angiogenesis is the proliferation of blood vessel through the injection of some external agents such as growth factors. Angiogenesis and neovascularization through the scaffold are important factors that determine the success of tissue engineering endeavor. Growth factor like VEGF is used in tissue engineering scaffolds to promote vascularization. Reactive oxygen species (ROS) can induce neovascularization through growth factor-mediated pathways. The use of metal oxide nanoparticles is a novel approach to induce vascularization through scaffolds.

13.1 Introduction

Blood vessel formation occurs through different mechanisms such as angiogenesis, vasculogenesis, and arteriogenesis [1]. These three mechanisms are interrelated and help for the vasculature formation of the body. In a developing embryo, the blood vessel formation is through vasculogenesis. In this process, endothelial progenitor cells combine together to form solid cords. These cords transform into parent vessels for the tubulogenesis [2]. Angiogenesis is different from vasculogenesis because it is the formation of a new blood vessel from an existing blood vessel. That is, it is the expansion of vascular bed formed through vasculogenesis [3]. Arteriogenesis helps the formation of functionally developed arteries. In the post-natal period, re-vascularization of undersupplied tissues happens mainly through angiogenesis and arteriogenesis [1].

Tissue engineering is a therapeutic strategy that uses a combination of cells, biomaterials, and substances which induce cell differentiation for the repair or replacement of tissue for the enhancement of tissue function [4]. Scaffold for tissue engineering is the three-dimensional backbone for the migration, proliferation, and differentiation of cells for the formation of extracellular matrix (ECM) and tissue [5, 6]. The most important issue in tissue engineering and regenerative medicine is the difficulty of formation of blood vessel through the scaffold [7]. For the integration of scaffold into the host tissue, a network of active blood vessels is necessary [8].

Angiogenesis is the process by which new blood vessels are formed in a part or an organ of the body, which is in need of oxygen-rich blood. This is normally a natural process happening in all organisms having ischemia, a condition in which the blood supply to a particular part of the body is decreased due to any obstruction or constriction of the blood vessels. Angiogenesis is a very complex process mediated by endothelial cells in the blood vessels [9]. At the time of angiogenic growth, some of the endothelial cells (EC) in the capillary wall are chosen for sprouting. They are known as the tip cells, which lead to the sprout growth [10]. Even though the growth of blood vessel to the scaffold will take place, the growth of the blood vessel will be very slow and that prevents it from providing an adequate amount of blood to the tissue construct lying inside the tissue-engineered construct [8]. Growth factors [11] such as FGF and VEGF are considered as very effective regulators for the induction of angiogenesis in tissue engineering scaffolds [12] (Figure 13.1). Because of their very less half-life, they have to overcome numerous challenges [13].

Clinical interests in the control of angiogenesis rouse from two different quarters. In one side, angiogenesis is suppressed for blocking the new blood vessel growth to inhibit the tumor growth or pathologies like diabetes. On the other side, angiogenesis is stimulated to increase the blood flow in patients with insufficient blood flow to particular parts of the body like ischemic heart disease or peripheral vascular diseases [14].

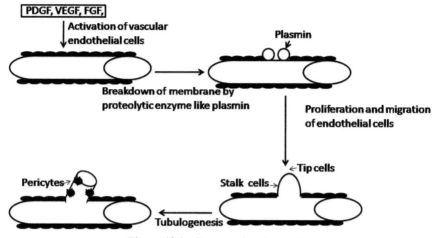

Figure 13.1 Process of angiogenesis.

13.2 Therapeutic Angiogenesis: Concept, Approaches, and Applications

Mainly angiogenesis occurs during embryogenesis [15]. But there are reports showing the existence of angiogenesis in adults [16–18]. Angiogenesis is a process which includes many factors such as growth factors which are released at adequate concentration and time. In growth factors related to angiogenesis, vascular endothelial growth factor (VEGF) is the most inspected one [19]. They are key regulators of angiogenesis because VEGF and their receptors entangled in remodeling and branching of the vasculature. They are involved in the migration of endothelial cells and maturation of nascent vessels [20]. In addition, VEGF and their receptors work as vascular permeability factors [21]. There are many additional angiogenic factors present in the body such as fibroblast growth factors (FGF), transforming growth factors (TGF), angiopoietins, and platelet-derived growth factors (PDGF) [19] (Table 13.1).

By definition, therapeutic angiogenesis is the process which stimulates the formation of new blood vessels to the ischemic parts or organs which need oxygen-rich blood [22]. The use of therapeutic angiogenesis is to stimulate the formation of new blood vessels into ischemic organs to raise flow and amount of oxygen-rich blood into these areas. It is based on delivery of external factors to enhance neovasculature formation. The use of genes, cells, and proteins exhibited their effectiveness in animal models. But the clinical

Table 13.1   Different types of angiogenic growth factors and their role in angiogenesis

Angiogenic Growth Factor	Location	Role in angiogenesis
VEGF-A	Secreted by circulating polymorphonuclear cells and platelets and expressed highly in lung, kidney, heart, and adrenal gland vascular smooth muscle cells	Stimulation of growth, endurance, relocation, proliferation, and differentiation of endothelial cells increases vascular permeability.
FGF	Skeletal muscle cells, intestinal enterochromaffin cells vascular smooth muscle cells, cardiac muscle cells	Production of collagenase and plasminogen activator stimulates endothelial cell survival, proliferation, and migration
Platelet-derived growth factor	Astrocytes, platelets, fibroblasts, keratinocytes, epithelial cells	VEGF upregulation, recruitment and pericytes, and smooth muscle cell proliferation.
Angiopoietin/Tie2	Adult: lung capillaries Embryo: vascular endothelium, angioblasts, endocardium	Regulates survival and integrity of endothelial cells, regulates sprouting and branching of vessels.
Hepatocyte growth factor	Secreted by biliary epithelial cells and polymorphonuclear cells in the liver	Promotes motility, survival, invasion, and morphogenesis of endothelial and epithelial cells.
Transforming growth factor-β	Secreted by cells including immune cells such as plasma cells and macrophages	Upregulation of angiogenic factors including VEGF and FGF, and proteinases promotes reformation of basement membrane, regulation of smooth muscle cell differentiation, and recruitment.

translation of them is challenging because of various reasons. Delivery of angiogenic factors is usually considered safe according to the trials. However, many obstacles must be surpassed before use in human therapy [23].

The delivery of angiogenic growth factors can be done by the administration of recombinant protein or by gene transfer. Recently, quinaprilat (angiotensin-converting enzyme (ACE)) inhibitor and DNA encoding the transcription factor for inducing hypoxia (hypoxia-inducible factor-1 (HIF-1)) have been exhibited to stimulate neovascularization [24]. The better preparation and delivery techniques for the therapeutic angiogenesis are now under clinical investigation. The advantage of gene therapy over recombinant protein therapy is it can maintain a high concentration in the local area over a long period.

The angiogenic growth factors have some important features in common. All act as the mitogens for the endothelial cells (EC). Similarly, all have exhibited to assist EC migration. Almost all angiogenic factors exhibit EC apoptosis inhibition. Other characteristic features of growth factors are upregulation of nitric oxide, matrix proteinases, matrix proteins, certain adhesion molecules, and a variety of cytokines [24].

13.2.1 Approaches

13.2.1.1 Direct VEGF administration

Vascular endothelial growth factor (VEGF) is a heparin-binding dimeric glycoprotein with a molecular weight of 45 KD. VEGF is different from other angiogenic growth factors in various features. VEGF can be secreted by all kind of cells. VEGF binding sites are present in endothelial cells, quiescent, and proliferating cells. But these are not present in other types of cells. Therefore, VEGF is considered as the endothelial cell-specific mitogen.

The direct VEGF delivery has been carried out using intracoronary and intravenous injections in human beings [25, 26]. But the amount of VEGF reached the target was very less and that failed to last more than one day [27]. So VEGF was administrated by incorporating into various polymers or encapsulated in micro or nanoparticles [28].

13.2.1.2 Cell-based therapy

Cell-based therapy is the most abundantly used method for therapeutic angiogenesis. Their contribution is either by direct participation in blood vessel formation or by paracrine signal secretion. Cells have the capability to act as hypoxia-responsive paracrine release vehicle by the advantage of their diverse cytokine contents. Most studied cell populations for therapeutic angiogenesis are bone marrow mesenchymal stem cells (BMMSCs) and endothelial progenitor cells (EPCs). EPCs have the capacity to move toward the growing vasculature area and bind with the growing vasculature [17, 29].

Related to ethical issues and immunogenicity, autologous stem cell therapy is more advantageous than any other methods. It is because of the absence of immunogenicity. That is, the cells are transplanted into the same body from which it is taken. As the cells are not taken from fetus but from adults, the ethical issues will be very less. Moreover, because of the limited differentiation potential of adult cells, the chances of aberrant tissue formation or cancer formation are very less. Nevertheless, cell-based therapies have several limitations. Because of the heterogeneous cell types, varied responses may occur. The increased cost, time, and regulatory concerns increase the barriers for clinical translation [30].

13.2.1.3 Regulation at genomic/molecular level

Thrombosporins (TSP-1 and TSP-2) are strong angiogenesis inhibitors. The inhibition is by the direct effect on endothelial cell survival, migration, proliferation, and apoptosis by opposing the activity of VEGF. Many membrane receptor systems and signal transduction molecules mediate the effect of thrombosporins [31]. TSP-1 has the antiangiogenic effect in the tumor microenvironment. TSP-1 counteracts VEGF in many important ways such as inhibition of the release of VEGF from the extracellular matrix, direct interaction of VEGF signal transduction inhibition. Rodriguez-Manzaneque et al. has studied the method of action of TSP-1 in transgenic mice. They compared tumor progression in wild-type, TSP-1-null, and TSP-1-overexpressing mice. They concluded that the action of TSP-1 is the inhibition of VEGF release from the extracellular matrix by suppressing MMP-9 (metalloproteinase-9) activity [32]. Since angiogenesis is regulated by change in oxygen tension, endothelial cells and smooth muscle cells are having a mechanism to sense the oxygen changes and also the angiogenesis. Hypoxia-inducible factor-1 (HIF-1) has a major role in cell's response to hypoxia through activating angiogenic agents such as vascular endothelial growth factor. HIF-1 is degraded due to the hydroxylation of specific prolyl residues by prolyl hydroxylase domain-2 (PHD2). So, the silencing of PHD2 gene by RNA interference (RNAi) might enhance the angiogenic growth factor expression and so angiogenesis [33].

13.2.1.4 Hypoxia-induced angiogenesis

Hypoxia is a well-recognized stimulant of angiogenesis, and hypoxia-inducible factor-1 (HIF-1) pathway is the most relevant regulator of oxygen homeostasis [34, 35]. The enzyme in this pathway: HIF-1 contains two subunits HIF-1α and HIF-1β, which are fundamentally expressed and HIF-1α being degraded under normoxia. But under hypoxic conditions, the degradation is prevented and HIF-1 continues its function. The function of HIF-1 is mainly the regulation of gene expression, particularly that of stromal cell-derived factor (SCDF-1) and VEGF. Even though hypoxia is viewed as an obstacle for tissue engineering, it has some advantages and advanced research require considering when constructing a scaffold, i.e., by a process called

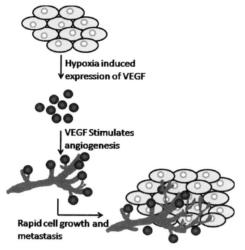

Figure 13.2 Hypoxia-induced angiogenesis. Hypoxia results in the expression of VEGF and this VEGF helps to stimulate angiogenesis.

"hypoxic preconditioning". In this, activation of Akt is happened due to the first exposure of BMMSC to hypoxic conditions; it in turn increased paracrine effects, reduced hypoxia related cell death, increased secretion of angiogenic factors, and increased proliferative potential [36, 37] (Figure 13.2).

For oxidative phosphorylation in ATP production, oxygen (O_2) is very essential for eukaryotes. For the proper division and development of cells and tissues, it is essential to maintain a constant supply of O_2 by vascular system [38]. Hypoxia is common in normal mammalian cells and in diseases when a change in oxygen supply occurs. The hypoxic condition can occur either by complete occlusion of blood vessels or by ruptures of blood vessels [39].

Judah Folkman first suggested the crucial role of angiogenesis in tumorigenesis. In avascular state, tumor mass cannot surpass 1 mm^3 due to the lack of oxygen and nutrients [40]. This avascular state in tumor because of low-oxygen tension (hypoxia) initiates angiogenesis due to the oxygen-sensing mechanism and following introduction of proangiogenic genes [41]. The most important gene is of vascular endothelial growth factor (VEGF) procure endothelial cells to hypoxic or avascular cites and invigorate proliferation. At first, oxygen deficiency is managed by hypoxia-inducible factor (HIF-1), a cellular oxygen-sensing transcription factor which influences the transcription of about 40 proteins like VEGF [42].

13.2.2 Applications of Therapeutic Angiogenesis

13.2.2.1 Wound healing

Rapid vascularization is very important to make sure fast wound healing. Current strategies are not meant for promoting angiogenesis and granulation. Limited knowledge about the process of vascularization in the graft is the main reason which prevents the development of skin substitutes having angiogenic properties [43]. By the incorporation of endothelial cells into the tissue engineered construct such as skin substitutes, scientists improved the vascularization into them and proved by preclinical and clinical trials [44].

During wound healing, the fibrin/fibronectin-rich wound clot is invaded by angiogenic capillary sprouts, and within a few days, it is arranged into a microvascular network throughout the granulation tissue. The density of blood vessels diminishes because collagen tends to accumulate in the granulation tissue for producing scar. An active interaction occurs between endothelial cells, angiogenic factors such as VEGF, FGF, angiopoietin, mast cell tryptase, TGF-β, and the extracellular matrix (ECM) environment. Specific ECM receptors of endothelial cell are critical for the blood vessel's morphogenetic changes during wound repair. Recent studies have shown that by modulating integrin receptor expression, wound ECM could regulate angiogenesis in part. Wound angiogenesis can also be regulated by interaction of endothelial cell with ECM environment in wound. By understanding the molecular mechanics to regulate wound angiogenesis, specifically how ECM regulates ECM receptor and angiogenic factor necessity, it is possible to provide new approaches for treating chronic wounds [45].

13.2.2.2 Bone development

The repairing of big bone defects is a serious clinical challenge. In bone, to maintain skeletal integrity, the spatial and temporal connection of bone cells and blood cells is very important. So angiogenesis plays an important role in bone development and fracture repair. The current methods for repairing the bone defects and providing mechanical support are grafts (autologous or allogenic) and implants (polymeric/metallic). These methods have a number of limitations due to potential disease transmission, insufficient supply, rejection, cost, and the inability of the grafts to integrate with the surrounding tissues. The bone tissue engineering is the novel method to heal musculoskeletal damages. Many scaffold constructs have been developed for bone tissue engineering but, for the survival and integration of the scaffold into host tissue, an active network of blood vessel is necessary. Combination of stem cells and polymeric scaffolds which release growth factors is under assessment to energize bone regeneration [46, 47].

13.2.2.3 Cardiac diseases

Cardiovascular disease is a major cause of death all over the world, and it is due to the partial or complete occlusion of the blood vessel in the nearby area. Restoration of blood supply is essential for the treatment of cardiovascular diseases. Therapeutic angiogenesis helps to stimulate the emergence of new blood vessels from the existing blood vessels. There are many methods for the delivery of angiogenic factors; however, using polymeric biomaterials give transitory and controlled release of growth factors. For targeted angiogenesis, it is possible to incorporate biomimetic signals into polymeric scaffolds; it will also allow environmentally responsive angiogenesis. Many recent progresses are emerged in the therapeutic angiogenesis such as use of genetically engineered stem cells and endogenous cell-homing mechanisms [48].

The recognition of angiogenic growth factors such as fibroblast growth factor and vascular endothelial growth factor has increased the interest in use of these factors for induction of therapeutic angiogenesis. The positive results from many animal studies and clinical trials proposed these new treatment modalities for ischemic diseases (diseases caused by restriction of blood supply to the tissues). Investigators and clinicians came to understand about the complexity of therapeutic angiogenesis by the increased understanding about cellular and molecular biology of blood vessel growth [14].

13.3 Growth Factors Needed for Angiogenesis

13.3.1 Fibroblast Growth Factor

Fibroblast growth factors (FGFs) are a group of growth factors involved in angiogenesis, endocrine signaling pathways, embryonic development, and wound healing. They are heparin-binding proteins, and for their signal transduction, the interaction with cell-surface-associated heparan sulfate proteoglycans is essential. They are also important in proliferation and differentiation of various cells and tissues [49]. The members of this family are with diverse biological effects and cellular targets [50]. Two members of this family, that is, FGF-1 (acidic fibroblast growth factor) and FGF-2, are having strong affinity to heparin, and their effects have been studied on vascular cells such as smooth muscle cells and endothelial cells. Both FGF-1 and FGF-2 are providing support for therapeutic angiogenesis as a stimulus *in vivo* [14].

13.3.2 Vascular Endothelial Growth Factor

Vascular endothelial growth factor (VEGF) is a fundamental regulator of angiogenesis, embryogenesis, reproductive functions and skeletal growth. Their biological effects are intervened by VEGFR-1, VEGFR-2, and two receptor tyrosine kinases (RTKs) but differ in signaling properties. VEGF inhibition is tested as a method to prevent angiogenesis and vascular leakage. Several VEGF inhibitors are developing to treat several malignancies [19]. VEGF-A is a homodimeric glycoproteins with the capacity to stimulate angiogenesis [51]. It can also increase vascular permeability than that of histamine (vasoactive substance) about 10,000 times. Initially, it was purified by this property and named as vascular permeability factor [52]. VEGF-A has a number of isoforms which biochemically distinct from each other. In mouse, three isoforms are present, and in humans, it is up to five, generated from alternative mRNA splicing in a single gene [53, 54]. The name of the isoforms is based on the number of amino acids present in the protein. The human isoforms of VEGF-A include VEGF121, VEGF145, VEGF165, VEGF189, and VEGF206 depending on the number of amino acids.

13.3.3 Platelet-Derived Growth Factor

Platelet-derived growth factor (PDGF) has also been involved in glioma angiogenesis. Initially, PDGF was discovered as a substance to increase the mitosis for fibroblasts in human serum and localized in the platelet's alpha granules [55, 56]. But subsequent studies proved that it is synthesized by a number of cell types and some of its target cite of action is capillary endothelial cells, osteoblasts, neurons, glia, and vascular smooth muscle cells [57–59]. It has pleiotropic effects and has roles in CNS development, embryonic development, tissue homeostasis, vascular system, and wound healing. An angiogenic role has also been demonstrated, even though it is weaker than the angiogenic effects of FGF and VEGF [59]. PDGF is a protein with molecular weight of 30 kDa consisting of dimers containing disulfide bonds in A and B chains. The A and B subunits have approximately 100 amino acids in length and share a 60% sequence analogy; the isoforms are functionally active only when dimerized as PDGF-AA, PDGF-BB, or PDGF-AB [58, 60]. The PDGF receptors are parts of the protein tyrosine kinase family receptors [61] and are activated by receptor subunits's ligand-induced dimerization. Two receptor subunits have been depicted: PDGFR-α (PDGF receptor-α) and PDGFR-β (PDGF receptor-β). PDGFR-α is a 170 kDa protein encrypted on chromosome 4q, while PDGFR-β is a 180 kDa protein coded on chromosome 5q [62–64].

13.4 Reactive Oxygen Species-Dependent Angiogenesis

Angiogenesis is an important process for tumor growth, wound healing, and embryonic development. It needs the processes such as cell multiplication, migration, and blood capillary formation in endothelial cells. In many cancer cells, reactive oxygen species (H_2O_2 and superoxide) is present in high levels. The function of ROS is mediating various growth-related responses such as angiogenesis by acting as signaling molecule (Figure 13.3). Thioredoxin and SOD, two endogenous antioxidant enzymes, can regulate ROS-dependent angiogenesis. In ECs, a NADPH oxidase is the major source of ROS which is having small G-protein Rac1, Nox (1, 2, 4 & 5), p22-phox and p47-phox. Growth factors such as VEGF and angiopoietin-1 can activate NADPH oxidase, and it can also be activated by the conditions such as hypoxia and ischemia. This oxidase can produce ROS which engage in autophosphorylation of VEGFR2 can also be involved in many redox signaling pathways to induce genes and transcription factors for angiogenesis. Dietary antioxidants are very effective factor for tumor angiogenesis treatment. Understanding the roles of NADPH oxidase and redox signaling events-derived ROS in

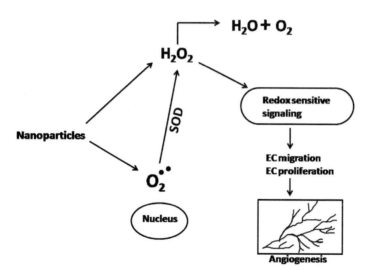

Figure 13.3   Role of reactive oxygen species in angiogenesis.

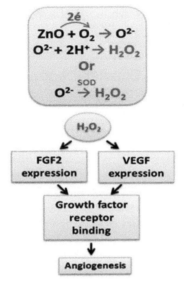

Figure 13.4 Mechanism of angiogenesis in PCL scaffolds containing ZnO nanoparticles.

angiogenesis will help to use it for the tumor therapeutic angiogenesis in the future.

Zinc oxide (ZNO) nanoparticle is a safe and widely used material for everyday applications. ZnO has the ability to form H_2O_2 by the chemisorption of oxygen from oxides to O^{2-} by taking electrons from them in the presence of oxygen. ZnO nanoparticle in scaffolds can produce reactive oxygen species (ROS) which may trigger angiogenesis in tissue-engineered scaffolds. ZnO nanoparticle in tissue-engineered PCL matrix can improve the angiogenic property of the scaffold. This property is due to the expression of FGF and VEGF by H_2O_2 produced by ZnO nanoparticle. This leads to the successful integration of the tissue into the scaffold. The capillaries help to carry oxygen and nutrients for the proliferating tissue and remove the waste materials [65] (Figure 13.4).

13.5 Metal Nanoparticle-Based Angiogenesis

Peptide-coated gold nanoparticles could activate or inhibit in vitro angiogenesis. Soumen Das et al. demonstrated the deliberate activation or inhibition of in vitro angiogenesis using functional gold nanoparticles coated with peptide. The peptides, anchored to oligo-ethylene glycol capped gold nanospheres,

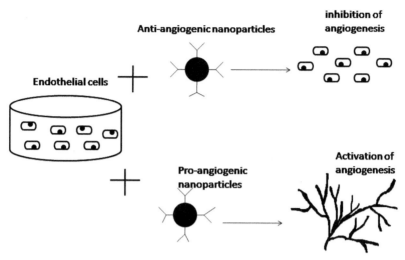

Figure 13.5 Metal nanoparticle induced angiogenesis. Anti-angiogenic nanoparticle inhibits the angiogenesis, and proangiogenic nanoparticle promotes the angiogenesis.

were designed to interact selectively with cell receptors responsible for activation or inhibition of angiogenesis (Figure 13.5). The functional particles are shown to influence the extent and morphology of vascular structures significantly, without causing toxicity. Mechanistic studies reveal that under various biological conditions, the nanoparticles have the ability to adjust the balance between naturally secreted pro- and anti-angiogenic factors. Nanoparticle-induced control on angiogenesis opens up new expectations in targeted drug delivery and therapy. The proangiogenic activity of metal nanoparticle critically depends on surface valence states of that metal. In particular, most of the reduced metal nanoparticles induce endothelial cell proliferation, tube formation in cell culture, and *in vivo* vascular sprouting. Particle size of the metal nanoparticle and ionic ratio are also very important for the proangiogenesis-enabling metal nanoparticles to act as oxygen modulators by catalyzing facile pathways for the replenishment and liberation of intracellular oxygen [66].

13.6 Stimulating Angiogenesis in Scaffolds by Therapeutic Angiogenesis

Angiogenesis is an important factor in tissue engineering for the successful incorporation of tissue-engineered scaffold [67]. The tissue-engineered matrix function as a reservoir of growth factors for the induction of new blood

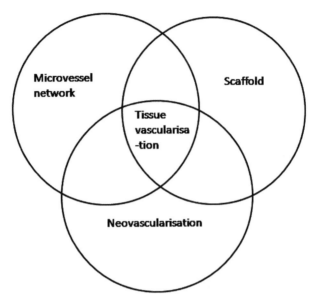

Figure 13.6   Triad of tissue vascularization.

vessel formation and as a scaffold for seeding the cells (EPCs) to participate in the blood vessel formation, i.e., arteriogenesis. Tissue vascularization includes microvessel network, neovascularisation, and artificial matrices and which constitute the triad of tissue vazcularisation (Figure 13.6). A successful vazcularisation is depending on the type of graft material and its arrangement. It should possess some important characteristics such as the material should be non-trombogenic, it should have similar compliance that of native blood vessel, and it should possess highly porous structure for the nutrient exchange at the capillary level [68].

Augustine et al. implanted the ZnO incorporated PCL scaffold and neat PCL scaffold in guinea pigs for histological evaluation. Within 5 days, they observed the initiation of angiogenesis in ZnO-incorporated PCL scaffold, i.e., the migration of a large number of RBCs to the membrane and arrangement of them in a circular manner (the first step of vascular sprouting) and also the migration of a large number of fibroblasts from the sides to the interior of the scaffold. But the vascular sprouting was not happened in the scaffold with neat PCL [69]. Here, ZnO nanoparticle acts as the stimulating agent for therapeutic angiogenesis (Figure 13.7). Such ZnO nanoparticle-incorporated scaffolds showed enhanced wound healing in guinea pigs [70].

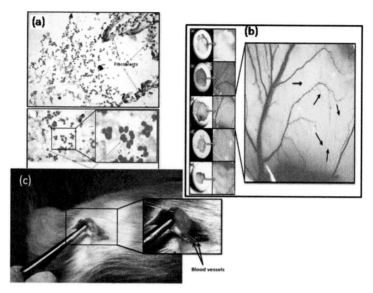

Figure 13.7 Stimulation of angiogenesis by using ZnO nanoparticle-incorporated PCL scaffold. (a) Initiation of sprouting of blood vessels through the scaffold, (b) blood vessel formation through the scaffold having different concentration of ZnO nanoparticle, (c) matured blood vessel through the scaffold after subcutaneous implantation.

Augustine and coworkers reported the formation of extensive vasculature when polyvinylidene fluoride-co-trifluoroethylene scaffolds incorporated with ZnO nanoparticles were implanted subcutaneously in rats. Blood vessel formation was further enhanced when human mesenchymal stem cells were seeded on the scaffolds prior to the implantation. Many metal oxide nanoparticles are showing angiogenic property and find application in modulating wound healing and tissue regeneration [71].

13.7 Challenges and Risks

The growth factors involved in the angiogenic process may be unstable in some solvents like organic solvents. The concentration of growth factors which apply for the angiogenesis is also a significant factor because the overexpression of VEGF may cause vascular tumors [72]. This is not occurred by angiogenesis but by a mechanism known as vasculogenesis. Low concentration of VEGF leads to angiogenesis, and high concentrations of VEGF may lead to vasculogenesis. So the regulated use of growth factors is very important for the controlled angiogenesis. Another major problem associated with VEGF therapy is that the newly formed blood vessels through therapeutic angiogenesis can be functionally abnormal, i.e., the formation of enlarged, thin-walled vessels with hyperpermeability, and lacking supporting pericytes [73].

The primary concern for tissue-engineered vascular network is to make sure that they are stable and functional in the scaffold within their *in vitro* and *in vivo* environment. It is also important to ensure that they will not merely produce capillaries but also helps for the formation of mature and stable blood vessels even after the absence of initial conditions [74]. In normal conditions, antioxidant systems of cell help to minimize the disturbances caused by ROS. But, if the ROS generation exceeds a certain level, it will overcome the activity of cellular antioxidants and results in oxidative stress. The prevention of oxidation is very influential in all aerobic organisms, because if the antioxidant protection is less, it may lead to carcinogenicity, cytotoxicity, and mutagenicity [75].

13.8 Conclusion

Large number of researches have been carried out to inhibit the angiogenesis to prevent the diseases such as cancer. But in the case of diseases which are occurring due to insufficient vascularization such as cardiovascular diseases, therapeutic angiogenesis is required. But many researches suggest that sustainable release of the angiogenic growth factors or agents is the major problem in the clinical trials of therapeutic angiogenesis. Nanotechnology plays a major role in providing materials with angiogenic ability and sustainable release at the specific target.

References

[1] Carmeliet, P., and Jain, R. K. (2011). Molecular mechanisms and clinical applications of angiogenesis. *Nature* 473, 298–307.
[2] Xu, K., and Cleaver, O. (2011). Tubulogenesis during blood vessel formation. *Semin. Cell Dev. Biol.* 22, 993–1004.
[3] Heil, M. 1., Eitenmüller, I., and Schmitz-Rixen, T., and Schaper, W. (2006). Arteriogenesis versus angiogenesis: similarities and differences. *J. Cell. Mol. Med.* 10, 45–55.
[4] Malafaya, P. B., Silva, G. A., and Reis, R. L. (2007). Natural–origin polymers as carriers and scaffolds for biomolecules and cell delivery in tissue engineering applications. *Adv. Drug Deliv. Rev.* 59, 207–233.

[5] Augustine, R., Hruda, N. M., Dinesh, K. S., Ayan, M., Dhruba, M., Nandakumar, K., et al. (2014a). Electrospun polycaprolactone/ZnO nanocomposite membranes as biomaterials with antibacterial and cell adhesion properties. *J. Polym. Res.* 21, 1–17.

[6] Recek, N., Resnik, M., Motaln, H., Lah-Turnšek, T., Augustine, R., et al. (2016). Cell adhesion on polycaprolactone modified by plasma treatment. *Int. J. Polym. Sci.* 2016, 354–366.

[7] Ennett, A. B., and Mooney, D. J. (2002). Tissue engineering strategies for in vivo neovascularisation. *Expert Opin. Biol. Ther.* 2, 805–818.

[8] Rouwkema, J., Rivron, N. C., and van Blitterswijk, C. A. (2008). Vascularization in tissue engineering. *Trends Biotechnol.* 26, 434–441.

[9] Daniel, T. O., and Abrahamson, D. (2000). Endothelial signal integration in vascular assembly. *Annu. Rev. Physiol.* 62, 649–671.

[10] Adams, Ralf. H., and Kari Alitalo, K. (2007). "Molecular regulation of angiogenesis and lymphangiogenesis." Nature. Reviews. Molecular. Cell Biology. 8, no. 6. 2007, 464–478.

[11] Madry, H., Rey-Rico, A., Venkatesan, J. K., Johnstone, B., and Cucchiarini, M. (2013). Transforming growth factor beta-releasing scaffolds for cartilage tissue engineering. *Tissue Eng. B Rev.* 20, 106–125.

[12] Nillesen, S. T., Geutjes, P. J., Wismans, R., Schalkwijk, J., Daamen, W. F., and van Kuppevelt, T. H. (2007). Increased angiogenesis and blood vessel maturation in acellular collagen–heparin scaffolds containing both FGF2 and VEGF. *Biomaterials* 28, 1123–1131.

[13] Richardson, T. P., Peters, M. C., Ennett, A. B., and Mooney, D. J. (2001). Polymeric system for dual growth factor delivery. *Nat. Biotechnol.* 19, 1029–1034.

[14] Ng, Y.-S., and D'Amore, P. A. (2001). Therapeutic angiogenesis for cardiovascular disease. *Curr. Control. Trials Cardiovasc. Med.* 2, 278–285.

[15] Risau, W., and Flamme, I. (1995). Vasculogenesis. *Annu. Rev. cell Dev. Biol.* 11, 73–91.

[16] Asahara, Takayuki, A., Toyoaki Murohara, T., Alison Sullivan, A., Marcy Silver, M., Rien van der Zee, R., and Tong Li, T. (1997). Bernhard Witzenbichler, Gina Schatteman, and Jeffrey M. Isner. "Isolation of putative progenitor endothelial cells for angiogenesis." *Science* 275, no. 5302. 1997, 964–966.

[17] Asahara, Takayuki, A., Tomono Takahashi, T., Haruchika Masuda, H., Christoph Kalka, C., Donghui Chen, D., Hideki Iwaguro, H., Yoko Inai, Marcy Silver, and Jeffrey M. Isner, et al. (1999). "VEGF contributes

to postnatal neovascularization by mobilizing bone marrow-derived endothelial progenitor cells." *The EMBO Journal.* 18, no. 14. 1999, 3964–3972.

[18] Urbich, C., and Dimmeler, S. (2004). Endothelial progenitor cells characterization and role in vascular biology. *Circ. Res.* 95, 343–353.

[19] Ferrara, N., Gerber, H.-P., and LeCouter, J. (2003). The biology of VEGF and its receptors. *Nat. Med.* 9, 669–676.

[20] Jain, R. K. (2003). Molecular regulation of vessel maturation. *Nat. Med.* 9, 685–693.

[21] Asahara, T., Chen, D., Tsurumi, Y., Kearney, M., Rossow, S., Passeri, J., et al. (1996). Accelerated restitution of endothelial integrity and endothelium-dependent function after phVEGF165 gene transfer. *Circulation* 94, 3291–3302.

[22] Isner, J. M. (1995). Therapeutic angiogenesis: a new frontier for vascular therapy. *Vasc. Med.* 1, 79–87.

[23] Chu, H., and Wang, Y. (2012). Therapeutic angiogenesis: controlled delivery of angiogenic factors. *Ther. Deliv.* 3, 693–714.

[24] Murohara, T., Asahara, T., Silver, M., Bauters, C., Masuda, H., Kalka, C., et al. (1998). Nitric oxide synthase modulates angiogenesis in response to tissue ischemia. *J. Clin. Invest.* 101, 2567–2578.

[25] Eppler, S. M., Combs, D. L., Henry, T. D., Lopez, J. J., Ellis, S. G., Yi, J. H., et al. (2002). A target-mediated model to describe the pharmacokinetics and hemodynamic effects of recombinant human vascular endothelial growth factor in humans. *Clin. Pharmacol. Ther.* 72, 20–32.

[26] Hendel, R. C., Henry, T. D., Rocha-Singh, K., Isner, J. M., Kereiakes, D. J., Giordano, F. J., et al. (2000). Effect of intracoronary recombinant human vascular endothelial growth factor on myocardial perfusion evidence for a dose-dependent effect. *Circulation* 101, 118–121.

[27] Cleland, J. L., Duenas, E. T., Park, A., Daugherty, A., Kahn, J., and Kowalski J. (2001). Development of poly-(D, L-lactide–coglycolide) microsphere formulations containing recombinant human vascular endothelial growth factor to promote local angiogenesis. *J. Control. Release* 72, 13–24.

[28] Simón-Yarza, T., Formiga, F. R., Tamayo, E., Pelacho, B., Prosper, F., and Blanco-Prieto, M. J. (2012). Vascular endothelial growth factor-delivery systems for cardiac repair: an overview. *Theranostics* 7, 2524–2536.

[29] Takahashi, T., Kalka, C., Masuda, H., Chen, D., Silver, M., Kearney, M., et al. (1999). Ischemia-and cytokine-induced mobilization of bone

marrow-derived endothelial progenitor cells for neovascularization. *Nat. Med.* 5, no. 4. 1999, 434–438.

[30] Fadini, G. P., Agostini, C., and Avogaro, A. (2010). Autologous stem cell therapy for peripheral arterial disease: meta-analysis and systematic review of the literature. *Atherosclerosis* 209, 10–17.

[31] Lawler, P. R., and Lawler, J. (2012). Molecular basis for the regulation of angiogenesis by thrombospondin-1 and-2. *Cold Spring Harb. Perspect. Med.* 2:a006627.

[32] Rodríguez-Manzaneque, J. C., Lane, T. F., Ortega, M. A., Hynes, R. O., Lawler, J., and Iruela-Arispe, M. L. (2001). Thrombospondin-1 suppresses spontaneous tumor growth and inhibits activation of matrix metalloproteinase-9 and mobilization of vascular endothelial growth factor. *Proc. Natl. Acad. Sci.* 98, 12485–12490.

[33] Wu, S., Nishiyama, N., Kano, M. R., Morishita, Y., Miyazono, K., Itaka, K., et al. (2008). Enhancement of angiogenesis through stabilization of hypoxia-inducible factor-1 by silencing prolyl hydroxylase domain-2 gene. *Mol. Ther.* 16, 1227–1234.

[34] Ceradini, D. J., Kulkarni, A. R., Callaghan, M. J., Tepper, O. M., Bastidas, N., and Kleinman, M. E. (2004). Progenitor cell trafficking is regulated by hypoxic gradients through HIF-1 induction of SDF-1. Nat. Med. 10, 858–864.

[35] Schioppa, T., Uranchimeg, B., Saccani, A., Biswas, S. K., Doni, A., Rapisarda, A., et al. (2003). Regulation of the chemokine receptor CXCR4 by hypoxia. *J. Exp. Med.* 198, 1391–1402.

[36] Uemura, R., Xu, M., Ahmad, N., and Ashraf, M. (2006). Bone marrow stem cells prevent left ventricular remodeling of ischemic heart through paracrine signaling. *Circ. Res.* 98, 1414–1421.

[37] Ma, T., Grayson, W. L., Fröhlich, M., and Vunjak-Novakovic, G. (2009). Hypoxia and stem cell-based engineering of mesenchymal tissues. *Biotechnol. Prog.* 25, 32–42.

[38] Semenza, G. L. (2007). Life with oxygen. *Science* 318, 62–64.

[39] Bertout, J. A., Patel, S. A., and Simon, M. C. (2008). The impact of O_2 availability on human cancer. *Nat. Rev. Cancer* 8, 967–975.

[40] Folkman, J., Merler, E., Abernathy, C., and Williams, G. (1971). Isolation of a tumor factor responsible for angiogenesis. *J. Exp. Med.* 133, 275–288.

[41] Giordano, F. J., and Johnson, R. S. (2001). Angiogenesis: the role of the microenvironment in flipping the switch. *Curr. Opin. Genet. Dev.* 11, 35–40.

[42] Semenza, G. L. (2002). HIF-1 and tumor progression: pathophysiology and therapeutics. *Trends Mol. Med.* 8, S62–S67.

[43] Lindenblatt, N., Platz, U., Althaus, M., Hegland, N., Schmidt, C. A., Contaldo, C., et al. (2010). Temporary angiogenic transformation of the skin graft vasculature after reperfusion. *Plast. Reconstr. Surg.* 126, 61–70.

[44] Black, A. F., Berthod, F., L'heureux, N., Germain, L., and Auger, F. A. (1998). In vitro reconstruction of a human capillary-like network in a tissue-engineered skin equivalent. *FASEB J.* 12, 1331–1340.

[45] Tonnesen, M. G., Feng, X., and Clark, R. A. (2000). Angiogenesis in wound healing. *J. Investig. Dermatol. Symp. Proc.* 5, 40–46.

[46] Kanczler, J. M., Ginty, P. J., Barry, J. J., Clarke, N. M., Howdle, S. M., Shakesheff, K. M., et al. (2008). The effect of mesenchymal populations and vascular endothelial growth factor delivered from biodegradable polymer scaffolds on bone formation. *Biomaterials* 12, 1892–1900.

[47] Kanczler, J. M., and Oreffo, R. O. (2008). Osteogenesis and angiogenesis: the potential for engineering bone. *Eur. Cell. Mater.* 15, 100–114.

[48] Deveza, L. (2012). *Therapeutic Angiogenesis for Treating Cardiovascular Diseases*. Ph.D. dissertation, Stanford, CA: Stanford University.

[49] Ay, I., Sugimori, H., & Finklestein, S. P. (2001). Intravenous basic fibroblast growth factor (bFGF) decreases DNA fragmentation and prevents downregulation of Bcl-2 expression in the ischemic brain following middle cerebral artery occlusion in rats. *Molecular brain research*, 87(1), 71–80.

[50] Powers, C. J., McLeskey, S. W., and Wellstein, A. (2000). Fibroblast growth factors, their receptors and signaling. *Endocr. Relat. Cancer* 7, 165–197.

[51] Leung, D. W., Cachianes, G., Kuang, W. J., Goeddel, D. V., and Ferrara, N. (1989). Vascular endothelial growth factor is a secreted angiogenic mitogen. *Science* 246, 1306–1309.

[52] Senger, D. R., Galli, S. J., Dvorak, A. M., Perruzzi, C. A., Harvey, V. S., and. Dvorak, H. F. (1983). Tumor cells secrete a vascular permeability factor that promotes accumulation of ascites fluid. *Science* 219, 983–985.

[53] Tischer, E., Mitchell, R., Hartman, T., Silva, M., Gospodarowicz, D., Fiddes, J. C., et al. (1991). The human gene for vascular endothelial growth factor. Multiple protein forms are encoded through alternative exon splicing. *J. Biol. Chem.* 266, 11947–11954.

[54] Shima, D. T., Kuroki, M., Deutsch, U., Ng, Y. S., Adamis, A. P., and D'Amore, P. A. (1996). The Mouse Gene for Vascular Endothelial Growth Factor genomic structure, definition of the transcriptional unit, and characterization of transcriptional and post-transcriptional regulatory sequences. *J. Biol. Chem.* 271 3877–3883.

[55] Kaplan, D. R., Chao, F. C., Stiles, C. D., Antoniades, H. N., and Scher, C. D. (1979). Platelet alpha granules contain a growth factor for fibroblasts. *Blood* 53, 1043–1052.

[56] Heldin, C.-H., Wasteson, Å., and Westermark, B. (1985). Platelet-derived growth factor. Mol. Cell. Endocrinol. 39, 169–187.

[57] Antoniades, H. N. (1991). PDGF: a multifunctional growth factor. *Baillières Clin. Endocrinol. Metab.* 5, 595–613.

[58] Antoniades, H. N., Scher, C. D., and Stiles, C. D. (1979). Purification of human platelet-derived growth factor. *Proc. Natl. Acad. Sci. U.S.A.* 76, 1809–1813.

[59] Heldin, C.-H., and Westermark, B. (1999). Mechanism of action and in vivo role of platelet-derived growth factor. Physiol. Rev. 79, 1283–1316.

[60] Bowen-Pope, D. F., Hart, C. E., and Seifert, R. A. (1989). Sera and conditioned media contain different isoforms of platelet-derived growth factor (PDGF) which bind to different classes of PDGF receptor. *J. Biol. Chem.* 264, 2502–2508.

[61] Claesson-Welsh, L. (1994). Signal transduction by the PDGF receptors. *Prog. Growth Factor Res.* 5, 37–54.

[62] Gronwald, R. G., Grant, F. J., Haldeman, B. A., Hart, C. E., O'Hara, P. J., and Hagen, F. S., (1988). Cloning and expression of a cDNA coding for the human platelet-derived growth factor receptor: evidence for more than one receptor class. *Proc. Natl. Acad. Sci.* 85, 3435–3439.

[63] Hart, C. E., Forstrom, J. W., Kelly, J. D., Seifert, R. A., Smith, R. A., Ross, R., et al. (1988). Two classes of PDGF receptor recognize different isoforms of PDGF. *Science* 240, 8, 1529–1531.

[64] Matsui, T., Heidaran, M., Miki, T., Popescu, N., Rochelle, W. L., Kraus, M., et al. (1989). Isolation of a novel receptor cDNA establishes the existence of two PDGF receptor genes. *Science* 243, 800–804.

[65] Augustine, R., Edwin, A. D., Indu, R., Balarama, K., Nandakumar, K., and Sabu, T. (2014C). Electrospun polycaprolactone membranes incorporated with ZnO nanoparticles as skin substitutes with enhanced fibroblast proliferation and wound healing. *RSC Adv.* 4, 24777–24785.

[66] Das, S., Singh, S., Dowding, J. M., Oommen, S., Kumar, A., Sayle, T. X., et al. (2012). The induction of angiogenesis by cerium oxide nanoparticles through the modulation of oxygen in intracellular environments. *Biomaterials* 33, 7746–7755.

[67] McGregor, I. A., and McGregor, A. D. (1995). *Fundamental Techniques of Plastic Surgery: And Their Surgical Applications.* Elsevier: WB Saunders Company.

[68] Cassell, C. S., Hofer, S. O., Morrison, W. A., and Knight, K. R. (2002). Vascularisation of tissue-engineered grafts: the regulation of angiogenesis in reconstructive surgery and in disease states. *Br. J. Plastic Surg.* 55, 603–610.

[69] Augustine, R., Edwin, A. D., Indu, R., Balarama, K., Nandakumar, K., and Sabu T. (2014b). Investigation of angiogenesis and its mechanism using zinc oxide nanoparticle-loaded electrospun tissue engineering scaffolds. *RSC Adv.* 4, 51528–51536.

[70] Augustine, R., Dan, P., Sosnik, A., Kalarikkal, N., Tran, N., Vincent, B., et al. (2017a). Electrospun poly(vinylidene fluoride-trifluoroethylene)/zinc oxide nanocomposite tissue engineering scaffolds with enhanced cell adhesion and blood vessel formation, *Nano Res.* doi: 10.1007/s12274-017-1549-8

[71] Augustine, R., Mathew, A. P., and Sosnik, A. (2017b). Metal Oxide Nanoparticles as Versatile Therapeutic Agents Modulating Cell Signaling Pathways: Linking Nanotechnology with Molecular Medicine, *Appl. Mater. Today* 7, 91–103.

[72] Springer, M. L., Chen, A. S., Kraft, P. E., Bednarski, M., and Blau, H. M. (1998). VEGF gene delivery to muscle: potential role for vasculogenesis in adults. *Mol. Cell* 2, 549–558.

[73] Pettersson, A., Nagy, J. A., Brown, L. F., Sundberg, C., Morgan, E., Jungles, S., et al. (2000). Heterogeneity of the angiogenic response induced in different normal adult tissues by vascular permeability factor/vascular endothelial growth factor. *Lab. Invest.* 80, 99–115.

[74] Heinke, J., Patterson, C., and Moser M., (2012). Life is a pattern: vascular assembly within the embryo. *Front. Biosci.* 4, 2269–2288.

[75] Mates, J. M. (2000). Effects of antioxidant enzymes in the molecular control of reactive oxygen species toxicology. *Toxicology* 153, 83–104.

14

Toxicity of Nanomaterials Used in Nanomedicine

Parvathy Prasad, Sunija Sukumaran, Nitheesha Shaji, V. K. Yadunath, Jiya Jose, Nandakumar Kalarikkal and Sabu Thomas

International and Inter University Centre for Nanoscience
and Nanotechnology, Mahatma Gandhi University, Kerala, India

Abstract

Nanotoxicology is an emerging subdiscipline of nanotechnology. Nanotoxicology refers to the study of the interactions of nanostructures with biological systems. Current advances in nanotechnology have led to the development of the new field of nanomedicine. Nanomedicine and Nanotoxicology are the two sides of the same coin, which includes many applications of nanomaterials and nanodevices for diagnostic and therapeutic purposes. The same unique physical and chemical properties that make nanomaterials so attractive may be associated with their potentially dreadful effects on cells, tissues, and environment. So the associated control requirements are essential, because their surface area and toxicity may be significantly greater than those of larger particles. Though, specific mechanisms and pathways through which nanomaterial may exert their toxic effects remain unknown. So the study of nanomaterials toxicity in nanomedicine is very essential.

14.1 Introduction

Nanomedicine and nanotoxicology are the two sides of the same coin. Nanotechnology involves the creation and manipulation of materials at nanoscale levels (1–100 nm) to make products that exhibit novel properties [1]. The application of nanotechnology to medicine, known as nanomedicine, concerns the use of exactly engineered materials at this scale to develop novel

365

therapeutic and diagnostic modalities. Nanomaterials have unique physico-chemical properties, such as ultrasmall size, large surface area to mass ratio, and high reactivity which are different from bulk materials (in microscale) of the same composition. These properties can be used to overcome some of the limitations found in traditional therapeutic and diagnostic agents. As the same time, it create toxic effects [2].

Nanotoxicology is developing as an important subdiscipline of nanotechnology. Nanotoxicology refers to the study of the interactions of nanostructures with biological systems revealing the relationship between the physical and chemical properties (e.g., size, shape, surface chemistry, composition, and aggregation) of nanostructures [3].

Several people can get exposed to nanostructures in a variety of manners such as researchers manufacturing the nanostructures, patients injected with nanostructures, or people using products containing nanostructures. In all the cases, there will be unique routes of exposure that will be directive to the specific fate of nanostructures [3]. The overall behavior of nanostructures could be illustrated below: (1) Nanostructures can enter the body via six principle routes: intra venous, dermal, subcutaneous, inhalation, intraperitoneal, and oral; (2) absorption can occur where the nanostructures first interact with biological components (i.e., proteins, cells); (3) they can distribute to various organs in the body and may remain the same structurally, be modified, or metabolized; and (4) they enter the cells of the organ and reside in the cells for an unknown amount of time before leaving to move to other organs or to be excreted.

The recent developments in nanotechnology and nanoscience have led to the discovery of the new field of nanomedicine, which comprises many applications of nanomaterials and nanodevices for diagnostic and therapeutic purposes. In specific, new approaches to site-specific drug targeting using nanoparticle drug carrier systems have been developed. Nanoparticle-based molecular imaging techniques have been introduced as revealing adjuncts in personalized treatment of patients. The extraordinary sensitivity of the physical characteristics of nanoparticle complexes with biomolecules has introduced a significant attention in the design of new sensors making use of a strong dependence of electron transfer and energy transfer on donor–acceptor distances. The distance-dependent devices offer the prospect to make use of the optical and electronic signals thus obtained for extremely sensitive and precise biomolecular recognition of bio-analytes of interest which is the creation of nanosensors.

The usage of nanodevices to facilitate molecular repair mechanisms of damage to macromolecules, particularly DNA, is another innovative venue of nanomedicine. Finally, recently created muscle-powered "Biobots" and nanosized excitable vesicles not only provide the possibility of crafting and manufacturing devices with the potential to substitute lost biological functions but also advance our understanding of the fundamentals of life and disease [4].

The current advancement in the areas of research in nanotechnology are undeniably important. The same distinctive physical and chemical properties that make nanomaterial so attractive may be associated with their potentially dreadful effects on cells and tissues. There is an emerging concern that nanosized particles merit a more difficult assessment of their potential effects on health and environment. So the associated control requirements are essential, because their surface area and toxicity may be significantly greater than those of larger particles. Though, specific mechanisms and pathways through which nanomaterial may exert their toxic effects remain unknown. A current report described unusual redox features of SWCNTs in the existence of physiologically relevant redox agents [5].

14.2 Nanomedicine

Nanomedicine is defined as the monitoring, repair, construction, and control of human biological systems at the molecular level, using engineered nanodevices and nanostructures. Our body is constructed from nanoscale building blocks such as DNA and proteins. Which have long been targeted by the pharmaceutical industry long before the beginning of nanotechnology. The category of drugs includes aspirin, cisplatin, and other anticancer agents as well as much more complex molecules like beta-blockers and anti-inflammatory agents. The difference between nanomedicine and conventional drugs is that, nanomedicine is entirely based on small molecule at nanoscale dimension [6]. It not only covers the therapeutic agents themselves, but also promise to combine the abilities to deliver the drugs to the targeted sites, ie to specific regions or tissues in the body, to specific cells, perhaps to a specific location within a cell, and also to make release of the therapeutic responsive to a physiological condition and perform definite task. The increased biological activity of nanoparticles can be either positive or required (e.g., antioxidant activity, carrier capacity for therapeutic penetration of blood-brain barrier, and the stomach wall or tumor pores), distributed throughout the whole body including entering the central nervous system, or negative and undesirable (e.g., toxicity, induction of oxidative stress, or cellular dysfunction) or a mix of both.

14.3 Nanomaterials Used for Nanomedicine

The development of nanotechnology leads the tremendous growth application of nanoparticles in different fields, which increase the risk of human exposure to these nanomaterials [7]. Nanomaterials generally show different, physical, chemical, and biological properties compared to fine or coarse particles of the same materials [8]. Nanomedicine is the application of nanotechnology in medical field. Development of nanocapsules and nanodevices might trigger a revolution in drug delivery, gene therapy, and medical diagnostics [9]. The potential advantage of nanotechnology in medicine is enormous ranging from novel approaches of designing artificial organs to nanorobotics, biosensors, diagnostic devices, bone grafting, and tiny vehicles for drug delivery [10]. But some insoluble nanoparticle carriers may penetrate biological membrane barriers and accumulate in tissues and organs of human body. Toxicity evaluations of these nanomaterials are necessary to facilitate integration into future medical applications. So it is important to study the toxicological effects of nanoparticles used in nanomedicine. There are many types of nanoparticles that are used as nanomedicines. Carbon-based nanoparticles are nanoparticles which are composed mostly of carbon. They have potential application in biomedical fields, including drug delivery, DNA and protein sensors, biocatalyst, and tissue scaffolds [11]. Metal-based nanoparticles are also used in biomedical fields. QD are small closely packed semiconductor vehicles that emits wavelength depended on their size. It makes them highly suitable for contrast agents for magnetic resonance imaging (MRI). They typically have a core made of inorganic element, but are generally coated with organic materials such as polyethylene glycol to enhance their biocompatibility or attach them to specific target molecule like proteins or DNA strands. If the target molecule is an indicator of disease, detection of that molecule may indicate a higher inclination for disease. An example is to use nanoparticles to bind to blood clots and to help make clots more visible by ultrasound [12]. Nanoshells are another nanodelivery system that is composed of copolymers, which are used in combination with specific wavelengths of light and heat technology used for cancer therapy. Gold nanoshells (GNSs) are particularly proper for use in the surgical arena as their outer shell is composed of a commonly used reduced inert gold. When activated by near infrared light, GNS can raise surrounding temperatures to levels sufficient for cellular ablation.

This strategy was recently used by Stern and Cadeddu et al., for the therapeutic ablation of urologic malignancies [13]. Nanopolymers have various applications in energy material science and medicine. Many synthetic and natural polymers are used as drug carriers in chemotherapy [14]. Figure 14.1 represents the different types of engineered nanoparticles.

Nanocomposites refer to the combination of NPs with other polymer. They have various applications in a packaging, automotive industry, nano-sized clay, etc. [15]. Microemulsion methods have also been used to deliver pharmaceuticals, specifically metal chelators to treat CNS related diseases [16, 17]. Veiseh et al. have reported that multifunctional nanoprobes which contained glioma cell targeting functions were also capable of being detected via MRI and fluorescent microscopy methods. Incorporation of cancer-killer genes into nanocapsules is being one tried out. One of the genes being investigated is the gene elaborate tumor necrosis factor, a protein that is toxic not only to cancer but also to healthy cells when injected in large doses. To avoid damage to normal tissue, the nanocapsule is coated with sensors that target only to the tumor cells [13]. Delivery of drugs to the brain always constitutes a big challenge. The usage of nanoparticles to deliver drug to the brain using the pathfinder technology is being examined. This technology uses nanoparticulate drug carriers in combination with the novel targeting principle of "differential protein adsorption" to cross the blood–brain barrier. As the nanoparticles are not efficiently scavenged by macrophages,

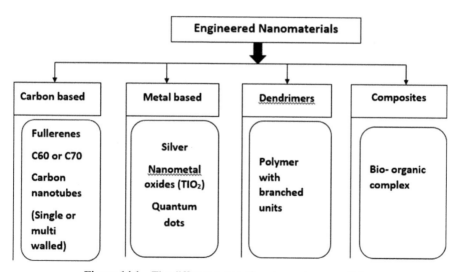

Figure 14.1   The different types of engineered nanoparticles.

the resulting increase in blood circulation time and therefore bioavailability are expected to extend the duration of controlled system drug delivery or to progress the diagnoses for nanoparticles to reach target sites by extravasation [12]. In biomedical science, the delivery of magnetic nanoparticles (MNPs) to or into various cell types has become an area of increasing research attention. Targeted delivery is used to deliver drugs or genes by attaching them to MNPs and locally concentrating the resulting complexes *in vivo* to the preferred location. Correspondingly, magnetic hyperthermia, the local concentration of MNPs and subsequent heating via magnetic fields, has shown as a promising candidate for potentially viable cancer therapy [18].

In summary, nanomateials have various applications in medical field, including drug delivery, molecular diagnostics, implants, tissue engineering, surgery, genomics, and nanodevice, but they also cause cytotoxic effects to human body.

14.4 Toxicokinetics of Nanoparticles

Toxicokinetics of nanoparticles includes different steps such as absorption, metabolism, distribution accumulation, and elimination. Nanoparticles have been found to be circulated to the colon, lungs, bone marrow, liver, spleen, and the lymphatic after intravenous injection. Distribution is followed by rapid clearance from the systemic circulation, predominantly by action of the liver and splenic macrophages. Clearance and opsonization, the process that prepares foreign materials to be more efficiently engulfed by macrophages, occur under certain conditions for nanoparticles depending on size and surface characteristics [19, 20]. Absorption is the first step of toxicokinetics. Nanoparticles may enter into human body through ingestion, inhalation, and dermal contact under the condition of occupational and environmental hazards. Inhalation is the primary route for nanoparticles to enter into the human body in occupational hazards. However, in the area of nanomedicine, intravenous and subcutaneous infection of nanoparticle carriers may represent a unique and more important exposure route than inhalation. The smaller thoracic particles and ultrafine particles with low density and Brownian motion show deposition and result in a deep penetration of nanoparticles in the lungs [21]. It causes adverse health effects in respiratory tract. The respiratory system acts as main port of entrance for airborne nanoparticles. Soluble nanoparticles captured or deposited in the lung may dissolve sudden and entered into the systemic circulation. The insoluble nanoparticles may be phagocytized by alveolar macrophages, migrate to the interstial space of the alveolar septa, or pulmonary lymphatics. The earlier studies found that carbon black (14 or 56 nm) and Tc-labeled monocolloids (less than 80 nm) cause lung inflammation and tissue damage [22]. The gastrointestinal tract (GIT) is an important route for nanoparticle absorption since drug carrier, food products [21]. In the field of nanomedicine, the GIT uptake of nanoparticles has been a subject of recent efforts to develop effective carriers that enhance the oral uptake of drug and vaccines [23]. These nanoparticles are then swallowed and subsequently ingested into the GIT. Then, NPs transport from the digestive tract into the lymphatic and blood circulation. GIT absorption increases with decreasing particle diameter. Dermal exposure represents an important nanoparticle absorption route in the field of nanomedicine [21]. Wound dressing contains silver nanoparticles. It acts as an antibacterial agent. The nanosilver-based dressing and surgical sutures have received approval for clinical application. Also, good control of wound infection is achieved; their dermal toxicity is still a topic of concern. Despite laboratory and clinical studies confirming the dermal biocompatibility of nanosilver-based dressings, several other researchers have established the cytotoxicity of these materials. Paddle-Ledinek et al. exposed cultured keratinocytes to extracts of several types of silver containing dressings. The results indicated that extracts of nanocrystalline-coated dressings are among those, which are the most cytotoxic. Keratinocyte proliferation was significantly inhibited, and cell morphology was affected [2]. So understanding potential epidermal and dermal penetration and toxicity of different nanoparticles will be important in the field of nanomedicine. Currently, there is no evidence that nanoparticles are destroyed in the body after being absorbed into the blood circulation from different exposure routes, but their chemical forms may be changed. The nanoparticles such as gold nanoparticles, silver nanoparticles, fullerences, and carbon nanotubes are unable to be metabolized by enzymes in human body. Some studies show that particles size, below 200 nm, gains access through the blood–brain barrier [24] and Au NPs, TiO$_2$ NPs are able to penetrate human red blood cells [25]. Nowadays, PLGA NPs are widely used for drug delivery. Semete et al. have administrated PLGA particles orally to mice. The result shows the accumulation of NPs in liver, kidney, brain, and spleen. Metabolism, tissue distribution, accumulation, and elimination of nanoparticles may be affected by various factors such as chemical form, particle size, and particle charge. Nanoparticles may also be distributed in different tissues or organs [26]. Elimination of nanoparticles may be through urine, bile, and feces [27]. It is not possible to eliminate all NPs from human body. As a result, deposition of NPs may take place in the human body.

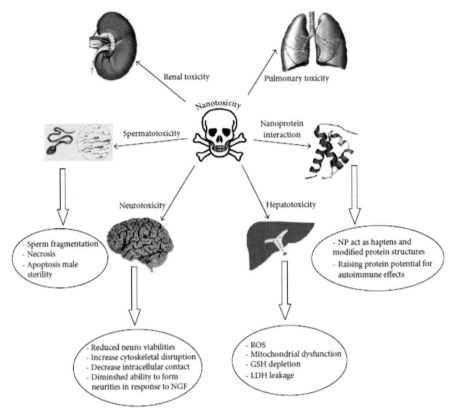

Figure 14.2 Important recorded toxic effects of therapeutically used nanoparticles (El-Ansary et al. 2009 [19]).

Figure 14.2 shows the important recorded toxic effects of therapeutically used nanoparticles [28].

14.5 Toxicity of Nanoparticles

The nanoscale dimension of particles is what makes nanotechnology so useful in medicine and industry. It is also one of the main factors that might make them potentially dangerous to human health as well as the environment. Researchers now presenting those smaller particles show more reactivity and toxic effects. It is because any intrinsic properties of particles will likely be emphasized with the increase in surface area per unit mass [29]. The potential risks inherent to any new technology are recognized and studied.

Though, the special concern with nanotechnology is the unique type of toxicity due to surface modification and small size. Improved endocytosis including a potential for inflammatory and pro-oxidant activity are shown to be mainly dependent on nanoparticles "surface chemistry (coating) and *in vivo* surface modifications." The increase in pulmonary toxicity (e.g., inflammation, granuloma formation) of carbon nanotubes when compared with that of the carbon black and carbonyl iron particles was seen in mice and rats [30]. Figure 14.3 shows the effects of nanoparticles into the organisms.

The large surface area of mesoporous SiO_2 NPs is enhanced by their surface pores and allows the particles to be filled with a drug. It makes SiO_2 NPs as a good drug carrier. Several studies show that crystalline SiO_2 NPs may induce carcinogenesis and produce inflammation, irritation, fibrosis, and other adverse health effects [31, 32]. The concept of targeted drug delivery shows selective accumulation of drug within a given tissue after systemic administration of the carrier-bound drug with the minimal possible side effects on the tissues and organs studied the toxicity of SiO_2 Nps in rats. They found that SiO_2 might be hepatoxic, and it also depends on its dose.

Oxidative stress caused by free radicals formed by the interaction of particles with cells may result in cell death. Indication of mitochondrial distribution and oxidative stress response after endocytosis of nanoparticles was noted. It was suggested that due to small size of the nanoparticles, it could act like haptens to modify protein structures, either by altering their function or rendering them antigenic, thus raising their potential for autoimmune effects. There is a primary finding that indicated that gold nanoparticles might move

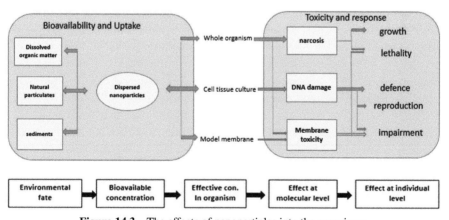

Figure 14.3 The effects of nanoparticles into the organisms.

through a mother's placenta to the fetus. Efficient uptake of nanoparticles via the gastrointestinal tract has also been well recognized in oral feeding studies and gavage studies. From all, these findings show that nanoparticles may potentially cause problems with body burdens and it is hypothesized that nanoparticles, because of their long retention in the body tissues, might repeat their highly catalytic activity with the host in cascade.

In recent years, Au NPs received significant attention in biomedicine because of their unique physical, chemical, and biological properties. Some *in vito* studies show that Au NPs induce cytotoxicity in different cells, involving apoptosis and necrosis [34]. Goodman et al. [34] found that cationic-functionalized Au NPs induce moderate toxicity when compared to anionic-functionalized Au NPs. Dosing or amount of therapeutic nanoparticles administered to the cells or tissues is critical to accessing the toxicity. Silver nanoparticles, the new group of antimicrobial agents, are becoming one of the progressively rising products in nanomedicine. Shavandi et al. checked the effect of commercial colloidal nanosilver on murine peritoneal macrophages by the MTT assay *in vitro*. A significant decrease in cell viability was detected at concentrations of 1–25 ppm of nanosilver compared to the control group after 24 h of cell culture. This acute dose-dependent cytotoxicity of silver nanoparticles on peritoneal macrophages needs attention about their use [35]. In recent years, fullerenes, a model of carbon-based nanoparticles, have attracted considerable interest due to their unique properties. The potential and the increasing use of fullerenes in nanomedicine have raised questions about their health safety. Different groups reported that nanoparticles of pristine fullerene C60 have no acute/subacute toxicity or genotoxicity in mice, rats, or guinea pigs [36–38]. However, some fullerene C60 derivatives, either covalently or noncovalently altered fullerenes, are highly toxic. Carbon nanotubes (CNT) including SWCNT and MWCNT recently developed as delivery vehicles for use in cancer treatment and gene therapy [39]. The potential toxicity of SWCNT and MWCNT in humans is also not known. Several researches show that both SWCNT and MWCNT induce cytotoxicity, oxidative stress, apoptosis, and necrosis *in vitro* [40, 41]. Han et al. studied the pulmonary response of mice to MWCNT and found that a single treatment of MWCNT is able to inducing a cytotoxic and inflammatory response in the lungs of mice [42]. CNT-induced fibrosis was due to direct stimulation of fibroblast proliferation and collagen production, rather than persistent inflammation as is the case with crystalline SiO_2 or asbestos. Schipper et al. [43] studied the toxicology of SWCNT when injected into the blood stream of mice. Survival as well as clinical and laboratory parameters revealed no evidence of toxicity over 4 ml/mg, but data recommended further confirmation with larger groups of animals was necessary. Therefore, the safety of CNT in drug delivery has been questioned. The literature has established that some of the nanomaterials used in nanomedicine may cause adverse health potentials on the human body [43]. To translate nanotechnology into nanomedicine, it is important to understand any potential toxicity produced by nanomaterials and to design methods to mitigate any detrimental effects [44]. TiO$_2$ nanoparticles are synthesized worldwide in large quantities for use in a wide range of applications including nanomedicine. The microscale titanium dioxide (TiO$_2$) is widely used in pharmaceutical and cosmetics industries. It is considered as biologically inert. Such that, there was no obvious lung toxicity in rats when a single instilled dose of TiO$_2$ was 5 mg/rat or 50 mg/kg. On the other hand, many studies have demonstrated that when TiO$_2$ particles size decreased to nanoscale dimension, they could produce more pulmonary toxicity than their bulk counterparts. In a comparative study done by Li et al., the acute pulmonary toxicity induced by 3 nm and 20 nm TiO$_2$ was investigated through measurement of selected biochemical parameters in bronchoalveolar lavage fluid (BALF). They observed that at 3-day post-exposure, the 3 nm TiO$_2$ induced significant increase in albumin, alkaline phosphatase (ALP), and acid phosphatase (ACP) concentrations in high-dose group (40 mg/kg) and also induced significant increase in ALP and ACP concentrations in mid-dose group (4 mg/kg). But it did not induce significant increase of total protein and LDH concentrations in any dose group. While 20 nm TiO$_2$ induced significant increase in all biochemical parameters in high- and mid-dose groups. At 3-day post-exposure, both TiO$_2$ particles did not induce obvious pulmonary toxicity in their low-dose (0.4 mg/kg) groups as evidence of no significant increase in all biochemical parameters. Arcinogenicity studies in rodents indicate that TiO$_2$ nanoparticles produced lung tumors when given by inhalation [45, 46]. Carbon black nanoparticles (CBN) are also carcinogenic to the lung of rats. Heinrich et al. [45] treated rats with CBN at a concentration of 7.5 mg/m^3 for 4 mo, then at a concentration of 12 mg/m^3 for 20 mo, and then by clean air for 6 ml. The occurrence of lung tumors (39%), including benign and malignant squamous cell tumors and bronchioalveolar cell tumors, was significantly increased compared to the clean air group (0.5%) [45]. Researches using rats by intratracheal instillation with CBNa at 1.5 to 6 mg/rat from 5 to 20 times. Animals were observed up to 30 mo. Total lung tumor incidence (56–80%) was significantly increased compared to the control (2%). The obtained *in vivo* data concerning the carcinogenic effects of TiO$_2$ nanoparticles, CBN,

SWCNT, and MWCNT have been reviewed. Of these, TiO$_2$ nanoparticles and CBN exhibited carcinogenicity in experimental animals. There are wide variety of nanoparticles which are used for drug delivery applications such as Doxil, nanoparticles coated by PEG, and liposomes. Kasid and Dritschilo [47] checked the toxicity of 467-nm-diameter cationic liposomes containing *raf* antisense oligonucleotide as well as the blank liposomes itself using multiple assay in multiple species. Although these loaded NPs do show promise for anticancer treatment, the liposomes themselves exhibit toxicity with significant increase in total neutrophil count and liver enzymes in mouse studies and changes in complement activity in monkey studies [47]. Hussain et al. found that cationic liposomes are significantly more toxic than anionic liposomes, and the concentration and size of liposomes do not depend on the toxicity. QDs are high quantum yield semiconductor NPs, used for *in vivo* and *in vitro* bio imaging because of its photostability and large florescence intensity [48]. The only limitations of these materials are its toxicity, stemming from the release of heavy metal ions from the QD core. To overcome the limitations, QDs are functionalized by using biopolymers so it prevent ion leeching and increase biocompatibility and bioavailability [37, 49]. Silva reviewed a list of fullerene (C60) derivatives studied both *in vitro* and *in vivo* for their neuroprotective ability. The model material responsible for providing neuroprotection is fullerenol which is hydroxyl-functionalized fullerene [50]. Yamawaki and Iwai [51] reported that the *in vitro* toxicity of fullerenols in human umbilical vein endothelial cells (ECs) that were treated with 1–100 µg/mL concentrations (average diameter 4.7–9.5 nm) for a day which induced cytotoxic morphological changes as well as showing cytotoxicity via LDH and WST assays in a dose-dependent manner [51]. Nowadays, the identification of cytotoxicity of nanoparticles toward mammalian germline stem cells has roused great concern over the biosafety of nanomaterial. In their study, they used a cell line with spermatogonial stem cell characteristics to test *in vitro* toxicity of numerous types of nanoparticles. The results illustrate that of all the tested materials (Ag, MoO$_3$, and Al), silver nanoparticles were the most toxic with manifestations such as drastic reduction of mitochondrial function, increased membrane leakage, necrosis, and induction of apoptosis. The results are of significant practical implications because silver nanoparticles are now able to access human sperms via a variety of commercialized products such as contraceptive devices and maternal hygiene items [52]. Based on these finding, fertility problems may occur. In addition to these, as a fair extrapolation, another question emerged is that what they will do to egg cells?

Liver appears to be the main accumulation site of circulatory silver nanoparticles. Like germ line stem cells, similar patterns of cytotoxicity of silver nanoparticles (decrease of mitochondrial function, LDH leakage, and abnormal cell morphologies) were detected with *in vitro* BRL 3A rat liver cells, but to a lesser degree. The same researchers in there another study, a neuroendocrine cell line (PC-12 cells) was exposed to silver nanoparticles as a control against Mn nanoparticles. The experimental results showed that silver nanoparticles were toxic to mitochondria than to Mn and Mn^{2+} [53]. All of these findings are of quite significant because considerable amount of silver could be detected in rat brain following inhalation of silver nanoparticles. The neurological toxicity of silver is not clinically ascertained. However, numerous seizures cases have been related to exposure to silver or silver compounds. Thus, it is evident that mitochondria seem to be sensitive targets of cytotoxicity of silver nanoparticles. Nevertheless, the mechanism of silver nanoparticles on mitochondria is yet to be explained. In the study with BRL 3A liver cell line, depletion of glutathione (GSH) level and an increased ROS were found in association with mitochondrial perturbation. This suggests that oxidative stress might mediate the cytotoxicity of silver nanoparticles. Based on these findings, a preliminary result can be formed that silver nanoparticles may interact with proteins and enzymes with thiol groups within mammalian cells. These proteins and enzymes such as glutathione, thioredoxin, SOD, and thioredoxin peroxidase are key components of the cell's antioxidant defense mechanism which is accountable to neutralize the oxidative stress of ROS largely generated by mitochondrial energy metabolism. As these effects of Ag+ could be completely blocked by sulfhydryl reagents as GSH, the surface modification of silver nanoparticles by phosphoryl disulfides was effective in improving silver biocompatibility and intracellular uptake. They prepared the phospholipid derivatives containing disulfide groups to modify silver nanoparticle surfaces. By adding sodium borohydride to reduce both disulfide bonds of the derivatives and silver ion simultaneously, the generated thiol group can be reacted with newborn silver atoms immediately to generate nanoclusters. The assemblies consisted of either phosphorylcholine (PC) or phosphorylethanolamine (PE) head groups, which made the silver clusters biocompatibile. In cell culture tests, the surfaces-modified nanoparticles were internalized into platelet and fibroblast cells in a short period of incubation without harming the cells [30]. It is reported that copper ions ingested are metabolized in liver and excreted via urine. If the intake of copper exceeds the range of the tolerance, it would cause toxic effects to hepatic and renal tissues, which is consistent with the finding of Meng et al. [54] that nanocopper possesses tremendously high bioavailability. Hence, the original safety limit may be modified to much lower level. From these finding, we can suggest that nano- and microcopper exhibit different biological behaviors *in vivo* via oral exposure routine. In terms of nanocopper particle, both copper overload and metabolic alkalosis contribute to their grave toxicity [54, 55]. High chemical reactivity of Ag nanoparticles was observed in the reaction with hydrochloric acid, that is, Ag (nanoparticles) + HCl $\rightarrow$ AgCl + H$_2$; the reaction product silver chloride was characterized by X-ray powder diffraction to give a direct evidence for the reaction. From the results, it is clear that which has been proved impossible for the bulk [56]. The nanoscale vanadium oxide (V$_2$O$_3$) potentiated toxicity compared to bulk material is demonstrated in human endo- and epithelial lung cells and might be due to the higher catalytic surface of the particles. Reduction in cell viability is almost ten times stronger and starts with the lowest concentrations of "nanoscaled" material (10 µg/mL). Vanadium oxide pointers to an induction of heme oxygenase 1 (HO-1), in a dose-dependent manner in ECV304 cells, whereas a reduction in protein levels can be observed for the epithelial cells (A549). Lipid peroxidation can be detected also for "nanoscaled" vanadium oxide to a much stronger extent in macrophages (RAW cells) than for bulk material. The observed properties cannot only be explained by oxidation from V$_2$O$_3$ to V$_2$O$_5$ as there are significant differences between the novel nano-vanadium and all used bulk materials (V$_2$O$_3$ and V$_2$O$_5$). It seems rather to be a nanoeffect of a high surface reactivity, here coupled with a yet-unknown toxicity potentiating effect of a technically important catalyst.

The challenge of creating new nanotherapeutics and making a redistic assessment of their toxicity will be pursured optimally with the collaboration of experts from materials science, chemistry, biomedical imaging, medicine, and toxicity with each expert performing the tasks at which they are experts. Some researchers recently reviewed a list of fullerene (C60) derivatives studied both *in vitro* and *in vivo* for their neuroprotective ability.

14.6 Effect of Nanoparticles in Some Aquatic Organisms

Various studies have been reported the effects of NPs on aquatic organisms. Lovern and Klaper found that fullerences C60 were toxic to Daphnia magna. They found that NPs cause death and behavioral changes to the fish. The toxicity of metal NPs is different in different group of organism [57]. It mainly depends on the type of NPs. For example, Cu and Ag NPs were highly toxic for D. magna. The concentration of Ag NPs increased the mortality rate and rate of abnormalities on zebrafish larvae [58]. Some researchers also checked the toxicity of metal NPs on green algae. Metal oxide NPs such as TiO_2, ZnO, and Al_2O_3 were inhibited the growth of algae and cause toxicity to embryos. G4 polyaminoamide (PAMAM) dendrimers with amine functional group cause reduction in growth and development of zebrafish embryos.

14.7 Discussion

Nanotoxicology mentions to the biokinetic assessment of engineered nanostructures and nanodevices. The necessity of this area for investigations became apparent after the intensive expansion of nanotechnology, which in the last two decades has been extensively used in the pharmaceutical industry, medicine, and engineering technology. Particle toxicology and the resulting adverse health effects of asbestos fibers and coal dust serve as a historical reference points to the development of nanotoxicological concepts. In the area of medicine, nanomedicine is defined as the monitoring, repair, construction, and control of human biological systems at the molecular level, using engineered nanodevices and nanostructures [59]. Macrophages as specialized host defense cells and endothelium as thin specialized epithelial cells that line the inner surface of lymph vessels and blood vessels serve as gate keeper to control passage of materials together. Then, tumors are the most common targets of nanoparticles. Within these biological targets, nanoparticles favor the formation of prooxidants particularly under exposure to light, ultraviolet light, or transition metals, thereby destabilizing the balance between the production of reactive oxygen species (ROS) and the biological system's ability to detoxify or repair the system. ROS can also be formed by the NADPH oxidase in phagocytic cells as target of nanoparticle devices. Nanoparticles can modify mitochondrial function as well as cellular redox signaling. Oxidative stress induced by nanoparticles is reported to improve inflammation through upregulation of redox-sensitive transcription factors including nuclear factor kappa B (NFkB), activating protein (AP-1), and extracellular signal regulator kinases (ERK) C-Jun, N-terminal kinases JNK, and p38 mitogen-activated protein kinases pathways [60].

14.8 Conclusion

Nanomaterials have unique applications in nanomedicine and other industries because of its interesting properties. The application of nanoparticles in medical field are drug delivery, molecular diagnostics, molecular imaging, dental implants, bone grafting, tissue engineering, etc., but these NPs cause some adverse health to human body. Mainly NPs enter into human body through inhalation, dermal contact, and intravenous subcutaneous injection. The metabolism, tissue distribution, and elimination of NPs depend on its chemical form, dose, and particle size. It is not able to remove all NPs from human body through elimination. In addition, they may also accumulate into human body. Some earlier studies in animals found that the NPs accumulated cause adverse health effects. TiO_2 and carbon black NPs cause carcinogenicity. Some studies show that carbon nanotubes including both SWCNT and MWCNT fullerenes C60, SiO_2, Au NPS, and Ag NPs also exhibit reproductive or developmental toxicity. So it is important to study the toxic effects of NPs which used for medicine.

References

[1] Hussain, S., et al. (2005). In vitro toxicity of nanoparticles in BRL 3A rat liver cells. *Toxicol. In Vitro* 19, 975–983.

[2] Oberdörster, G., et al. (2005). Principles for characterizing the potential human health effects from exposure to nanomaterials: elements of a screening strategy. *Part. Fibre Toxicol.* 2:8.

[3] Oberdörster, G., Stone, V., and Donaldson, K. (2007). Toxicology of nanoparticles: a historical perspective. *Nanotoxicology* 1, 2–25.

[4] Wang, J. (2005). Nanomaterial-based electrochemical biosensors. Analyst 130, 421–426.

[5] Rajendra, J. and Rodger, A. (2005). The binding of single-stranded DNA and PNA to single-walled carbon nanotubes probed by flow linear dichroism. *Chem. A Eur. J.* 11, 4841–4847.

[6] Miller, J. (2003). Beyond biotechnology: FDA regulation of nanomedicine. *Columbia Sci. Technol. Law Rev* 4:E5.

[7] Kisin, E. R., et al. (2007). Single-walled carbon nanotubes: geno-and cytotoxic effects in lung fibroblast V79 cells. *J. Toxicol. Environ. Health Part A* 70, 2071–2079.

[8] Zhao, J., et al. (2009). Titanium dioxide (TiO2) nanoparticles induce JB6 cell apoptosis through activation of the caspase-8/Bid and mitochondrial pathways. *J. Toxicol. Environ. Health Part A* 72, 1141–1149.

[9] Oberdörster, G. (2010). Safety assessment for nanotechnology and nanomedicine: concepts of nanotoxicology. *J. Intern. Med.* 267, 89–105.

[10] Shvedova, A., and Kagan, V. (2010). The role of nanotoxicology in realizing the 'helping without harm'paradigm of nanomedicine: lessons

from studies of pulmonary effects of single-walled carbon nanotubes. *J. intern. Med.* 267, 106–118.

[11] Koyama, S., et al. (2006). Medical application of carbon-nanotube-filled nanocomposites: the microcatheter. *Small* 2, 1406–1411.

[12] Hardman, R. (2006). A toxicologic review of quantum dots: toxicity depends on physicochemical and environmental factors. *Environ. Health Perspect.* 165–172.

[13] Müller, R. H. and Keck, C. M. (2004). Drug delivery to the brain–realization by novel drug carriers. *J. Nanosci. Nanotechnol.* 4, 471–483.

[14] Piddubnyak, V., et al. (2004). Oligo-3-hydroxybutyrates as potential carriers for drug delivery. *Biomaterials* 25, 5271–5279.

[15] Schwarz, F., et al. (2006). Healing of intrabony peri-implantitis defects following application of a nanocrystalline hydroxyapatite (Ostim^{TM}) or a bovine-derived xenograft (Bio-Oss^{TM}) in combination with a collagen membrane (Bio-Gide^{TM}). A case series. *J. Clin. Periodontol.* 33, 491–499.

[16] Cui, Z., et al. (2005). Novel D-penicillamine carrying nanoparticles for metal chelation therapy in Alzheimer's and other CNS diseases. *Eur. J. Pharm. Biopharm.* 59, 263–272.

[17] Veiseh, O., et al. (2005). Optical and MRI multifunctional nanoprobe for targeting gliomas. *Nano Lett.* 5, 1003–1008.

[18] Liang, H., and Blomley, M. (2014). The role of ultrasound in molecular imaging. *Br. J. Radiol.* 2, S140–S150.

[19] El-Ansary, A. and Al-Daihan, S. (2009). On the toxicity of therapeutically used nanoparticles: an overview. *J. Toxicol.* 2009:754810. doi:10.1155/2009/754810

[20] Chen, F. (2014). Toxicology and cellular effect of manufactured nanomaterials. U.S. Patent No WO 2007094870 A2. Washington, DC: U.S. Patent and Trademark Office.

[21] Hagens, W. I., et al. (2007). What do we (need to) know about the kinetic properties of nanoparticles in the body? *Regul. Toxicol. Pharmacol.* 49, 217–229.

[22] Nemmar, A., et al. (2003). Size effect of intratracheally instilled particles on pulmonary inflammation and vascular thrombosis. *Toxicol. Appl. Pharmacol.* 186, 38–45.

[23] Hillyer, J. F., and Albrecht, R. M. (2001). Gastrointestinal persorption and tissue distribution of differently sized colloidal gold nanoparticles. *J. Pharm. Sci.* 90, 1927–1936.

[24] Tsuji, J. S., et al. (2006). Research strategies for safety evaluation of nanomaterials, part IV: risk assessment of nanoparticles. *Toxicol. Sci.* 89, 42–50.

[25] Rothen-Rutishauser, B. M., et al. (2006). Interaction of fine particles and nanoparticles with red blood cells visualized with advanced microscopic techniques. *Environ. Sci. Technol.* 40, 4353–4359.

[26] Semete, B., et al. (2010). In vivo evaluation of the biodistribution and safety of PLGA nanoparticles as drug delivery systems. *Nanomedicine* 6, 662–671.

[27] Lasagna-Reeves, C., et al. (2010). Bioaccumulation and toxicity of gold nanoparticles after repeated administration in mice. *Biochem. Biophys. Res. Commun.* 393, 649–655.

[28] Griffitt, R. J., et al. (2007). Exposure to copper nanoparticles causes gill injury and acute lethality in zebrafish (Danio rerio). *Environ. Sci. Technol.* 41, 8178–8186.

[29] Warheit, D. B., et al. (2004). Comparative pulmonary toxicity assessment of single-wall carbon nanotubes in rats. *Toxicol. Sci.* 77, 117–125.

[30] Donaldson, K., et al. (2001). Ultrafine particles. *Occup. Environ. Med.* 58, 211–216.

[31] Castranova, V., and Vallyathan, V. (2000). Silicosis and coal workers' pneumoconiosis. *Environ. Health Perspect.* **108**(Suppl 4):675–684.

[32] Castranova, V., et al. (2002). Effect of inhaled crystalline silica in a rat model: time course of pulmonary reactions. 234/235, 177–184.

[33] Galagudza, M. M., et al. (2010). Targeted drug delivery into reversibly injured myocardium with silica nanoparticles: surface functionalization, natural biodistribution, and acute toxicity. *Int. J. Nanomed.* 5, 231–237.

[34] Goodman, C. M., et al. (2004). Toxicity of gold nanoparticles functionalized with cationic and anionic side chains. *Bioconjug. Chem.* 15, 897–900.

[35] Shavandi, Z., Ghazanfari, T., and Moghaddam, K. N. (2011). In vitro toxicity of silver nanoparticles on murine peritoneal macrophages. *Immunopharmacol. Immunotoxicol.* 33, 135–140.

[36] Gharbi, N., et al. (2005). [60] fullerene is a powerful antioxidant in vivo with no acute or subacute toxicity. *Nano Lett.* 5, 2578–2585.

[37] Kolosnjaj, J., Szwarc, H., and Moussa, F. (2007). Toxicity studies of fullerenes and derivatives. *Adv. Exp. Med. Biol.* 620, 168–180.

[38] Mori, T., et al. (2006). Preclinical studies on safety of fullerene upon acute oral administration and evaluation for no mutagenesis. *Toxicology* 225, 48–54.

[39] Firme, C. P., and Bandaru, P. R. (2010). Toxicity issues in the application of carbon nanotubes to biological systems. *Nanomedicine* 6, 245–256.

[40] Casey, A., et al. (2008). Single walled carbon nanotubes induce indirect cytotoxicity by medium depletion in A549 lung cells. *Toxicol. Lett.* 179, 78–84.

[41] De Nicola, M., et al. (2009). Effects of carbon nanotubes on human monocytes. *Ann. N. Y. Acad. Sci.* 1171, 600–605.

[42] Han, S. G., Andrews, R., and Gairola, C. G. (2010). Acute pulmonary response of mice to multi-wall carbon nanotubes. *Inhal. Toxicol.* 22, 340–347.

[43] Schipper, M. L., et al. (2008). A pilot toxicology study of single-walled carbon nanotubes in a small sample of mice. *Nat. Nanotechnol.* 3, 216–221.

[44] Liu, M., Zhang, H., and Slutsky, A. S. (2009). Acute lung injury: a yellow card for engineered nanoparticles? *J. Mol. Cell Biol.* 1, 6–7.

[45] Heinrich, U., et al. (1995). Chronic inhalation exposure of Wistar rats and two different strains of mice to diesel engine exhaust, carbon black, and titanium dioxide. *Inhal. Toxicol.* 7, 533–556.

[46] Rittinghausen, S., Mohr, U., and Dungworth, D. (1997). Pulmonary cystic keratinizing squamous cell lesions of rats after inhalation/instillation of different particles. *Exp. Toxicol. Pathol.* 49, 433–446.

[47] Kasid, U., and Dritschilo, A. (2003). RAF antisense oligonucleotide as a tumor radiosensitizer. *Oncogene* 22, 5876–5884.

[48] Hussain, S., et al. (2006). Chemosensitization of carcinoma cells using epithelial cell adhesion molecule–targeted liposomal antisense against bcl-2/bcl-xL. *Mol. Cancer Ther.* 5, 3170–3180.

[49] Michalet, X., et al. (2005). Quantum dots for live cells, in vivo imaging, and diagnostics. *Science* 307, 538–544.

[50] Silva, G. A. (2005). Nanotechnology approaches for the regeneration and neuroprotection of the central nervous system. *Surg. Neurol.* 63, 301–306.

[51] Yamawaki, H., and Iwai, N. (2006). Cytotoxicity of water-soluble fullerene in vascular endothelial cells. *Am. J. Physiol. Cell Physiol.* 290, C1495–C1502.

[52] Limbach, L. K., et al. (2007). Exposure of engineered nanoparticles to human lung epithelial cells: influence of chemical composition and catalytic activity on oxidative stress. *Environ. Sci. Technol.* 41, 4158–4163.

[53] Wright, J. B., et al. (2002). Early healing events in a porcine model of contaminated wounds: effects of nanocrystalline silver on matrix metalloproteinases, cell apoptosis, and healing. *Wound Repair Regen.* 10, 141–151.

[54] Meng, H., et al. (2007). Ultrahigh reactivity provokes nanotoxicity: explanation of oral toxicity of nano-copper particles. *Toxicol. Lett.* 175, 102–110.

[55] Vega-Villa, K. R., et al. (2008). Clinical toxicities of nanocarrier systems. *Adv. Drug Deliv. Rev.* 60, 929–938.

[56] Ferin, J., and Oberdürster, G. (1985). Biological effects and toxicity assessment of titanium dioxides: anatase and rutile. *Am. Ind. Hyg. Assoc. J.* 46, 69–72.

[57] Lovern, S. B. and Klaper, R. (2006). Daphnia magna mortality when exposed to titanium dioxide and fullerene (C60) nanoparticles. *Environ. Toxicol. Chem.* 25, 1132–1137.

[58] Lee, K. J., et al. (2007). In vivo imaging of transport and biocompatibility of single silver nanoparticles in early development of zebrafish embryos. *ACS Nano* 1, 133–143.

[59] Kurath, M., and Maasen, S. (2006). Toxicology as a nanoscience?– Disciplinary identities reconsidered. *Part. Fibre Toxicol.* 3:6.

[60] Levi, N., et al. (2006). C 60-Fullerenes: detection of intracellular photoluminescence and lack of cytotoxic effects. *J. Nanobiotechnol.* 4:14.

Index

A

Angiogenesis 37, 146, 351, 358
Antibacterial efficacy of AgNPs
263
Antimicrobial activity 65, 85,
263, 272
Anti-psychotic drug 95

B

Biocompatibility 23, 124,
252, 377
Biodegradable 10, 44, 60, 143
Biomedical application 22, 57,
157, 184
BMNPs 157
Breast cancer 2, 33, 227, 246

C

Calmodulin (CaM) 219, 222
Cancer 2, 30, 182, 374
Carbon nanotube (CNT)
219, 228
Chemotherapy 23, 33, 76, 369
Chitosan 23, 57, 61, 75
Ciprofloxacin 83, 207,
212, 265
Clinical study 183, 293
Curcuma longa Linn 290
Curcumin 256, 289, 290, 305
Curcuminoids 289, 291, 301, 308
Cureit^{TM} 294, 303, 305, 307

Cyto-compatibility 122
Cytotoxic activity 198, 207, 211

D

Daphnia magna 378
Drug delivery 2, 34, 121, 138

E

Extraction 66, 69, 73, 291

F

FeCo 168, 173, 175, 184
Formulation development 95
Fullerenes 374, 380

G

Gold nanoparticles 3, 38,
258, 373
Growth factors 343, 346,
351, 358

M

Magic bullets 7, 28
Magnetic Nanoparticles 3, 35,
158, 162
MCF-7 cell lines 198, 213
Molecular dynamics (MD)
219, 220
Mucoadhesion 57, 65, 77, 82
Multi-walled carbon nanotube
(MWNT) 34

N

Nanogel 121, 123,
    128, 133
Nanomedicine 1, 10, 365, 368
Nanoparticles 1, 23, 30, 37
Nanorobot 3, 39, 42, 368
Nanoscaffold 3, 4
Nanoscale zero valent iron
    (nZVI) particles 197, 203,
    208, 213
Nanotechnology 1, 9, 19, 29
Nanotoxicology 365, 366, 379

O

Olanzapine 95, 97, 103, 110

P

Permeation enhancement 57,
    65, 82
pH sensitive 36, 148
Phytolacca Berry (*Phytolacca
    Decandra*) 198, 203, 205
Phytosynthesis 249, 251,
    253, 272
Proliferation 83, 207,
    307, 374

Q

Quantum dots 3, 19, 143, 253

R

Raloxifene hydrochloride 227

S

Scaffold 4, 84, 343, 357
Silver nanoparticles (AgNPs)
    249, 271, 371, 377
Single-walled carbon nanotube
    (SWNT) 32
Solid lipid nanoparticles 31, 98,
    117, 293   **T**

Tissue engineering 3, 75,
    160, 355
Toxicokinetics 370

W

Wound healing 3, 83,
    343, 353

Z

Zebrafish 379

About the Editors

Jince Thomas is currently working as Senior Research Fellow at the International and Inter University Centre for Nanoscience and Nanotechnology, Mahatma Gandhi University, Kerala, India. His research focuses on Polymer Blends, Polymer Nanocomposites, Conductive and Insulating Polymer composites and biomaterials. He has two master's degrees (M.Tech. in Polymer Science and Technology and M.Sc. Chemistry). He also worked in University of Tennessee Knoxville, USA. He has authored three book chapters.

Sabu Thomas is the Pro vice chancellor of Mahatma Gandhi University, Kottayam, Kerala, India. Prof. Thomas has (co-)authored 700 papers in international peer-reviewed journals in the area of polymer science and nanomaterials. He has organized several international conferences and workshops. His research group focuses on specialized areas of polymers and has extensive collaborative exchange programs with various industries and institutions throughout the world. He is in the list of most productive researchers in India. He has 5 patents to his credit, and edited 73 books. The H-index of Prof. Thomas is 79. He has been awarded MRSI, CRSI and Nanotech medals. Prof. Thomas has been conferred Honoris Causa DSc from University of South Brittany, France and University of Lorraine, France.

Jiya Jose is a Post-doctoral fellow of International and Interuniversity Centre for Nanoscience and Nanotechnology at Mahatma Gandhi University, Kottayam, Kerala, India. Her Ph.D. is from CSIR-National Institute of Oceanography under the guidance of Dr. Shanta Achuthankutty. Her research mainly focuses on Nano medicine, Drug delivery, nano materials for Water purification, Polymer scaffolds for wound healing applications, Interaction studies on Nanoparticles and Microorganisms. She has authored more than eight publications in peer reviewed journals. She has also authored seven book chapters.

Nandakumar Kalarikkal is an Associate Professor at School of Pure and Applied Physics and Joint Director of International and Inter University Centre for Nanoscience and Nanotechnology of Mahatma Gandhi University, Kottayam, Kerala, India. His research activities involve applications of nanostructured materials, laser plasma, phase transitions, etc. He is the recipient of research fellowships and associateships from prestigious government organizations such as the Department of Science and Technology and Council of Scientific and Industrial Research of Government of India. He has active collaboration with national and international scientific institutions in India, South Africa, Slovenia, Canada, France, Germany, Malaysia, Australia and US. He has more than 130 publications in peer reviewed journals. He has also co-edited 9 books of scientific interest and co-authored many book chapters.